Economics and Management of
Competitive Strategy

Economics and Management of
Competitive Strategy

Daniel F Spulber
Northwestern University, USA

World Scientific

NEW JERSEY · LONDON · SINGAPORE · BEIJING · SHANGHAI · HONG KONG · TAIPEI · CHENNAI

Published by

World Scientific Publishing Co. Pte. Ltd.

5 Toh Tuck Link, Singapore 596224

USA office: 27 Warren Street, Suite 401-402, Hackensack, NJ 07601

UK office: 57 Shelton Street, Covent Garden, London WC2H 9HE

Library of Congress Cataloging-in-Publication Data
Spulber, Daniel F.
 Economics and management of competitive strategy / Daniel F. Spulber.
  p. cm.
 Includes bibliographical and indexes.
 ISBN-13 978-981-283-846-9
 ISBN-10 981-283-846-5
 1. Strategic planning--Economic aspects--Textbooks. 2. Business planning--Economic aspects--Textbooks.
3. Competition--Textbooks. 4. Managerial economics--Textbooks. 5. Strategic planning--Economic aspects--Textbooks.
6. Business planning--Economic aspects--Textbooks. 7. Competition--Textbooks. 8. Managerial economics--Textbooks.
9. Strategisches Management. 10. Theorie.
 HD30.28 .S643 2009
 658.4012

                                                         2012406257

British Library Cataloguing-in-Publication Data
A catalogue record for this book is available from the British Library.

First published 2009
Reprinted 2013

Typeset by Stallion Press
Email: enquiries@stallionpress.com

Printed in Singapore by World Scientific Printers.

CONTENTS

PREFACE

After completing the text, managers will be able to perform a strategic analysis. The process of strategy making includes five basic steps: goal formulation, external analysis of the company's markets and internal analysis of the organization, choosing methods to achieve competitive advantage, devising competitive strategies that anticipate rival actions, and designing the organization to execute the strategy. Understanding these concepts prepares the manager for business leadership.

The text presents an integrated approach to management strategy by identifying both market opportunities and organizational abilities. Managers can apply the concepts to understand how companies gain competitive advantage by effectively matching organizational abilities with market opportunities. Managers consider sources of competitive advantage based on costs, product differentiation, and creating innovative transactions. Further, managers examine strategies for entering markets, creating new markets, and handling regulation.

The book is action-oriented: It presents a set of imperatives for managers and entrepreneurs for achieving success, and perhaps more importantly, coping with success. The market-focused strategy presented in this text helps managers understand market dynamics, anticipate competitor reactions, and build a responsive organization. The text emphasizes that any particular competitive advantage cannot be sustained indefinitely. Managers are urged to look ahead, taking into account technological innovation and potential competition. The book illustrates the concepts with many practical applications including mini-business cases drawn from both domestic and international business. CEOs, managers, entrepreneurs, corporate directors and venture capitalists should find the book's market-focused framework to be useful in evaluating strategic alternatives.

The book is primarily intended for the introductory course in Management Strategy at the Masters level. The book may also be used for more advanced undergraduate business courses in management and strategy. The book contains economic reasoning but does not attempt to impose economic analysis on the reader. Instead, management considerations guide the discussion. Economic reasoning is integrated within the fundamental management strategy topics.

The text does not require training in economics or mathematics and can be approached readily by first-year management students. The text features numerical examples that can be solved using very simple calculations or spreadsheets.

I am grateful to Dean Dipak Jain of Northwestern's Kellogg School of Management for his great interest in research and scholarship. I also thank Dean Dipak Jain for encouraging faculty to create textbooks as a means of communicating research to students, faculty and business leaders. I thank my outstanding students and colleagues at Kellogg with whom I have discussed many of the ideas in this text. I thank my students, Ofer Azar and Francisco Ruiz-Aliseda, for valuable research assistance and corrections on the manuscript. I thank my wife, Susan, and my children Rachelle, Aaron and Benjamin for their great intelligence, enthusiasm and delightful company. Without them, I could not have successfully completed this work.

Take-away points of Economics and Management of Competitive Strategy

Setting goals	• Choice of markets the firm will serve • Choice of markets maximizes the value of the firm • Choice of markets matches organizational abilities with market opportunities
External analysis and internal analysis	• The market compass: customers, suppliers, competitors, partners • The organization grid: structure, performance, abilities, resources
Competitive advantage	• Competitive advantage requires creating value: Advantage is often temporary • Value chain: Vertical structure based on transaction costs and need for coordination
Competitive strategy	• Price leadership strategy • Product differentiation strategy • Transaction coordination strategy • Entrants need to overcome perceived advantages of incumbent firms

---

PART I

MANAGEMENT STRATEGY

CHAPTER 1

DESIGNING MANAGEMENT STRATEGY

Contents

Scenario #1.   You are the CEO of an entrepreneurial start-up company that is being established to capitalize on a technological innovation. Working with the owner, you have assembled an accomplished advisory board, and a high-quality management team as well as some talented engineers and marketing professionals. You have already made initial contacts with some impressive potential clients and begun discussions with technology partners. An investment bank has scheduled meetings with a group of venture capital investors in two weeks. To launch the venture, the advisory board, management group, a representative from the investment bank and other personnel are meeting in a conference room at a downtown hotel. You have distributed a basic business plan to the participants. After introductions have been made around the table and a set of presentations given during the morning meeting, a question arises during the afternoon session: What is our strategy? A heated discussion breaks out and the group turns to you for leadership.

Scenario #2.   You have just been made head of a division of a rapidly growing regional business. The company's brands are gaining national recognition and the firm's financial performance to date has been outstanding. The board went to great lengths recruiting you as the leading prospect among over a dozen attractive candidates. Since you are new to the organization, you are being deluged with information about the company as you get up to speed. Although you have always been a quick study, what information should be the focus of your attention? Should you continue the successful policies of your predecessor or chart out a new direction? What is the relationship of the division to the rest of the company? Taking the helm is both a thrill and a challenge, but what do you do now?

Scenario #3.   You have just completed business school and joined a consumer products division of a major corporation. Your job description is fairly specific and the immediate tasks that you need to accomplish are spelled out when you join the team. A few weeks after your arrival, the senior manager of your division convenes the division personnel for a briefing. He makes a rather long speech, outlining the company's mission and specific goals. The senior manager provides some details about the company's overall strategy and the challenges that the division faces. The senior manager also points to some changes in the firm's market position and the possible impact on the organization. It all sounds vague and rather distant from your day-to-day activities. You wonder whether the briefing is really necessary and if all the discussion of the company's strategy will affect your job description, not to mention your stock options. How important is the briefing for your assigned tasks?

Scenario #4.   You have just joined a top international consulting company. You are sent with a team to meet with an important corporate client in Europe. The client company is under siege, with entrepreneurial European firms introducing low-cost alternatives and international competitors attracting its customers with superior products and services. The client's company is in an industry that, until now, has been sheltered from competition by trade restrictions and other government controls. A combination of technological change and market deregulation have altered the landscape considerably. You wonder whether consumer characteristics and market conditions in Europe should affect your recommendations, or whether you can give management some generic strategic advice. As you meet with senior personnel at the client company, they urgently seek your guidance on a course of action.

These four scenarios are based on actual experience. Whatever your position — CEO, division manager, employee, consultant — similar questions arise. What are you trying to achieve? What information do you need? What actions should you take? How should you deal with competitors? What are your responsibilities and those of other people in the organization?

Strategic analysis provides a method for answering these crucial questions. By the time you have gone through this text, you will know the basic principles of strategic analysis. You will have in mind all the steps needed to create your own strategy. You will know how to start from your position in the company, how to gather the necessary information, and how to design your strategy. You will have a systematic framework to help you decide on a course of action and to get others on board. Understanding the strategy-making process set out in this book will prepare you to become a successful business leader.

Formulating strategy is the primary responsibility of the company's managers. As a manager, you must prepare a strategy, regardless of whether you are leading the whole company or a division of a company. If you are an entrepreneur, you must devise a strategy to prepare a complete business plan and to guide the company. As a consultant, you need to understand the strategy process since you have to assist clients in formulating strategy. Even if you are a company employee who is not a part of the management, it is often necessary to understand the strategy process because your actions must help carry out the company's strategy.

What is the purpose of having a strategy? Simply put: *Management strategy* is a broad *plan of action* to achieve the company's goals. As a manager, you must formulate a strategy to prepare your company for competition. You also need a strategy to lead and coordinate the members of your organization. This chapter outlines the main steps of strategic analysis. After you have completed the chapter, you will know the key concepts which you will need for strategy-making.

Management strategy has five main components which form the outline of the book. (1) The manager begins the strategic analysis by selecting the *goals* of the company. (2) The manager performs a comprehensive *external analysis* of market conditions and a careful *internal analysis* of the characteristics of the company's organization. The manager adjusts the choice of the company's goals taking into account information about both the company's potential markets and its organizational abilities. (3) The manager identifies the critical factors that distinguish the firm from its competitors and allow the firm to attain a *competitive advantage*. (4) The manager formulates a *competitive strategy* that anticipates the strategies of competitors and chooses market actions to outperform competitors. (5) Finally, the manager turns to the design of an *organizational structure* that conforms to the company's overall strategy.

Even though market conditions are changing constantly and business practices are evolving rapidly, managers will always benefit from the guidance provided by the basic principles of management strategy. However, strategy-making is an ongoing process. As market conditions shift and organizations develop, it is often necessary for the manager to start the process again. The basic steps of strategy-making covered in this chapter can be applied repeatedly to respond to changing market conditions.

Chapter 1: Take-Away Points

Managers formulate the company's strategy by following the five steps of strategic analysis — select goals, perform external and internal analyses, identify competitive advantage, devise competitive strategy, and design the organization:

- The manager selects the company's *goals* to make the best match between organizational abilities and market opportunities. Information from the manager's external and internal analyses guides the manager's choice of goals and in turn, the company's goals serve to refine and update the manager's external and internal analyses.
- The manager performs an *external analysis* to examine what types of markets the firm will encounter as the strategy unfolds. The manager performs an *internal analysis* to determine the characteristics of the organization and its potential to realize market opportunities.
- The manager selects activities to obtain a *competitive advantage* by emphasizing factors that lead to superior market performance, including costs, products, and transactions.
- The manager devises a *competitive strategy* by anticipating rival strategies and choosing actions to outperform rival firms, including choosing the types of moves, timing of moves, defensive actions, and methods of market entry.
- Having set the goals of the business and chosen the competitive strategy to achieve them, the manager designs an *organizational structure* that conforms to the company's overall strategy.

1.1 *Strategic Analysis and the Goals of the Firm*

Strategic analysis is a management decision-making process. After applying the five steps of strategy making, the manager is ready to guide the business organization in competition. Accordingly, strategy is forward-looking — the manager asks what the company should do. Moreover, strategy is externally-focused — the manager asks what will succeed in the marketplace. Also, strategy has an internal-perspective — the manager asks how the organization will carry out the company's strategy. There is no universal strategic prescription; strategy depends on context and what works in one type of market may not work in another. However, the method of strategic analysis is sufficiently general that it works in many different market situations. The basic steps of strategic analysis are illustrated in Figure 1.1.

Every journey begins with a single step. But where do we begin? Just as the traveler needs a destination, the company needs a goal. The goal is what the company is trying to achieve. If goals are the company's destination, strategies are the route to the destination — strategies are the means to achieve the company's goals. Goal setting is the crucial first step in strategy-making.

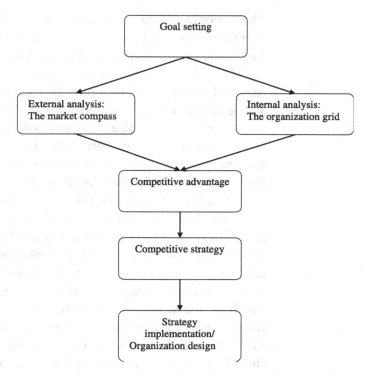

Figure 1.1: The manager's process of strategic analysis.
The process includes feedback between goal setting and the external and
internal analyses. Managers apply the process repeatedly to address changes
in the company's markets and its organization.

The manager's first responsibility in the strategy process is goal
selection. The company's *goals* often are framed in terms of *serving
specific markets*. The manager poses the key question: "What business
should we be in?" Answering this question can require continuing to
serve the company's existing markets. The company may need to enter
promising new markets and to exit from unfavorable markets. The goal
might look like one of the following:

The company will produce sportswear for sale in its own outlets in
    North America.

The company will operate hair care salons in Japan.

The company will conduct basic research in biotechnology for pharma-
    ceutical companies.

The company will provide a complete line of financial services
    throughout Europe.

The company will serve the market for electric power generation in
    Brazil.

The company will create and supply specialized designs of micro-
    processors.

The company will operate supermarket chains in many countries
    around the world.

These goals are fairly simple. Generally, defining the goal requires a careful definition of the business and the market that it serves. Often, companies want to be the leading firm in their chosen markets.

How should a manager choose the company's goals? The process cannot be accidental — a dart thrown at a chart on the wall will not do. The goal can change over time, but that does not mean that managers can wing it, hoping that the goal will emerge over time as the company experiments with different activities. The manager needs to follow a systematic procedure so as to take advantage of available information and to increase the chances that the company will be successful.

Understandably, the question of choosing goals is hotly debated in strategy making. Some argue that the company should choose goals based only on the best market opportunities. Market-driven goals are based on discovering original market opportunities that are characterized by growing customer demand and relatively limited competition. However, in some markets, the company may find that competitors are much better at satisfying customer needs. The video game market may look attractive but other companies might be better at designing or marketing games. Then, the company's goals must adapt to the abilities of the organization relative to its competitors.

Others argue that the company should engage in those tasks that reflect the company's unique skills and competencies, particularly those that are hard for others to copy. Organization-driven goals are based on recognizing unique organizational abilities and resources that will help the company prevail over its competitors. However, in some cases, there is little demand for the things the company is best able to do. The company may be very good at designing wooden tennis rackets when the market has switched to composite materials.

This text focuses on *value-driven strategy*. The manager chooses goals and strategies that maximize the total value of the firm. The manager chooses *value-driven goals by making the best match of organizational abilities with market opportunities*. This means that a compromise may be necessary. The company may not necessarily chase the most attractive market opportunity or employ the best skills of its organization. The path to success is choosing the best *combination* of market opportunities and organizational strengths. The company might target an apparently less attractive opportunity that fits its skills or develop some secondary skills to meet an opportunity. This is why the manager must integrate information from both the external analysis and the internal analyses when choosing the company's goals.[1]

The goals of the company are not only to serve particular markets but also to serve them well. Companies strive for winning performance in their markets. Kenichi Ohmae observes that "[i]n the real world of business, 'perfect' strategies are not called for. What counts ... is not performance in absolute terms but performance relative to competitors".[2]

Corporate strategy is the overall strategy of a multi-business company. The corporation's senior management sets its goals by choosing the collection of individual businesses, thus determining the scope of the firm's activities. Each business serves a specific set of markets. Top management chooses a policy of diversification if entering multiple businesses increases the total value of the firm. Companies should operate multiple businesses only if those businesses are worth more when operated together than if they were operated separately. Corporate strategy involves the selection and coordination of the multiple businesses. Thus, the goals of corporate strategy are expressed in terms of the set of businesses the company wishes to operate. For example, Groupe Danone chooses to operate three global businesses: dairy products, biscuits, and beverages.

Business strategy is the overall strategy of a business unit of a large corporation or that of a standalone business. Business strategy refers to the plans of an established firm for serving existing markets or new markets. Business strategy also refers to the plans of an entrepreneur contemplating entry into a market. Many dot-coms failed because they followed a policy of ready-fire-aim. A start-up company certainly needs to have a business plan before it is established. The goals of a business are expressed in terms of the set of markets that the business wishes to serve. For example, one of the goals of Danone's dairy products business is to continue providing the highest-selling brand of yogurt worldwide.

This text uses the general term *management strategy* to include both corporate strategy and business unit strategy. The term *manager* is used generally here to refer to a business leader. Management strategy refers to the manager's decisions and plans that are necessary to guide the company. As already emphasized, management strategy results from the process of selecting goals, performing external and internal analyses, identifying competitive advantage, devising competitive strategy and designing the organization.

Most organizations have well-defined goals. Many types of organizations exist, including profit-maximizing businesses, educational establishments, religious institutions, informal communities, non-profit enterprises, governmental agencies, and the military. Objectives and decision-making procedures vary across organizations, making generalizations difficult. Different types of organizations might have different measures of success. A non-profit organization might have public service goals. A private club is operated for the benefit of its members. A government agency may serve political or social interests. To carefully define the main issues, this text concentrates on *profit-maximizing* businesses, whether privately held or publicly traded.

After the manager completes the strategic analysis, the process of strategy implementation begins. *Implementation* means that the manager must *act through others* to execute the strategy. The manager communicates the strategy to the members to the organization. The manager exercises leadership to guide the company's employees and motivates employees to cooperate with each other and to act in the interests of the company. After monitoring the performance of the organization in executing the strategy, managers make the necessary adjustments both in organizational design and in the strategy itself. Implementation brings a strategy to life, but implementation cannot be effective without a good strategy. Because the focus of this book is on strategy analysis, the important but extensive problem of strategy implementation is beyond the scope of the discussion.

Taking some time to choose goals is worthwhile because a brilliant strategy in pursuit of the wrong objectives can be of limited value. The company's top managers must be involved with defining its goals. Strategy in this text is concerned with *rational actors*, that is, managers seeking to make explicit goals and choosing the best means to achieve these goals. In contrast, some analysts emphasize the importance of evolutionary, non-rational approaches to forming strategy. Henry Mintzberg speaks of managers crafting strategy, which is the outcome of rational deliberations and of responses that emerge as reactions to unforeseen events.[3] Many managers form their ideas based on experience or substantially change their plans over time.

Strategies always involve risk and conjecture. The failure or success of the firm's actions provides valuable feedback. Companies continually encounter technological innovations, changes in customer demand, and creative competitors. Companies and their managers update their goals and strategies as new information becomes available. Strategies are not set in stone, but, instead, anticipate and respond to competition. Accordingly, the strategy process in this text can be applied to develop dynamic strategies that adapt to changing market conditions.

Exhibit 1.1   Jack Welch at General Electric

Jack Welch, the legendary chairman and CEO of General Electric, spelled out his strategy clearly with his "number one, number two" concept. He pledged to continue to operate or to acquire only those businesses that would be number one or number two in their market. He took over a company in 1981 that had 350 different businesses in its portfolio clustered in 43 Strategic Business Units.[4] Those businesses that did not perform or could not be improved to meet the goal would be divested. The corporate winners would only be those who are the "leanest", the "lowest-cost worldwide producers of quality goods and services or those who have a clear technological edge, a clear advantage in a market niche".[5]

(Continued)

Exhibit 1.1 *(Continued)*

Welch put forward a "fix, close, or sell" policy.[6] He drew three circles representing services, high technology and core businesses, and placed the names of 15 businesses inside the circles. The businesses outside the circles would be divested if they could not be improved. Welch stated, "I'm looking at the competitive arena. Where does the business sit? What are its strengths vis-à-vis the competition? And what are its weaknesses? What can the competition do to us despite our hard work that can kill us a year or two years down the road? What can we do to them to change the playing field?" He continued: "If you have a game that's vulnerable, somebody can move fast, get you. And you don't have a checkmate play or another move. You've got to get out of that game".[7]

In addition to the "number one, number two" policy, Jack Welch reoriented the company toward services. When Jack Welch began managing GE, the company was composed of 15 percent services and 85 percent products. As the company began the 21st century, it had a mix of 25 percent products and 75 percent services, including financial services and medical systems.[8]

While focusing on strengths and shifting to services, GE under Welch embraced globalization. Operating in more than 100 countries, the company has about 45 percent of its 293,000 employees outside the United States and earns approximately the same share of revenues outside the country. The company was named the World's Most Admired Company by *Fortune* and the World's Most Respected Company by the *Financial Times*. Welch's choices of what businesses to maintain, improve or to exit; the decision to enter into international markets; and orientation toward services established *goals* for General Electric.

1.2 *External Analysis and Internal Analysis*

The manager's choice of the company's goals and strategy depends critically on information. Managers are responsible for gathering information about the company's markets and the company's organization. The manager's external analysis evaluates market opportunities both to suggest the goals of the firm and to evaluate the potential economic returns after goals are selected. The manager's internal analysis evaluates the organization's abilities both to help in the selection of goals and to determine how the organization should be structured to achieve those goals.

External Analysis

The company's goals are generally framed in terms of the markets that the company intends to serve. Accordingly, the strategy process requires an external analysis of the characteristics of the markets the company proposes to serve. The manager evaluates how the firm's markets are changing and how those markets will look in the future. The manager generally knows well who the firm's customers, suppliers, competitors, and partners are today. Choosing the company's goals requires thinking about how the company's markets will look in the future. Because the external analysis looks at the future of the firm's markets, it is outward-directed and forward-looking. Scanning the company's actual and potential markets is a critical step in the strategy-making process.

Effective strategy requires competing in markets that highlight a company's strengths relative to its potential competitors.[9] Webvan, an Internet-based grocery delivery service, failed in trying to take on the major grocery chains. Despite its large warehouses, Webvan was not able to achieve lower costs than the supermarkets because it did not generate enough orders to benefit from scale, and customers preferred shopping at supermarkets to having their groceries delivered.[10] Toshiba abandoned the manufacture of commodity memory chips, where it did not have an advantage over low-cost producers, choosing instead to focus on custom-designed chips where its technological strength might provide an edge over competitors.[11]

Context matters a great deal when few, if any, universal policy prescriptions are available. The ultimate success or failure of strategic actions depends on the markets in which the company operates. What type of products and services are customers seeking? What type of products and services do suppliers offer? What kind of competitors does the company face? Who are the company's potential partners? Selling computers in China is not the same as selling computers in Europe. The manager must tailor strategy to the market situation, as defined by potential customers, suppliers, competitors, and partners.

Other factors also affect the company's strategy. The geographic scope of the market is significant: Is the market local, national, regional, or global? The pace of technological change and the sources of innovation also have considerable impact: Is the firm operating in an industry with a rapid or slower rate of technological change? The social context affects available strategies by influencing relationships between employers and workers and between firms and customers. The legal and regulatory environment affects strategy by setting constraints on the actions of firms and creating opportunities.

The manager's external analysis looks at how the firm's markets will change as its competitive strategy is implemented. The external

Customers

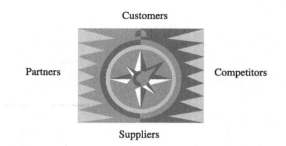

Partners

Competitors

Suppliers

Figure 1.2: The market compass.

analysis entails information gathering, data analysis and informed prediction. Strategy-making guides external analysis since the manager's strategic choices require gathering specific types of information. The information that is obtained leads to refinements in strategy.

The external analysis looks at the firm's markets. I introduce a concept that guides the external analysis called "the market compass". There are four main players: customers, suppliers, competitors and partners. The external analysis seeks to identify the firm's *prospective* customers, suppliers, competitors and partners. What will the company's markets look like at the time that strategies will be implemented? How will the company's strategies change the market conditions that it faces? Figure 1.2 shows the market compass.

Internal Analysis

Managers also perform an internal analysis as part of the strategy process. The purpose of the internal analysis is to support the decision-making process by determining whether the company's goals and strategies are feasible for the organization and whether the design of the organization should be modified to adapt to the company's strategy. Do the company's goals make the best match between the abilities of the organization and market opportunities? Can the organization implement the required strategies to attain the goals?

The internal analysis examines the company's organizational structure, performance, abilities, and resources. The company's *organizational structure* refers to its boundaries, divisions, lines of authority, management practices, and incentives. The company's *performance* is evaluated in terms of the total value of the firm, which depends on the present discounted value of the company's economic profit over the long term. The company's *abilities* include the capabilities and competencies of the company's employees as they work together to achieve the company's goals. Finally, the company's *resources* encompass tangible assets such as plant and equipment, inventories, and accounts receivable as well as less tangible assets

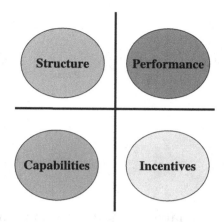

Figure 1.3: The organizational grid.

such as intellectual property, technological knowledge, product brands, and goodwill.

As part of the strategy-making process the manager evaluates how the company's organization should change as a result of the company's strategy. The market opportunities identified through the process of setting goals and examining market dynamics can suggest the need for changing the company's organizational structure.

I introduce a concept that guides the internal analysis called "the organizational grid." This concept refers to the organization's structure, performance, abilities, and resources. In computer science, virtual organizations draw upon a "grid" to bring together resources.[12]

A manager can use the concept of the organizational grid to identify the organization's abilities and resources, as well as its structure and performance. Then, the organization's abilities and resources can be organized to implement to manager's strategy. Figure 1.3 shows the organizational grid.

1.3 *Competitive Advantage and Competitive Strategy*

To succeed in attaining its goals, the firm must perform better than its competitors. This means that the company must identify a *competitive advantage* that distinguishes the business from its competitors. A company with a competitive advantage is able to create greater economic value for its shareholders, customers, and suppliers than its competitors. The company attempts to sustain its competitive advantage by engaging in continual innovation in production processes, product features, and transaction methods.

A competitive advantage is not in itself a guarantee of success. The company must put its competitive advantage into practice by devising and executing a *competitive strategy*. The company takes actions in the marketplace that anticipate the strategies of established competitors and potential entrants. The company exits some of its existing markets and enters new markets. The company further extends its advantage by creating and operating markets. In addition, the company pursues non-market strategies to address law, regulation, and public policy.

Creating Value and Competitive Advantage

Competition is the fundamental challenge of business strategy. Good products and an efficient organization are rarely enough. To survive and succeed, the company must outperform competitors in many different ways. Its products and services must offer customers greater value through higher quality and convenience or lower prices than rivals offer. The company must be more attractive to critical suppliers and distributors. It also must compete for the attention of investors. The company's cost efficiencies must exceed market benchmarks. Capturing market share can boost profits if the company takes advantage of its sales to achieve recognition, market power and cost economies.

To win markets, companies seek competitive advantage, an edge that will differentiate the company and allow it to outperform other companies. The purpose of strategy is to enhance the firm's competitive chances in the market by choosing actions that yield the greatest expected value for the company. The manager's external analysis identifies attractive businesses. The manager's internal analysis identifies effective organizational activities, resources, and competencies. To distinguish itself from its competitors, the company creates innovative matches between organizational abilities and market opportunities.

A company must create greater value than its competitors to continue attracting customers, suppliers and investors. Companies create value for customers by offering them better product features, lower prices, or both. Companies create value for suppliers by lowering their costs through cooperation, giving them a greater share of earnings, or both. Companies create value for shareholders by capturing a greater share of total value created, resulting in a higher net present value of earnings over time.

Thus, to create greater value for customers, suppliers and shareholders, the company's *total* value created must be greater than that of its competitors. *Competitive advantage* is defined as the difference between the value created by the company and the value created by competitors. There are three main sources of competitive advantage. First, the firm has a *cost advantage* if it operates at lower cost than competitors. Second, the firm has a *differentiation advantage* if the firm's product creates greater customer benefits than those of its competitors. Third, the firm has a *transaction advantage* if the company has lower transaction costs or creates innovative combinations of buyers and sellers in comparison with competitors, thus generating greater value than competitors.

A number of strategic actions can achieve a transaction advantage. As intermediaries, companies bring buyers and sellers together, acting as agents for their customers and suppliers, monitoring the performance of contractual partners, brokering transactions, and communicating market information to customers and suppliers. Intermediaries earn rent by improving transactional convenience for their suppliers and customers. Acting as entrepreneurs, companies create new combinations of buyers and sellers through new types of transactions, novel goods and services, and involvement of new customers and suppliers.

As *market makers*, companies manage market transactions and create the institutions of exchange. Markets are mechanisms or institutions that bring buyers and sellers together. The firm's four primary market-making actions include price setting, market clearing, coordinating buyers and suppliers, and allocating goods and services. By creating markets, companies not only gain a home-court advantage, they become the focal point of their industry. Companies make markets work by forming relationships with suppliers and customers. Through information gathering, communication, and price-setting activities, firms provide competitive outlets for suppliers and competitive sources of products for their customers, discerning new opportunities for purchasing and selling goods and services. By coordinating their selling and buying activities companies create markets. Managers that are market-focused are better prepared to observe and respond to changes in customer needs and supplier abilities.

The market-making firm stands between its suppliers and customers, creating innovative transactions that coordinate their economic activities. By creating and operating markets, companies remain at the center of economic activity even as the characteristics of products and the way they are produced continue to change. Advantages from creating markets can be more durable than those founded on specific product designs or production processes.

Strategic analysis helps managers identify sources of competitive advantage. The relentless give-and-take of competition suggests that it is difficult to find sources of competitive advantage. The great complexity of business decision-making implies that opportunities are hard to find and still harder to exploit. Turbulent markets create risk and uncertainty for decision makers. Yet it is precisely these difficulties that make strategic analysis a necessity. Because of complexity and uncertainty, decision makers are unlikely to choose the same strategies. Managers must sort through many strategic alternatives and work with diverse sources of information. Strategic analysis is valuable because it helps managers process and organize information, pose the necessary questions, and sort through solutions, effectively.

Is Competitive Advantage Sustainable?

Although competitive advantage is attainable through business strategy, can success be sustained? A theme of this text is that *no specific competitive advantage can be sustained indefinitely*. Continual innovation surpasses existing products, processes, and transaction methods. Once a competitive advantage is attained through innovative processes, products, or transactions, market forces are likely to erode the returns. Strategy-making is not a one-shot deal — companies cannot hope that successful actions will remain so indefinitely. Maximizing the present value of profit requires long-term strategies, not short-term fixes. Strategy-making is an inherently *dynamic* process, requiring continual updating of company goals and the plans to achieve them.

Creating greater value than competitors requires innovation. Three main types of innovation correspond to the three main types of competitive advantage. *Process innovation* is the employment of new methods of producing and delivering goods and services. Process innovations include such significant changes as the assembly line, industrial robots, and computer-aided design and manufacturing, as well as specific management changes that consistently yield operating efficiencies. *Product innovation* refers to the introduction of new goods and services. Product innovations involve both major changes such as the creation of the automobile and enhancements such as closed auto bodies, steel frames, and anti-lock brakes. *Transaction innovation* is the use of new methods of buying, selling and contracting for goods and services. Transaction innovations include credit cards, bar-coding merchandise, Internet commerce, and other specific changes that enhance customer convenience. Finally, entrepreneurial companies engage in transaction innovation by introducing new combinations of products, processes and transactions into the marketplace.

The key to contemporary strategy is *technological change*. Because technological change is not particular to any specific industry or sector of the economy, it is fundamental to strategic analysis. The accelerating pace of technological change calls for more strategic analysis — to process information effectively, to adjust goals frequently, to recognize market opportunities, to act decisively and to organize with greater flexibility.

Management experts emphasize the concept of change in many colorful ways. Joseph A. Schumpeter described competition as gales of creative destruction. More recently, Peter Drucker speaks of managing in a time of great change. Richard D'Aveni refers to escalating market challenges as hypercompetition. Tom Peters tells managers to thrive on chaos. Charles Fine speaks of managing at clockspeed in an environment of temporary advantage. Philip Evans and Thomas S. Wurster suggest that some existing business approaches will be blown to bits by e-commerce.[13]

Tremendous technological change sweeps through the economy leaving no industry or business intact. Advances in computers and communications affect the production processes of practically every company, whether manufacturer, service provider, wholesaler, or retailer. The Internet changes the shape of retailing and reconfigures business-to-business transactions. Progress in science and engineering, including biology, chemistry, physics, electronics and materials sciences, impacts specific industries.

Technological change creates winners and losers. Occupations that existed at the start of the 20th century are rare at the start of the 21st century. There were over 100,000 carriage and harness-maker workers in 1900 and over 238,000 blacksmiths in 1910, whereas there are only a few thousand of each today. Railroad employees have fallen from 2 million in 1920 to about 230,000, and farm workers have declined from 11.5 million in 1910 to about 850,000 today. In contrast, there are well over 1.38 million medical technicians; 1.3 million computer programmers and operators, 864,000 auto mechanics; 3.3 million truck, bus and taxi drivers; and 200,000 airline pilots — professions that did not exist at the start of the 20th century.[14]

In addition, individual firms come and go. In the industries shown in Table 1.1, the number of establishments has fallen considerably, giving an indication of the attrition of individual firms. A pattern of growth, shakeout, and leveling off has been observed in many industries including compressors, electrocardiographs, gyroscopes, jet-propelled engines, lasers, outboard motors, nylon, paints, ballpoint pens, photocopy machines, radar, heat pumps, nuclear reactors, transistors, and zippers.[15] Shakeouts in wholesaling occurred in over a dozen industries including flowers, woodworking machinery, locksmith, specialty tools and fasteners, sporting goods, wholesale grocers, air conditioning and refrigeration, electronic components, wine and spirits, waste equipment, and periodicals.[16]

Enhanced computer power and improvements in communication have increased the information available to managers and changed their focus from data processing to strategic decision-making. These developments continue to reshape organizations, flattening corporate hierarchies and increasing outsourcing.

Technological change is the greatest force changing the market landscape but markets change in many other ways. Customer preferences for goods and services are likely to fluctuate. The characteristics of suppliers and the goods they offer vary over time. The actions of competitors are likely to change significantly as they pursue innovative strategies and discover new market opportunities. The company's actual and prospective partners change their objectives and seek new alliances.

Technological change has far-reaching social effects that translate into further market change. The transformations set in motion by the

Table 1.1: Selected shrinking and expanding industries.

	Number of Establishments	
	1970	**1996**
Selected Shrinking Industries		
Fur goods	980	133
Barber shops	24,577	4,499
Asbestos products	133	30
Drive-in theaters	1,567	408
Leather and leather products	3,430	1,938
General merchandise stores	25,032	14,797
Glass containers	128	78
Brooms and brushes	449	278
Trailer parks and campsites	6,419	3,984
Bowling centers	9,215	5,735
Concrete block and brick	1,332	901
Manufactured ice	800	578
Variety stores	14,439	10,848
Radio and television repair	7,953	6,212
Labor organizations	20,376	19,536
Selected Expanding Industries		
Videotape rental	0	20,816
Computer and data processing services	6,517 (1975)	88,911
Carpet and upholstery cleaning	816	8,879
Prepackaged software	1,522 (1975)	9,084
Vocational schools	1,188	6,816
Movie production and services	2,922	14,680
Semiconductors and related devices	291	1,052
Amusement parks	362	1,174
Chocolate and cocoa products	51 (1975)	165
Car washes	4,624	13,334
Political organizations	928	2,579
Office and computing equipment	923	2,112
Eating and drinking places	233,048	466,386
Colleges and universities	1,855	3,663
Florists	13,865	26,728
Tour operators	2,464 (1988)	4,725
Dental offices	63,817	113,054
Internal combustion engines	162	277
Passenger car rental	2,556	4,231
Pharmaceuticals	1,041	1,637
Aircraft	163	255
Plastic bottles	280 (1988)	437
Aircraft engines and parts	247	355
Physical fitness facilities	7,723 (1990)	10,720

(Continued)

Table 1.1: *(Continued)*

	Number of Establishments	
	1970	**1996**
Hotels and motels	34,674	45,252
Travel agencies	22,609 (1988)	28,735
Space vehicle equipment	39 (1975)	45
Beauty shops	70,967	81,872

Source: W. Michael Cox and Richard Alm, "The Churn Among Firms," *Southwest Economy*, Federal Reserve Bank of Dallas, January/February 1999, pp. 6–9, based on data from U.S. Bureau of the Census County Business Patterns, various years. Establishments are classified based on their major activity.

automobile and air travel are well recognized. In turn, political and social changes impact markets. Regulatory and legal changes also have far-reaching effects. Deregulation of transportation, energy, telecommunications, and financial transactions has had significant effects in many countries. Lowering or raising barriers to trade alters the volume and direction of world trade.

Accordingly, attaining and sustaining competitive advantage demand continual creativity. No matter how innovative they are, the firm's strategies are eventually subject to imitation or counteraction by competitors. Indeed, the more successful the strategy, the greater the returns to competitors seeking to emulate or surpass it. With the increasing pace of technological change, companies can no longer rely on building castle walls for long-term protection.

The notion that all advantages are temporary is both good news for entrants and bad news for established companies. On the other hand, others argue that technological change enhances first-mover advantages, thus harming entrants and helping established companies. Which view is correct? Does the entrant always win and the incumbent always lose, or vice versa? As one might expect, the outcome of competition is not so predictable. Advantages may be temporary, and companies can only develop or sustain a winning performance by finding new sources of competitive advantage.

Competitive Strategy and Anticipating Rival Strategies

Competitive strategy refers to the actions of the firm that are best responses to the observed or anticipated actions of competitors. Competitive strategy is a critical component of the company's overall strategy because it specifies the company's market moves. The manager's competitive strategy spells out the types of moves and their timing. Should the firm try to move before or after competitors? Should the firm emphasize prices or distinctive product features? How should the firm carry out market entry? What market segments should be targeted? Managers must understand that competitive strategy is difficult because they are playing a game against clever opponents.

Strategic analysis by managers must be suited to the market context. Accordingly, the choice of strategic moves depends on industry conditions such as the number of competing firms and their market power, the extent of product differentiation and the rate of technological change. Moreover, companies must anticipate the potential for entry of new competitors. The manager chooses competitive strategy by building on market information from the external analysis.

Competitive strategy can depend strongly on the number of competitors the firm faces. There are two important competitive situations: competition *in the market* and competition *for the market*. Companies competing *in* the market face a known set of rivals. The outcome of the competition is defined in terms of strategic moves. Companies pay close attention to the number and size of competitors. The number and size of firms in the market are referred to as *market structure.*

In contrast, when companies compete *for* the market, the number of competitors changes. New competitors may choose to enter the industry or established companies may choose to exit the industry. Established companies plan for the entry or exit of competitors or the possibility that they may be forced from the industry. Moreover, potential entrants make plans to establish their companies in anticipation of the reaction of incumbent firms.

Competitive strategy requires anticipating the future actions of competitors and preparing the best response. Companies need to develop market intelligence regarding their competitors so as to respond effectively to their strategies. The manager chooses prices, production, and products that are best responses to expected strategies of competitors. The company competes to win markets by delivering superior performance. Entrepreneurial companies devise entry strategies to attempt to win new markets. There are no guarantees of success in a competitive environment. However, it is useful to understand the strategy-making process.

Managers seek to understand factors that affect rivals' profits. For example, by understanding how competitor revenues and costs vary with sales, it is possible to make some predictions about future pricing and other strategic behavior. Companies try to identify the types of strategies available to their rivals. Are competitors most likely to vary their prices, productive capacity, or product features? The choice of strategic instruments can change the outcome of the game significantly.

In addition to the number of competitors, their payoffs, and the types of strategic moves that are available, *timing* of moves is fundamental to interaction with competitors. When should a new venture be launched?

When should new products be announced and when should they be introduced? When should a price change be put in place? When should a promotion begin? Should the company respond immediately to a competitor's price cuts or a targeted marketing campaign? Should the company introduce products to the market before their rivals or try to leapfrog over competitors' products after they are introduced? Competitive strategy takes account of the trade-offs between first-mover advantages and second-mover technological improvements.

Companies tend to be better informed than their competitors about their own costs, revenues, and strategic options. Competitive strategy operates in a world with information asymmetries. The manager must make predictions on the basis of limited knowledge of the motivations and competencies of competitors. Competitive analysis can be useful in narrowing the possible outcomes and pointing out effective moves.

Competitive strategy requires choosing key market segments to be contested and other segments to be conceded to competitors. An attempt to dominate all segments of the market can strain the company's limited resources and increase costs, putting the company at a disadvantage resulting in losing the overall strategic objective. Competitive strategy spells out the specifics of pricing, products and technology needed to surpass competitors. Companies choose their battles carefully, entering market segments that promise the highest return first, and avoiding battles in other segments that do not contribute directly to the company's overall strategy.[17] The reason for this is that any company, whether long established or a start-up, has limited resources. It faces constraints on productive capacity, technical capability, market knowledge, or availability of managers and employees. Thus, the need to attack from strength can require focusing attention on a few market segments.

Competitive strategies must be flexible and are subject to far less planning than the company's overall strategy. Some competitive decisions must be delegated to personnel with the best information on the actions and strategic intent of competitors, but they require constant monitoring to determine their effectiveness. Competitive strategies directly involve many functional areas of the company, particularly marketing and sales personnel who are in the trenches and can closely track the actions of rivals and customer responses. Pre-emptive actions or competitive reactions such as product promotions are designed to address the actions of close rivals. Competitive strategy must adapt rapidly to market change.

Competing for the Market and Entry Strategies

Market structure is far from stable. Although the number of competitors may seem fixed in the short run, companies enter and exit industries all the time. Innovative entrants seek advantage over incumbents. Companies with weaker sales and profitability exit the business. Competitive strategy takes place in the context of ever-changing rivalries. Companies compete with established firms and entrants to be the market leader.

Competition is a dynamic process in which companies seek to win their markets. Being the leading firm confers additional advantages such as customer recognition and cost economies. Companies that become complacent, however, are often displaced by energetic rivals. Entrants attack incumbent firms where the entrants' strength is greatest in relation to established firms. Entrants offer customers better value through lower prices or higher quality, and they can be expected to go all out in terms of marketing, sales efforts, management attention, customer service and innovation.

Established companies must be constantly prepared to defend market segments against the entry of competitors, but it may not be possible to defend all market segments simultaneously. The incumbent needs a game plan to discern which segments will draw the attention of entrants.

There are many asymmetries between incumbent companies and entrants. Incumbents operate at a greater scale than entrants by virtue of being established in the market. Most significantly, incumbent firms have already made irreversible investments in plant and equipment, marketing, and research and development (R&D), while entrants have yet to make commitments to the market. Incumbents often have established brands and customers may be accustomed to using their products and services. In contrast, entrants often introduce new brands and products with unfamiliar features. Incumbents may have greater knowledge of customer preferences and production technology than start-ups.

These and other differences between incumbents and entrants have been termed *barriers to entry*. Entrants perceive that there are barriers to entry if incumbents have a competitive advantage over entrants. Incumbents may have cost advantages, differentiation advantages or transaction advantages over entrants. Such advantages can give market power and economic rents to incumbents as long as they are sustainable.

In a static environment, barriers to entry would appear to be important competitive weapons. However, in a dynamic environment, such advantages are likely to be temporary. Technological change erodes potential differences between entrants and incumbents. Entrants can outperform incumbents by adopting efficient production processes, introducing innovative products, or employing new transaction techniques. Changes in consumer preferences also create opportunities for entrants to provide new products and services. Changes in international trade restrictions and other government regulations further open markets to new competitors.

Entrants can overcome incumbent advantages in a variety of ways. For example, if an incumbent has a cost advantage, entrants can counter with a differentiation or transaction advantage. Entrants can address the costs of building production facilities by joining with others in joint venture agreements, sharing risks by contracting with customers, or outsourcing to suppliers. Entrants can build distribution networks, partner with established distributors, or sell through independent wholesale and retail companies. By varying their mode of entry, entrants reduce the costs and risks of market commitment. Companies that identify significant weaknesses relative to their competitors pursue *indirect strategies*, serving niches or entering markets overlooked by incumbent firms.

Exhibit 1.2   Fox Television and Entry Strategies

Fox Television, a subsidiary of Rupert Murdoch's News Corporation, became the fourth broadcast television network in the 1990s. Its battles with the big three traditional networks, NBC, CBS and ABC, illustrate successful use of strategic maneuvers. Rather than attempting to conquer every segment of the TV market, Fox has avoided some battles and joined others with its full strength. For example, it outbid other networks for rights to broadcast National Football League games and even hired CBS announcers John Madden and Pat Summerall. A key area of focus for Fox was the children's television segment.

Children's television was a fast-growing and important segment of the television marketplace. For a time, that segment had been conquered by Fox's superheroes, the Mighty Morphin Power Rangers. Six happy, well-adjusted teenagers "morph" or transform into a team of superheroes. A marketer's dream, their outfits came in a rainbow of colors: white, red, blue, black, yellow, and pink. While they preferred karate chops, each hero had a different weapon that fit together with the others to form a superweapon, all sold separately along with action figures. Every ranger had a mechanical dinosaur ally, called a "dinozord", that he or she summons whenever they needed assistance. These friendly machines also fit together to form the ultimate weapon, the "megazord". Each weekday, the rangers fought and defeated a new extraterrestrial monster to the tune of "go, go, Power Rangers".

While saving the universe from evil, the Power Rangers won another battle — they topped all other children's TV shows. Moreover, they led the Fox Children's Television Network to the lead position in children's TV by its fifth season. Of the top 10 children's

(Continued)

Exhibit 1.2 *(Continued)*

TV shows, Fox had the top four (Power Rangers, Animaniacs, Batman, and Tiny Toons) and the number 10 show (X-Men). Fox led not only in ratings but in advertising revenue as well, with $200 million out of a market segment total of $700 million in that year.[18] The Fox Children's Television Network, along with the cable channel Nickelodeon, initially dominated the segment while ABC and CBS curtailed their offerings and NBC ceded the segment altogether.[19]

In the children's television market, individual shows and their assignment to time slots were of less importance than broader maneuvers. Chase Carey, chairman of Fox Television, cautioned that the comprehensiveness of Fox president Margaret Loesch's strategy was more important than the network's ratings. Carey noted that in a very short period of time Loesch "pulled together the support of the children's creative community, of the affiliates, of the advertisers — really all the aspects of the business you'd want in place — she was able to win over".[20]

The start-up network also was strong in the youth market with a variety of innovative shows. The tension showed at the other networks. Howard Stringer, president of the CBS broadcast group, groused "sometimes I think that youth is wasted on Fox".[21] Fox had successfully entered the market by concentrating its efforts on a market segment where it was relatively stronger than incumbent firms — the youth market.

1.4 *Strategy and Organizational Structure*

Having established the company's goals and strategies to achieve the goals, the manager next makes sure that the organizational structure conforms to the company's strategy. The reason that the company's organizational structure must follow its strategy is that the organization is responsible for putting strategy into practice. After senior managers have completed the strategy process, including redesign of the organization, they assign tasks to the members of the organization. For the company's strategy to be carried out effectively, the organizational design must facilitate the assignment and completion of the necessary tasks by managers and employees.

Organizational structure has two main aspects: horizontal and vertical. A company's *horizontal structure* refers to the scope of the company's product and service offerings and the divisions of the organization. For example, Pepsico has three principal divisions: Frito-Lay Company, the largest manufacturer and distributor of snack chips; Pepsi-Cola Company, the second largest soft drink business; and

Tropicana Products, the largest marketer and producer of branded juice.[22]

A company's *vertical structure* refers to the types of functional activities the organization performs and the degree of vertical integration between them. For example, Nike focuses its attention on product design, product development, marketing and distribution. The company forms Category Product Teams, consisting of its own designers, developers and marketing specialists, to develop an athletic shoe as well as a marketing plan, a process that the company says takes up to a year and a half. The company then puts together a technical package consisting of designs, patterns, lasted uppers, and model shoes. The company does not produce a shoe but instead ships the technical package to manufacturing subcontractors that operate factories in Europe and Asia. Then, after the shoes are produced, they are shipped to Nike distribution centers and finally to independent retailers. Thus, Nike is involved in design, development, contracting, marketing and distribution, but is not vertically integrated significantly into manufacturing or retailing.[23]

The Horizontal Structure of the Firm

The manager's choice of goals specifies the company's target markets. As a result, the goals of the company determine what businesses the company wants to begin operating, continue operating, and cease operating. The organization must conform to these objectives. If the company plans to enter a new market, it should establish a corresponding business unit, adapt an existing business unit, or acquire an existing business. If the company plans to exit from a market, it must close the corresponding business unit, adapt the business unit to focus on other activities, or divest the unit.

The manager's choice of goals thus helps to specify the scope of the firm. As already stated, the company's strategy specifies the plan of action to achieve the goals, whether the firm will emphasize a cost advantage, a differentiation advantage, or a transaction advantage. In addition, the firm will choose market actions that anticipate the strategies of competitors. To implement the company's strategy, the organization must have sufficient personnel to perform the necessary tasks. Moreover, the strategic tasks should somehow be divided among the members of the organization. Deciding how to allocate strategic tasks across the organization guides the process of organizational design.

The manager should choose the organizational form that best implements the company's strategy. Companies operate in the realm of the possible. Therefore, management cannot choose the organizational form arbitrarily because the company is limited by many constraints, including the availability of qualified personnel, the costs of travel, telecommunications and information systems, and legal and regulatory restrictions. If the company is already established, the manager may not be able to redesign the organization from scratch because of the significant costs of adjustment. These costs of adjustment are a source of inertia and explain why some organizations are slow to adapt to market change. The manager must compare the costs of organizational change with the benefits of improved strategy implementation.

A business organization can be structured in many different ways. One way to structure an organization is along *functional* lines, by dividing the organization into units responsible for R&D, finance, human resources, purchasing, marketing, sales, operations, and information systems. Such an organization is often appropriate for a company operating a single business or a closely related collection of businesses. A functional structure tends to favor central control of the organization's activities by its managers.

For a company operating multiple businesses, another way to structure the organization is to divide it into individual *business units*. The company can further create strategic business units that combine multiple related lines of business. The company's divisions then contain groupings of lines of business based on related products and services or based on the provision of products and services to particular target customers. An organizational structure based on business units tends to make the company more responsive to market forces. In choosing the design of the organization, managers consider the trade-offs between central control of organizational activities and overall market responsiveness.

Organizational design matches organization to the company's strategy. The critical design issue is defining the company's markets and then forming an organizational structure that helps the company serve those markets most effectively. Managers creating organizations dedicated to retail, wholesale, manufacturing, or R&D will select distinct structural forms. Managers operating closely related business lines or significantly diversified business lines will configure the organization's divisions differently. A manager pursuing a cost advantage, differentiation advantage or transaction advantage will adjust the organization appropriately.

Exhibit 1.3   Microsoft Reorganizations

Prior to 1999, Microsoft had been organized in three technology-oriented divisions: operating systems for personal computers, applications such as word processing and spreadsheet programs, and Internet-related businesses. Then, in March of that year, Microsoft reorganized the company into five units to better reflect its five core businesses.

(Continued)

Exhibit 1.3 (*Continued*)

Microsoft identified the company's five core businesses as the Windows operating system for consumers, applications for small business and home office knowledge workers, software for information technology use by large organizations, tools for software developers, and e-commerce applications including Internet portals and access service.[24] The goals of the company were to maintain or achieve a winning position in each of these businesses. In particular, the company sought to maintain the position of Windows as the leading operating system for personal computers in the world. After the company's reorganization, the five main divisions of the reorganized company mirrored its five core businesses.[25] They included

1. The Consumer Windows Division, which provided Windows software.
2. The Business Productivity Division, which offered business applications such as Microsoft Office; server applications including BackOffice, Small Business Server, and Exchange; and software for appliances (Windows CE).
3. The Business and Enterprise Division, which dealt with the information technology needs of large organizations and development of the Windows product line.
4. The Developer Group, which offered products for software developers that included SQL Server, COM+, Visual Basic, and Visual C++.
5. The Consumer and Commerce Group, which offered Commerce Server and WebTV products and operated the Microsoft network.[26]

The Microsoft reorganization in 1999 showed how a company's vision of its core businesses are reflected in its organizational divisions.

With the development of the Internet, Microsoft's vision changed again, going beyond its Windows-based approach. The emphasis on software for use within business enterprises led to the formation of the company's ".Net" strategy. The company's increased focus on the Internet and on software for use within business enterprises was then reflected in a new configuration of Microsoft's businesses. The company's three main product segments were as follows: (1) Desktop and Enterprise Software and Services, (2) Consumer Software, Services, and Devices, and (3) Consumer Commerce Investments.

(Continued)

Exhibit 1.3 *(Continued)*

According to Microsoft, its business groups and its product segments were once again closely aligned:

1. Desktop and Enterprise Software and Services. Associated with this product segment were two divisions: the Platforms Group and the Productivity and Business Services Group. The Platforms Group focused on the Windows platform and software for enterprise servers. According to the company, the Productivity and Business Services Group "drives Microsoft's broad vision for productivity and business process applications and services," including Microsoft Office and other business applications.
2. Consumer Software, Services, and Devices. Associated with this product segment were three divisions of the company: The MSN Business Group developed television network programming. The Personal Services Group provided platforms and software for consumers using the Internet and mobile devices, and extended the company's ".Net" strategy to consumer devices. The Home and Retail Division was responsible for learning and entertainment software, and the Xbox game player.
3. Consumer Commerce Investments. Finally, according to the company, associated with this product segment are various products and services including Expedia Inc., the HomeAdvisor online real estate service, and the MSN CarPoint online automotive service.[27]

As Microsoft's activities and market vision changed, its organizational structure followed. The company's corporate strategies were reflected in the configuration of business groups it operated. The company's businesses strategies were reflected in the activities of the business units within each business group.

The Vertical Structure of the Firm

The boundaries of the firm are defined both by the company's product and service offerings and its functional activities. The manager considers the company's target markets and its product and service offerings in creating the divisions of the firm. The manager next considers what functional tasks are required to execute the company's strategy and chooses which of those tasks the organization will perform and which of those tasks the company's suppliers will perform. The choice of what tasks the organization will perform is a crucial determinant of the vertical structure of the organization.

Practically any of the company's functional tasks — R&D, finance, human resources, purchasing, marketing, sales, operations, and information systems — can be outsourced. Transaction costs are an important component of the manager's decision. The manager examines not only the direct costs of procuring a good or service from the company's suppliers, but also the indirect costs of creating market transactions. The manager then examines the costs of producing that good or service within the organization, taking into account not only the direct costs but also the indirect costs of expanding management responsibilities. The activities of the organization must be selected to minimize the combined costs of managing the organization and conducting market transactions.

The manager must then allocate functional tasks between the company's central office and its divisions. Tasks such as marketing, sales and operations can be centralized or decentralized, that is, split up between divisions. These decisions depend on the trade-off between the benefits of scale and coordination when functions are centralized and the benefits of market responsiveness when functions are decentralized. Finally, the organization must delegate authority to its employees and provide them with incentives to implement the strategy.

Having chosen the firm's market boundaries, management assigns tasks to members of the organization. Senior managers often retain a major share of strategy-making functions. They devise the company's strategy, design the organization, and define projects for employees. Managers monitor the performance of employees to make sure that the organization is executing the company's strategy and achieving its goals. For employees to carry out their tasks, management must delegate authority to employees and specify responsibility for satisfactory performance.

1.5 *Overview*

Business strategy continues to be increasingly sophisticated and the speed of competition continues to accelerate. Managers face challenges from technological developments such as electronic commerce and market developments such as the growth of global competition. Managers and entrepreneurs cannot expect to succeed with enthusiasm and guesswork. Effective management depends on knowing how to perform a strategic analysis, and then applying the strategy to face market competition and to lead the company's organization. The five basic steps of strategic analysis set out in this chapter are the main topics of the textbook.

Designing strategy requires formulating goals that specify what markets the company should serve. Thus, goals are forward-looking and outward-oriented. Part I of this text shows how to start the process of strategic analysis. This chapter introduces goal setting and the basic outline of the book. Chapter 2 explains value-driven strategy and examines how managers should choose goals.

Managers examine the company's actual and potential markets and the existing and desired characteristics of the organization. Part II introduces the concept of the market compass and considers the manager's external analysis. Chapters 3 and 4 provide basic tools for conducting the external analysis of the firm's actual and potential markets, including customers, suppliers, competitors, and partners.

Part III of the book introduces the concept of the organizational grid and examines the manager's internal analysis. Chapters 5 and 6 present concepts used in conducting the internal analysis of the organization, including structure, performance, abilities, and resources.

Part IV considers the company's sources of its competitive advantage. The manager considers what the company needs to do to win its markets or, if the company is already successful, continue to maintain its leadership position. Chapter 7 introduces the pivotal concept of creating value and obtaining competitive advantage. Chapter 8 considers the value chain and applies transaction cost analysis to the choice of the vertical boundaries of the firm.

Part V turns to the manager's formulation of the firm's competitive strategy. The manager formulates competitive strategies that are targeted to market segments and anticipate competitor actions. Chapter 9 looks at the firm's price leadership strategy when the firm builds on a cost advantage. Chapter 10 sets out the firm's product differentiation strategy when the firm builds on a differentiation advantage by creating and operating markets. Chapter 11 introduces the transaction coordination strategy when the firm builds on a transaction advantage. Chapter 12 examines entry strategies and their relationship to barriers to entry and mobility.

Questions for Discussion

1. Select a company and examine its annual report or visit its corporate website. Does the company specify goals or objectives? Does the company describe its target markets? Does the company provide a mission statement? Does the company's management articulate a vision of the industry?
2. What is the difference between a company's goal and its overall strategy?
3. How does the manager's choice of company goals affect the external and internal analyses? How would information from the

manager's external and internal analysis affect the manager's choice of a goal? How is the process of selecting goals connected with the external and internal analyses?

4. Which is more important for the manager's decision-making process, the external analysis or the internal analysis, or are both equally important?

5. Why does the strategy process consider the sources of competitive advantage? Should managers be concerned about the differences or similarities between the company and its competitors? Would it be sufficient to focus only on customers?

6. How are competitive advantages different from competitive strategies?

7. Why does the manager's strategy process take into account the design of the organization?

8. Select a specific company and list the businesses that the company is involved in. Consider the organizational structure of the company and list its divisions. Compare the two lists.

9. Consider the first two scenarios given at the beginning of the chapter, the entrepreneurial start-up and the growing regional business. How will the strategy process change in each case? Will you as a manager be choosing different types of goals for the company? Will your process of external analysis and internal analysis be similar or different? How might your analysis of competitive advantage and competitive strategy be similar or different in the two companies? Would you expect the two types of companies to have similar or different organizational structures?

10. Consider the fourth scenario at the beginning of the chapter. Suppose that you are not familiar with the countries that the client firm operates in. How might that affect your strategic analysis? What types of information would you want to obtain about the client firm and its markets?

Endnotes

1. This key insight originates with Kenneth R. Andrews, *The Concept of Corporate Strategy* (Homewood, IL: Irwin, 1971).

2. Keniche Ohmae, *The Mind of the Strategist: The Art of Japanese Business* (New York: McGraw-Hill, 1982). Ohmae emphasizes the distinction between strategy, which involves competitive actions that improve the firm's position relative to other companies, and management actions that improve profitability, streamline the organization, or increase the effectiveness of management.

3. Henry Mintzberg and J. A. Waters, "Of Strategies, Deliberate and Emergent," *Strategic Management Journal* 6 (1985), pp. 257–272, and

Henry Mintzberg, "Crafting Strategy," *Harvard Business Review* 65 (July–August 1987), pp. 66–75.

4. Robert Slater, *The New GE: How Jack Welch Revived an American Institution* (Homewood, IL: Irwin Business One, 1993), p. 80.

5. Ibid., p. 78.

6. Ibid., p. 84.

7. Ibid., pp. 84–85.

8. General Electric Annual Report, 1998.

9. Similar issues arise in military strategy where the terrain — landscape, forest, rivers, plains, and mountains — can have profound effects on the outcome of conflict. The relative strengths of fighting forces will differ as physical conditions vary; tanks may have an edge over infantry in flat plains, whereas infantry with anti-tank weapons has the advantage in the woods or hilly terrain. Strategy, maneuvers and tactics must be appropriate to the terrain. As the military strategist Liddell Hart observed, the essence of military strategy is concentration of strength against the opponent's weakness.

10. See Miguel Helft, "What a Long, Strange Trip It's Been for Webvan," *The Industry Standard*, July 23, 2001, and Saul Hansell, "Webvan Hits Technological Barriers and Ingrained Habits of Consumers," *New York Times*, February 19, 2001, p. C1.

11. See Ken Belson and Don Kirk, "Toshiba Will Abandon Commodity Chip Business," *New York Times*, December 19, 2001, p. W1, and "Toshiba to Drop Low-End Chips, Sell U.S. Plant to Micron," *The Wall Street Journal*, December 19, 2001, p. B6.

12. See Ian Foster and Carl Kesselman, *The Grid: Blueprint for a New Computing Infrastructure* (Morgan Kaufmann Publishers, 1999), and Pawel Plaszczak and Rich Wellner, *Grid computing*, (San Francisco: Elsevier/Morgan Kaufmann, 2005).

13. Josef A. Schumpeter, *History of Economic Analysis: Capitalism, Socialism and Democracy* (New York: Harper, 1942); Peter F. Drucker, *Managing in a Time of Great Change* (New York: Truman Talley Books/Dutton, 1995); Richard A. D'Aveni with Robert Gunther, 1994, *Hypercompetition: Managing the Dynamics of Strategic Maneuvering* (New York: Free Press, 1994); Thomas J. Peters, *Thriving on Chaos: Handbook for a Management Revolution* (New York: Knopf, 1987); Charles H. Fine, *Clockspeed: Winning Industry Control in the Age of Temporary Advantage* (Reading, MA: Perseus Books, 1998); Philip Evans and Thomas S. Wurster, *Blown to Bits: How the New Economics of Information Transforms Strategy* (Cambridge MA: Harvard Business School Publishing, 1999).

14. Michael W. Cox and Richard Alm, *Myths of Rich and Poor* (New York: Basic Books, 1999).

15. See Michael Gort and Steven Klepper, "Time Paths in the Diffusion of Product Innovations," *Economic Journal* 92 (September 1982), pp. 630–653; and Steven Klepper and Elizabeth Graddy, "The Evolution of New

Industries and the Determinants of Market Structure," *Rand Journal of Economics* 21 (Spring 1990), pp. 27–44.

16. See Adam J. Fein, "Understanding Evolutionary Processes in Non-Manufacturing Industries: Empirical Insights from the Shakeout in Pharmaceutical Wholesaling," *Evolutionary Economics* 8 (1998), pp. 231–270.

17. Competitive strategy has similarities to maneuver in military strategy. *Maneuver* is an important planning level between so-called grand strategy and tactics. According to Robert R. Leonhard, strategy is the plan of the war and tactics are chosen to win individual battles, but maneuvers use battles as building blocks to carry out strategy, see Robert R. Leonhard, *The Art of Maneuver* (Novato, CA: Presidio Press, 1991). Leonhard observes that "the greater part of maneuver warfare takes place at the operational level". In military strategy, "operations" is the corresponding intermediate level of planning between strategy and tactics. Strategy forms the plan of the war and tactics are the individual battles and engagements. Operations constructs campaigns, which use battles as building blocks. Leonhard points to the Battle of Bull Run (or Manassas Junction) as an example of a battle that should never have been fought, as part of a misdirected campaign to capture Richmond. The North's strategy was the Anaconda plan to close off Southern ports and isolate the Confederacy by interdicting the Mississippi and other routes of transportation and communication. The Anaconda plan was the strategy that eventually would win the war for the Union. The problem was not simply that the North lost the Battle of Bull Run; the fault was that the battle and the campaign for Richmond were not in the service of the overall strategy. Maneuvers are designed to carefully select the battles to be won and those to be avoided: "history abounds with examples of battles that should not have been fought, that were irrelevant to the outcome of the campaign" (ibid.).

18. Lawrie Mifflin, "Fox Powers Way to Top of Children's TV," *New York Times*, July 3, 1995, p. C1.

19. Ibid.

20. Ibid.

21. Bill Carter, "CBS-Fox War of Words Extends to the Top Ranks," *New York Times*, July 25, 1994, p. C1.

22. See http://www.pepsico.com/.

23. See http://www.nikebiz.com/story/pr_make.shtml.

24. John Markoff, "Microsoft Will Reorganize Into 5 Units," *New York Times*, March 30, 1999, p. C2.

25. Microsoft also had a Streaming Media Division to develop and market Windows Media Technologies. In addition, the company had three centralized functional units: research, worldwide sales and support (including its enterprise, education and organization customer units and a Home and Retail Products Group that offered games, input devices and

reference products such as the Encarta Encyclopedia), and operations (which includes finance, administration, human resources, and information technology).

26. John Markoff, "Microsoft Will Reorganize Into 5 Units," *New York Times*, March 30, 1999, p. C2, and www.microsoft.com/presspass/cpOrg.htm.

27. The discussion of the 2001 organization is based on company statements in its 2001 Annual Report, microsoft.com.

CHAPTER 2

VALUE-DRIVEN STRATEGY

Contents

Value-driven strategy is a method for choosing the appropriate goals and strategies. The manager evaluates alternative goals and strategies based on whether they increase the total value of the firm. This chapter sets out the basic framework for value-driven strategy. After completing the chapter, you will have some of the necessary tools for choosing goals and strategies for competitive advantage.

The *total value of the firm* is the present value of the firm's economic profits over the long term. Maximizing the total value of the firm provides the manager with a consistent and reliable way to evaluate goals and strategies. The company chooses its goals, that is, the markets it wishes to serve, based on their contribution to the total value of the firm. In addition, the strategies to achieve those goals are chosen to maximize the total value of the firm.

The manager's choice of goals and strategies should make the *best match* between the company's organizational abilities and its market opportunities. The manager integrates the information obtained from the external analysis and the internal analysis. The interplay between market opportunities and organizational abilities is the basis of effective strategy.[1]

The strategy process often requires changes in the company's direction. Making the best match between organizational abilities and market opportunities involves two types of choices. The first type of choice is external: Should the company continue to serve its existing markets or should it enter new markets? The second type of choice is internal: Should the company rely on its existing organizational abilities and structure or should it develop new resources and

37

competencies and change the firm's organizational structure? This chapter extends the basic framework for choosing goals to dynamic strategy making.

Chapter 2: Take-Away Points

Managers should follow value-driven strategy by selecting goals and strategic actions to maximize the total value of the firm:

- The total value of the firm is the present value of the firm's economic profits over the long term.
- The company's goals and strategies should make the best match between the organization's abilities and market opportunities.
- The attributes of the organization should be evaluated in the context of the company's market opportunities.
- Market opportunities should be evaluated in terms of their contribution to the total value of the firm, given the organization's abilities to attain those opportunities.
- The 4R framework should guide the manager in dynamic strategy making: The manager chooses whether to remain on course, reposition the company, redesign the organization, or restructure the company.
- Managers should choose goals and strategies respecting ethical and social norms as well as legal, regulatory, and social concerns.

2.1 *Selecting Goals*

Management strategy celebrates the diversity among companies. No two organizations are exactly alike; they differ in terms of administrative structure, management and employee abilities, knowledge, corporate culture, traditions and routines. Companies seeking competitive advantage emphasize and create differences by emulating and improving on what others are doing and setting off in new directions. The manager should select the company's goals and strategies to make the match between organizational abilities and market opportunities.

How are new directions chosen? Which market opportunities should a company pursue? The organization may be particularly qualified to serve certain markets better than others. The company wishes to extract the greatest return from its existing resources and competencies. Choosing what markets to serve depends on which opportunities the company is best suited for and the organization's capacity for change, so the existing organization imposes constraints on the choice of objectives.

Maximizing the Total Value of the Firm

Maximizing the total value of the firm gives managers a consistent way to evaluate the company's goals. At the corporate level, managers should assemble a set of businesses that maximizes the total value of the firm. At the business level, managers should choose to serve those markets that yield the greatest value for the firm. The manager chooses business strategies to achieve the company's goals in a way that yields the greatest total value for the firm.

Maximizing the value of the company is not the goal in itself. Instead, the value of the company is a way of *evaluating* alternative goals — measuring potential success. Thus, "making money" is not a goal — establishing and operating a successful business that serves customers is a goal. Goals are specific business objectives. Consider for example, Atlas Air, the largest air cargo outsourcer in the air freight industry. In determining whether to enter into a long-term contract to supply freight outsourcing services to a major international airline, Atlas Air must consider whether or not the contract adds to the value of the firm.[2] The company's goal is to provide air freight services. Managers evaluate alternative contracts based on their contribution to the value of the firm.

The *total value of the firm* is the present value of the stream of profits over the long term, as Figure 2.1 illustrates. Maximizing the value of the firm requires obtaining the greatest present value of economic profit. Economic profit refers to the revenues of the firm net of all costs, including the costs of labor services, resources such as land and energy, services, manufactured inputs, technology, capital equipment, and

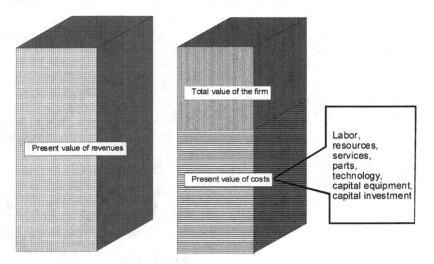

Figure 2.1: **The total value of the firm equals the present value of the stream of revenues over time minus the present value of the stream of costs over time.**

capital investment. The firm's value or long-run economic profit is calculated by discounting the company's cash flow at the appropriate rate of interest, thus taking into account the time-value of money. The rate of interest used to discount earnings should be the company's cost of capital, which reflects the riskiness of its business.

Management strategy is of course consistent with finance. The total value of the firm is the total of the net present value (NPV) of the firm's investment projects. The net present value of an investment project is the present value of operating profits minus the capital expenditure. Another way to look at the total value of the firm is that it is equal to the value of the firm's debt and equity (see Figure 2.2). For publicly-traded firms, the value of the firm's shares is equal to the total value of the firm minus the value of its debt. The value of the firm's shares is referred to as shareholder value. For privately held firms, the value of the firm to its owners is equity, the total value of the firm minus its debt obligations.

Value-driven strategy applies whether the company is privately-held or publicly-traded. To the extent that the capital markets function efficiently, the market value of the firm's debt and equity adjusts to equal the total value of the firm. The value of debt and equity do not determine the underlying value of the firm, they reflect it. The market's expectations about the performance of the firm's businesses and the effectiveness of the company's strategy determine how investors will value the firm's debt and equity. Corporate finance is a bit like owning a house. The value of the house is determined by its location, the size of the yard, and the quality of the building. The homeowner's equity is the value of the house minus the mortgage debt.

The market value of the equity of a firm's owners is the total value of the firm minus debt. So, with fixed debt obligations, maximizing the

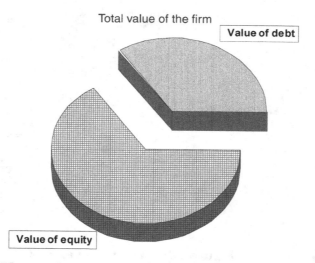

Figure 2.2: **The market value of the firm's debt and equity will reflect the underlying total value of the firm.**

total value of the firm also maximizes the value of the owner's equity. If the firm is publicly-traded, the stock market values the firm's equity equal to the long-term value of the firm, net of debt obligations. The price of a share of stock is simply the firm's equity divided by the number of shares. If the firm is privately-held, the market value of the owner's equity would be the total value of the firm minus its debt. Debt refers to corporate bonds, bank debt, and trade credit from suppliers.

Why is the total value of the firm the right way to evaluate the company's goals? Companies that fail to be profitable go out of business, so that activities that do not maximize the company's total value are likely to result in bankruptcy. Also, managers of publicly-traded companies that do not maximize shareholder value will often be replaced either by boards of directors or through the market for corporate control. Managers of privately-held companies that do not maximize value for their owners are also likely to be replaced by company's owners.

Evaluating goals based on profits means that the company makes the best use of its resources, including investment capital. The company must take into account the cost of natural resources, manufactured inputs, and labor services to promote productive efficiency and avoid wasting resources. Also, maximizing the value of the firm means that the company will serve those markets at which it earns the greatest profit, thus allowing the firm to be guided by customer needs. Companies that try to maximize profit necessarily respond to market forces, only serving those markets in which the company adds value relative to its competitors.

Profit-maximization does not mean earning the highest possible profit in any given year, but rather obtaining the highest present value of the stream of profits going forward. The need to take a long-term view is exemplified by William W. George, chairman and chief executive officer (CEO) of Medtronic, a leading medical technology company. He joined the company in 1989 when it was worth about $1 billion, and retired as CEO when the value of the company's equity exceeded $60 billion. George was fond of telling the company's shareholders and Wall Street analysts that "Medtronic is not in the business of maximizing shareholder value. We are in the business of maximizing value to the patients we serve". George pointed out that "a corporation's ability to survive and prosper is directly correlated with how well it serves its customers". According to George, companies should "pursue a worthy mission with a passion that inspires your employees, stay true to your purpose, practice your values with an unyielding consistency, and employ an adaptable business strategy to match changing market conditions". George concludes that by following these rules, "the long-term increases in shareholder value will far surpass the highest expectations of those who would seek only to maximize shareholder value".[3] George's observations emphasize that an increase in a company's stock price is not the company's goal, only a measure of the company's success in serving its markets.

Closely related to choosing goals, also referred to as purpose or objectives, managers set out a *mission* that articulates the nature of the business and its general commitment to performance in terms of product quality, customer service, and innovation. For example, Microsoft's expressed mission was "to create software that empowers and enriches people in the workplace, at school and at home". In electronic commerce, Microsoft's Ballmer sought to provide "a very rich infrastructure — stable, secure and enterprise-ready, for supporting the kinds of applications you want to deploy on the Internet".

Managers also articulate a *vision* that describes the future products, technology, and organization of the industry, and implicitly specifies the company's anticipated contribution. Top managers are charged not only with choosing the company's goals, but also with clearly communicating them to the organization. Defining and communicating the company's goals is an essential management tool. The strategy process creates opportunities for employees' and managers' participation. It is also a coordination device that motivates employees to strive for the common goal. Scott McNealy, chairman of Sun Microsystems, notes, "The bigger the organization, the more you need a vision and an architecture to allow people to make decisions that are all compatible".[4]

Exhibit 2.1   Lou Gerstner at IBM

When Lou Gerstner took the helm at IBM he said, "The last thing IBM needs right now is a vision". Instead, he emphasized the need to cut costs and focus on the company's markets.[5] Many criticized him for this statement but other leaders in the industry said similar things. Andy Grove, CEO of Intel, said: "Nobody in this business can have a sophisticated technological vision. This is a big-time fishing expedition. You cast a bunch of hooks in the water; some get nibbled on, some catch fish, some you reel in, some you let go".[6] Bill Gates of Microsoft agreed. "Being a visionary is trivial. Being a CEO is hard. All you have to do to be a visionary is to give the old 'MIPS to the moon' speech — everything will be everywhere, everything will be converged. Everybody knows that. That's different from being the CEO of a company and seeing where the profits are".[7] After his first year at IBM, Lou Gerstner introduced his remaking of IBM to its

(Continued)

Exhibit 2.1 (*Continued*)

shareholders with the observation, "Some call it mission. Some call it vision. I call it strategy".[8]

Gerstner observed that the company must do "three simple things" to succeed: win in the marketplace, execute its plans with a sense of urgency, and work more as a team. "This is just like *Coke* versus *Pepsi*, or *McDonald's* versus *Burger King*",[9] he said, introducing the basic outline of a difficult plan.[10]

Gerstner set goals for IBM by choosing what markets to enter and exit. IBM attempted to shift from mainframes and minicomputers to *client-server computing*. The "clients" are personal computers linked in a network to "servers", powerful computers that provide data and application programs to the desktop PCs. In addition, IBM would focus on its fastest growing segment, selling services to business. The goals might be summarized as servers and services.

Having set the goals, Gerstner outlined the broad strategy to achieve them. Providing client-server computing would require selling both the hardware and software to companies building networks of PCs, a particularly challenging prospect because of the host of strong competitors in client-server software, including Microsoft. On the hardware side, IBM faced intense rivalry from leading server makers Compaq, Hewlett-Packard, and others. Finally, any early success in this segment would come at the expense of its own mainframe business.

Providing business services would require not only supplying IBM equipment and software but also creating supplier networks so that IBM's consultants could solve customer problems rather than pushing IBM products. In this area, IBM would take on a retailer role by acting as a value-added reseller of networking products. The company set out to create high-speed data networks for companies and to develop hardware and software to enable its clients to enter into electronic commerce.

In terms of organizational structure, IBM would continue to be a vertically integrated company developing much of its own technology. Thus, IBM would use its core technology in more product lines, and sell more components such as microprocessor chips to outsiders. This means that the company had to produce components that were competitive with those of rivals in the marketplace rather than relying on internal adoption of its own components to give these divisions a competitive edge. It would replace mainframe computers, whose sales had fallen drastically,

(*Continued*)

Exhibit 2.1 (*Continued*)

with microprocessor-based machines. Most significantly, IBM sought to continue its vertical integration, rather than to rely heavily on outside suppliers. IBM tried to place the PowerPC chip, developed jointly with Apple and Motorola, throughout its product line, but remained heavily dependent on Intel. Not only would it compete with Intel and others in the chips market upstream, IBM would try to take on personal computer makers such as Compaq, Hewlett-Packard, Dell, and others.

Gerstner's strategy took years to bear fruit, but it seems particularly prescient. By the turn of the century, IBM's PowerPC chip had little success in computers, being largely confined to Apple's Macintosh line, but it had been chosen by Nintendo for the next generation of high-powered video game machines. IBM's sales of microchips for cellular phones and other wireless communication devices were impressive. IBM's servers made significant headway against competitors. IBM had become a leader in the provision of hardware and software for electronic commerce. Finally, IBM's service offerings were growing successfully.

Looking back, Gerstner observed that stage one in his turnaround plan had been simply survival. Stage two was putting the company's e-commerce strategy in place. Not content with the company's apparent success, Gerstner embarked on a broad reorganization of IBM that represented the "third turn of the wheel". Gerstner observed that "I've been absolutely convinced that you've got to blow things up and start over again every few years, and that puts a whole new face on people's jobs. It gets people focused externally rather than internally". The change in the company's organizational structure was targeted at implementing the company's Internet strategy in the most effective way. Gerstner emphasized the "linkage between structure and strategy, and that you really don't have an effective strategy without a structure that reinforces and complements that strategy". The management shakeup that accompanied the reorganization prepared for a successor; as Gerstner noted: "a new generation of leaders is starting to emerge at I.B.M".[11]

The Need for a Consistent Measure of Value

Maximizing the total value of the firm allows the manager to select the company's goals in a consistent manner. The company cannot pursue every market opportunity. Companies encounter constraints on their resources that limit the number of markets they can serve at any given time. These limits require managers to make *choices* between markets that the company could serve. Management strategy always involves recognizing *trade-offs* and choosing the best markets to serve.

The process of goal selection often will yield multiple alternatives. How does the manager choose among competing goals? Managers need a *consistent* way to compare goals. The manager must be able to compare very different goals; for example, the company might have to choose between expanding an existing business and establishing a new business. The company might have to choose between supplying a manufactured product or supplying a customer service. Whatever the choices are, the criterion used must be consistent over time so the manager can evaluate the company's actual performance relative to its anticipated performance.

To select the company's goals, the manager must determine the best markets to be served by the company. Managers cannot make a decision based on multiple factors, say for example profits and revenue, because choices can only be ranked based on a single measure of performance. Suppose that serving the North American market for a particular product would yield low revenue but high profit, while serving the South American market for that product would yield high revenue but low profit. What should be the manager's choice? If both profit and revenue are factors in the manager's choice, the manager cannot make a choice between these two alternatives.[12] Choosing between the two markets based on these two factors means that the manager would have to assign some weights to the factors. However, managers should avoid arbitrary weighted averages, such as some combination of profits and revenues, or even very complicated scoring systems. The manager still faces the problem of choosing the weights.

Even if the manager were to select a goal without spelling out the underlying objective, he or she would have implicitly chosen an objective. The manager that chooses between two goals based on a gut feeling has still compared the two goals. In the previous example, choosing to serve the North American market would indicate that the manager placed greater weight on profit whereas choosing the South American market would indicate that the manager placed greater weight on revenue. Choices reveal information about the manager's objectives, even if these objectives are not made explicit.

Some management recommendations suggest that managers simultaneously use multiple criteria to evaluate the company's goals. As Michael Jensen observes, the problem with such systems is that they do not yield a score to measure performance, but rather a dashboard or instrument panel that tells managers many interesting things about their business.[13] Of course, real-time information is highly valuable. Still, the manager must integrate the information to distinguish between alternative goals.

A single criterion for evaluating performance is also needed to motivate managers and employees of the company. When given multiple tasks, decision makers tend to pursue those that yield the greatest rewards and neglect the other tasks. The tasks that are chosen may not reflect the interests of the company. Properly aligning the interests of managers and employees with those of the company requires a clear and unified performance measure. Accordingly, managers should consistently follow a single overall objective.

The main point about measuring value is that alternative measures of performance tend to conflict. Managers cannot serve every market because the firm's resources are necessarily limited. Each market offers the firm different types of opportunities. The company might earn greater expected profits but face higher risks in some markets than in others. Some markets may offer the firm a high rate of growth whereas other markets will yield steady earnings. The firm may expect to have a higher market share in some markets and it may serve a high-quality niche in other markets. The firm might be a technology leader in some markets and a cost leader in others. Maximizing the total value of the firm provides a single criterion that cuts through these difficult trade-offs.

2.2 *Matching Organizational Abilities with Market Opportunities*

The manager chooses *value-driven strategy* to maximize the total value of the firm. Value-driven goals and strategies make the best *match* between organizational abilities and market opportunities. The manager integrates information generated by both the internal analysis of the organization and the external analysis of the company's potential markets. Using this information, the manager identifies combinations of organizational abilities and market opportunities that can create competitive advantages for the company.

Value-Driven Strategy: The Basic Framework

The manager's external analysis yields information about market opportunities. An automobile manufacturer might be interested in the growth of the markets for sports cars or for sport utility vehicles. The manager's internal analysis yields information about organizational abilities. The automobile manufacturer might observe the ability of its organization to design stylish automobile bodies or to improve on engine performance. Value-based strategy involves putting the two sources of information together to effectively choose the firm's goals and the strategies to achieve those goals. Thus, the manager of the auto company tries to determine if the firm can provide the market with a competitively styled vehicle with the necessary engine performance.

The company's goals are the targeted market opportunities, say the United States market for sports cars. The company's strategies for achieving those opportunities include decisions about prices, products, technologies, and input purchases. The company's goals and strategic choices require evaluating the company's organization. For example, does the company have the ability to produce the necessary body style and engine to provide the sports car? The organizational abilities are inputs needed to produce the products used to realize the market opportunities.

The manager often must make difficult choices, selecting some market opportunities over others and emphasizing some organizational abilities over others. The value of market opportunities depends on the organization's abilities. The value of organizational abilities depends on the market opportunities. The company evaluates alternative goals and strategies based on their potential for increasing the total value of the firm (see Figure 2.3).

The basic framework of value-based strategy can be summarized by a grid showing the payoffs in terms of the value of the firm in dollars. Suppose that the manager must choose between two market opportunities and between two types of organizational abilities. The numerical example in Table 2.1 makes two important points. First, notice that neither of the market opportunities is inherently best. Market B is better for a firm with Ability 1 and Market A is better for a firm with Ability 2. This implies that market opportunities cannot be evaluated in the absence of the firm's organizational abilities. Second, notice that neither of the organizational abilities is inherently best. Ability 1 is better for a firm serving Market B and Ability 2 is better for a firm serving Market A. This implies that organizational abilities cannot be evaluated without the context of the market opportunity. Putting it all together, the best match is for the firm to serve Market B using Ability 1, since that maximizes the value of the firm. The information in Table 2.1 is shown as a *value map* in Figure 2.4.

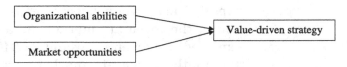

Figure 2.3: **Value-driven strategy: Choosing goals and strategies to create competitive advantage.**

Table 2.1: Value-driven strategy.

The manager makes the best match of organizational abilities and market opportunities. The payoffs are the value of the firm with different combinations of organizational abilities and market opportunities.

	Market Opportunities	
Organizational Abilities	**Market A**	**Market B**
Ability 1	50	*100*
Ability 2	65	40

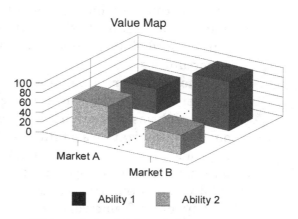

Figure 2.4: Value-driven strategy.

A value map shows the contribution to the firm's value of matches between organizational abilities and market opportunities.

Example 2.1 Products and Services

A manager has a group of highly skilled and technically trained personnel in the information technology (IT) field. The manager must decide whether to deploy the company's personnel as a sales force for IT products or as an IT consulting group providing services. To obtain the benefits of specialization and to avoid conflicts of interest, the manager must choose between sales and services. The personnel could operate either as a sales force selling equipment wholesale to resellers or as a sales force serving corporate customers, but not both. Alternatively, the personnel could provide consulting services to resellers or they could provide consulting services to corporate customers. In sales, the company's expected net revenues would be $75 in the wholesale market and $60 in the corporate market. As a services group, the company's expected net revenues would be $40 in the wholesale market and $90 in the corporate market. Accordingly, the manager would choose to provide consulting services in the corporate market.

Companies often use *multiple* organizational abilities to seize a set of market opportunities. Intel combined its organizational abilities with its identification of an original market opportunity by designing a new generation of computer chips targeted at mobile communications. Intel's chip design drew on the company's innovation and manufacturing expertise by combining three functions: communications tasks, multimedia capabilities, and extended battery life. The company used the three combined functions to address an original market opportunity: a multi-function processor allowing the convergence of mobile phones and handheld computer devices. Devices based on the Intel chips can handle Internet access as well as audio, video and gaming applications. According to Intel, the company has an advantage over its competitors in developing such solutions for a host of original equipment manufacturers, including Acer, Casio, Compaq, DaimlerChrysler, Hewlett-Packard, Hitachi, NEC, Symbol Technologies, and Toshiba and hundreds of software applications developers such as Adobe, Macromedia, PacketVideo and RealNetworks.[14]

Matching organizational resources and market opportunities involves greater strategic complexity than actions based on either organizational attributes or market features alone. This makes value-based strategy more difficult for the company to implement but also difficult for rivals to copy. Competitors must first identify the critical combinations of organizational abilities and market opportunities. Then, competitive attempts to equal or surpass these combinations require coordinated organizational processes and market actions.

Value-driven strategy then follows the following steps:

- Determine the firm's key abilities through the internal analysis.
- Determine the firm's main market opportunities through the external analysis.
- Choose goals that make the best match between organizational abilities and market opportunities based on maximizing the total value of the firm.
- Choose strategies to attain the firm's goals that maximize the value of the firm.

Evaluating Organizational Abilities

Evaluating organizational abilities depends on market opportunities; the manager cannot evaluate the organization's resources and competencies in isolation. Appraising the company's resources and competencies requires knowing their net contribution to the firm's value. Even if the company is the world's best maker of horseshoes, such an ability has little value if most people drive cars. The company's abilities must be suited to providing goods and services that customers want.

The manager should evaluate organizational resources in terms of their potential revenue-generating ability and the costs of employing those resources. This means that the manager cannot evaluate the company's resources and abilities just on the basis of the internal analysis. The company's external analysis is necessary to help determine the value of organizational attributes.

Organizational abilities and resources could be simply inputs into production, such as the services of capital equipment owned by the firm, or they could be much more complex intellectual property, organizational processes or production technologies. Strategy researchers have emphasized the importance of having organizational resources that are scarce, difficult to imitate, and without close substitutes.[15] It is not possible to know the value of the assets to the company without specifying how the company intends to use the asset. The question the manager faces is how to evaluate the strategic contribution of these resources.

The value of the assets the company owns is based on the best use of those assets. The manager should examine the incremental profits that can be generated using the assets, as Figure 2.5 shows. The cost of using the assets is not what the company originally paid for them, but rather the opportunity cost, which is measured by the current market value of the assets. If the incremental operating profits from using the assets are greater than the market value of the assets, then the firm should use the asset. The firm should employ the asset to target the market opportunity that yields the greatest value for the firm. Companies that own petroleum or natural gas deposits, such as BP or Shell, compare the operating profits they can realize from developing those resources with the market value of those resource deposits.

Critical strategic decisions about deploying resources are easy to observe when firms face bankruptcy. For example, Global Crossing established a unique international fiber optic telecommunications network, which was the firm's critical asset. After the firm went bankrupt,

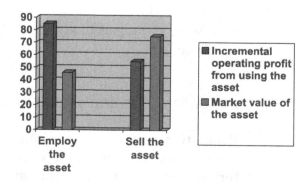

Figure 2.5: Asset value based on usage.
In deploying assets owned by the firm, the manager compares the incremental profit from using the asset to the market value that can be realized by selling the asset.

creditors debated whether the greatest value would be realized by operating the network or by liquidating the asset.[16]

The decision to employ resources also depends on the market value of the firm's alternative activities. Consider a restaurant that owns a high-quality pizza oven in good working condition. Suppose, moreover, that personnel at the restaurant are skilled in making pizza. This does not imply that the restaurant should make pizza. The restaurant should offer pizza only if its customers prefer to eat pizza and if the restaurant can earn more making pizzas than it could by selling off the oven and employing its personnel in other activities.

A high-quality resource will yield economic rent only when employed in the right market context. A cement company should not make music even if it were to own a Stradivarius violin. Even though the instrument is rare, it would not confer a competitive advantage on the cement company. In contrast, the Chicago Symphony Orchestra, which does own a Stradivarius violin, makes superb use of the violin in its performances. Resources must be complementary to the firm's activities to increase the value of the firm.

The manager should evaluate the organization's abilities and resources in terms of their market potential. Unique resources and competencies can generate returns for some period of time. However, reliance on the company's existing resources and competencies does not guarantee competitive advantage. Such advantages can be eroded in many ways. The manager should apply measures of expected costs and revenues.

The company may own popular brands that distinguish the company's products and generate customer demand. To obtain the greatest total value for the firm, the manager should compare the returns from employing the brands to sell products to the alternatives of retiring the brands or selling them off. Changes in consumer tastes can reduce the appeal of brands. Competitors can imitate brands or create more attractive products. Procter & Gamble has substantial expertise in developing and marketing brands of household items, with over 300 brands and 25,000 patents. Only 10 percent of the company's patents are used in its products. P&G's CEO Durk Jager announced the company's intention to give away or license many of its patents to combat the company's slow rate of innovation and development. For example, although P&G spent many years developing Febreze, a product that reduces odors in fabrics, Johnson & Johnson and Clorox offered competing products within a year of the introduction of Febreze, thus reducing the returns to P&G's product innovation.[17]

The company might have unique technological knowledge that is not available to competitors and is difficult to copy. Special skills in product design, R&D, manufacturing, customer service, and management attract customers and yield operating efficiencies. Some organizational abilities are the result of training, experience and skilled personnel that are difficult for rivals to imitate. Some organizational resources and abilities are difficult to observe due to organizational complexity, so competitors cannot know precisely what is responsible for the company's successful performance.[18] Critical patents preserve unique features of a company's products and services. Patents also cover production processes that improve the company's products or lower production costs. The value of the company's technical knowledge depends on the incremental profits that the company can obtain producing products and services using that knowledge. Technological change, however, can render patented production processes obsolete, and competitors can invent around patents. Organizational resources and competencies can be learned or surpassed by competitors through creativity or trial and error. The manager should consider market opportunities in determining the strategic value of the company's technology.

Consider the choice between producing a high-quality product that cannot easily be matched by competitors or a more standard product. The high-quality product makes full use of unique abilities of the company's employees and unique technological resources available to the company. Producing the standard product depends less upon the abilities of employees and technological resources of the company. Which option should the manager choose? If the manager is driven only by organizational considerations, the high-quality product might appear more attractive. Value-driven strategy depends on a consideration of expected revenues and costs.

Example 2.2   The Benefits of Standard Products

A company can produce two types of computer chips. The high-quality type costs $40 to produce per unit and the standard type costs $25 to produce per unit. The high-quality computer chip can be sold for $70 per unit while the standard type of computer chip can be sold for only $60. If the company can produce only one type of computer chip, which type should it produce? The profit per unit is $30 for the high-quality chip and $35 for the standard type of chip. In this case, the company should produce the standard type of chip. Should the company change its strategy if high-quality chips can be sold for $80?

The company's technological skills should be directed at satisfying *the needs of customers* rather than the interests of the company's engineers and product designers. An example of misdirected technological expertise is Motorola's Iridium project — a system of satellites covering the globe to enable use of a worldwide phone. The global communications system cost $5 billion but was sold for a mere $25 million. Before it found a buyer, Motorola engineers planned to let the satellites burn up on re-entry to earth. The company would have needed 1 million customers to break even, but the worldwide phone was bulky and expensive and appealed to very few users. "This was a technology that worked", according to a Motorola spokesman Scott Wyman. "But Iridium management pursued a marketing plan that in retrospect may not have been targeted at the most attractive available market".[19]

Thus the manager should choose between alternative technologies in terms of their performance in meeting customer needs. This requires considering expected costs and revenues. Suppose that the manager must choose between developing a cutting-edge technology that cannot easily be imitated or surpassed by competitors or using an off-the-shelf technology that is widely available. The manager knows that the cutting-edge technology draws on the abilities of the company's employees and unique company resources including patents and proprietary information. Employing the off-the-shelf technology does not draw extensively on these company resources. Which technology should the manager choose? If the manager is driven only by organizational considerations, the cutting-edge technology might be more appealing. The value-driven solution requires consideration of expected revenues and costs.

Example 2.3   Choosing the Appropriate Technology

A company can choose between two alternative technologies for delivering a service to its customers. The company may develop a cutting-edge technology that delivers a superior service to its customers. The cutting-edge technology is difficult for competitors to imitate and draws on unique engineering and scientific know-how within the company's organization. The cutting-edge technology has a development cost of $2,000 and an operating cost of $10 per customer. The company expects to price the service at $15 and to attract 1,000 customers with the resulting superior service. The company can use an alternative off-the-shelf technology that requires no development costs and has an operating cost of $5 per customer. The company expects to price the generic service at $10 and to attract 500 customers. Which technology should the company choose? In this case, the value of the cutting-edge technology is $3,000 while that of the off-the-shelf technology is $2,500, so the company should choose the cutting-edge technology. Would your answer change if the off-the-shelf technology could be used to serve 800 customers?

When purchasing inputs, the manager should compare the market inputs with organizational alternatives. Does the organization have resources that would substitute for the purchased inputs? Is it more effective to rely on market services or organizational resources? The market alternative may be superior to organizational solutions when the firm requires flexibility and when it can rely on innovative and efficient suppliers. Organizational alternatives may be preferable when there are high transaction costs of using markets or when the organization has innovative and efficient internal functions.

Exhibit 2.2   Ford and the Palladium Problem

Ford Motor Co. paid a heavy price when it relied on the firm's purchasing department rather than its internal research capabilities. Ford used the precious metal palladium in its catalytic converters, a part of the exhaust system that reduces air pollution. The control of air pollution was essential for Ford to comply with Federal regulatory standards and to maintain its environmentally-friendly publicity campaign. When the Russians held up shipments of palladium, the market price rose substantially and Ford began to stockpile palladium and to lock in long-term contracts for the metal. According to *The Wall Street Journal*, estimates were that Ford had stockpiled over two million ounces of the stuff. The market price of palladium hit $1,094 per ounce in January 2001. General Motors was less affected since it had locked in supplies before the price spike. Perhaps more importantly, GM executive David Andres had given "the Armageddon speech" to that company's engineers about the need to find ways of using less palladium.

Ford's researchers were able to reduce the company's dependence on palladium by the use of rare-earth elements to improve the effectiveness of precious metals such as palladium and by improvements in the way that metals are spread within the converter. Ford's research breakthroughs occurred just as increased supplies and falling market demand lowered the price to about $350 per ounce. Suppliers in South Africa increased production in response to high market prices. Demand fell due to the business cycle and similar research breakthroughs at other auto companies. Ford was forced to write off over $1 billion in the value of its palladium stockpile. The company might have avoided some of its stockpiling of palladium by better coordinating market purchasing decisions with the company's research efforts.[20]

Evaluating Market Opportunities

The manager's external analysis provides information about potential customers, suppliers, competitors, and partners. A market might appear *attractive* for many reasons. The manager may learn that in a particular market, customer demand is growing rapidly and that customers have a high willingness to pay for certain types of products. The company might have access to skilled, low-cost suppliers. There may be few competitors with high costs or obsolete products. There may be potential partners with high-quality products and loyal customers. Perhaps surprisingly, these market features are not sufficient to make the market attractive.

Industry attractiveness requires comparing the organization's abilities with competitors. The firm examines whether it has competitive advantages based on production costs, product differentiation, or transaction costs. The firm determines what industry to enter not on the general attractiveness of the industry but on the basis of the industry that generates the highest value of the firm.

Determining industry attractiveness requires integrating information from the external analysis and the internal analysis. Does the organization have the resources and abilities to compete for the market? Although an industry such as computer software may be experiencing high rates of growth, the manager may find that the company does not have a competitive advantage in that industry.

A market with high demand might appear to be more attractive than a market with low demand. In the high-demand market, there may be more customers or customers may have a greater willingness to pay for a particular product. However, each firm should consider the relative costs of serving the two markets. Different costs may be associated with serving the two markets, depending on the firm's abilities and resources. The firm may know more about customers in one market than in the other. Depending on its characteristics, the firm may have greater transaction costs of serving one market than the other. As a result, it is not possible to know whether or not the high-demand market is more attractive without additional information about the costs to the firm of serving that market.

Example 2.4

A manager in Japan who is contemplating exporting the firm's product must choose between two foreign markets, one in Europe and one in South America. In the European market, customers are willing to pay $75 for a certain type of product and in the South American market, customers are willing to pay $50 for a similar

(*Continued*)

Example 2.4 (Continued)

product. Suppose that the firm has the same costs of production and transportation in serving either market. However, because the firm has greater knowledge of the South American market, it will encounter transaction costs of $60 to serve the European market and transaction costs of $30 to serve the South American market. Accordingly, the manager should serve the South American market, since the additional value for the firm is $15 for the European market and $20 for the South American market. The South American market is more attractive because of the combination of market demand and the firm's ability to transact there.

The company's goals and its strategies depend on a comparison of the organization with the company's competitors. Although a global power in fast food, McDonald's experienced difficulties in entering the Philippines. A first mover in fast foods, McDonald's had the benefit of its worldwide economies of scale in marketing and extensive franchising experience. However, local rival Jollibee produced foods better suited to local tastes, and became the fast-food market leader in the Philippines.[21] Jollibee dominated the market with such products as Honey Beef Rice, Amazing Aloha (pineapple and bacon on a burger), Palabok Fiesta (shrimp, pork and smoked fish over rice noodles), and a desert of Peach Mango Pie.[22] The rapidly growing Filipino market became less attractive to McDonald's because its ability to adapt to local tastes did not compare well with that of Jollibee.

Executing strategies to attain the company's goals entails costs. The company's goals must be feasible, that is, the revenues generated by successfully meeting the goals should cover the company's costs. Some organizations make strategic plans without considering their feasibility. For example, the U.S. Congress spelled out a set of national educational goals but did not evaluate whether the goals were attainable and did not provide the resources that were necessary to achieve the goals (see Exhibit 2.3).

Exhibit 2.3 Education Goals 2000

Backed by President Bill Clinton, the U.S. Congress adopted a set of education goals for the year 2000. The goals extended six goals set out by President George Bush and the nation's governors in 1989 and were inspired by a report under President Ronald Reagan

(Continued)

Exhibit 2.3 *(Continued)*

titled *A Nation at Risk*. The goals were passed by Congress in 1994 with the strong support of both political parties.

Under the *Educate America Act* (HR 1804), the U.S. would achieve eight goals by the year 2000:

1. All children in America will start school ready to learn.
2. The high school graduation rate will increase to at least 90 percent.
3. All students will leave grades 4, 8, and 12 having demonstrated competency over challenging subject matter including English, mathematics, science, foreign languages, civics and government, economics, arts, history, and geography, and every school in America will ensure that all students learn to use their minds well, so they may be prepared for responsible citizenship, further learning, and productive employment in the nation's modern economy.
4. The nation's teaching force will have access to programs for the continued improvement of their professional skills and the opportunity to acquire the knowledge and skills needed to instruct and prepare all American students for the next century.
5. United States students will be first in the world in mathematics and science achievement.
6. Every adult American will be literate and will possess the knowledge and skills necessary to compete in a global economy and exercise the rights and responsibilities of citizenship.
7. Every school in the United States will be free of drugs, violence, and the unauthorized presence of firearms and alcohol and will offer a disciplined environment conducive to learning.
8. Every school will promote partnerships that will increase parental involvement and participation promoting the social, emotional, and academic growth of children.

The goals, while commendable, were at best wildly optimistic and at worst seriously misleading. Why would Congress adopt such unrealistic goals? Cynics might observe that politicians would promise the moon and the stars to get elected; idealists might suggest that setting high goals is the duty of elected officials. Looking ahead from the year 1994, the year 2000 must have seemed far enough in the future to lend some credence to the goals. Writing in the *New York Times*, Richard Rothstein observed that some of the Goals 2000 "were ridiculous in the first place".[23]

(Continued)

Exhibit 2.3 *(Continued)*

A progress report by the National Educational Goals panel evaluated the nation's performance. Some goals were within reach. According to the report, although high school graduation rates went down, more 18- to 24-year olds got General Education Development (GED) degrees so 15 states attained the 90 percent graduate rate. In other cases things simply got worse, as in the case of the goal to eliminate drugs and violence in schools.

In the report, Harvard's Professor Richard Elmore noted that "clear goals may not be good goals. Becoming first in the world in math and science is a clear goal but one that is not even remotely within range for the foreseeable future. This wording may focus attention on the failure to meet the goal, rather than monitoring and promoting actual progress that is being made".

It is not clear whether setting such unrealistic goals is helpful or causes harm. The report suggests that the goals helped to "stimulate reforms". However, in many cases, public resources needed to attain the goals were not provided. The announcement of the goals may have obscured the lack of funding. If the goals were not appropriate, then striving to achieve them may have misdirected education resources. The difficulty in measuring performance for some of the goals raises doubts about their value.

The experience with Goals 2000 holds some important lessons for managers. Choose goals that can conceivably be attained. Avoid pressures for a wish list. Limit the number of goals since too many can disperse efforts. Define goals such that performance is measurable to avoid a wild goose chase. Avoid excessive specification of performance objectives to keep from "teaching to the test". Make sure that strategies are chosen to achieve the goals. Appropriate resources to tackle the tasks at hand. Managers choosing goals should note that they may be held accountable for the company's performance.

Education Goals 2000 were not well defined because they far outran the resources and competencies of the educational system without sufficient provisions for organizational change. The U.S. Congress did not choose goals that matched abilities with opportunities.

To execute strategy, the manager prepares a detailed road map that associates specific functional tasks to the company's strategies. The strategic plan associates means with ends. The company sets out both revenues it expects to earn and the costs of its activities. The strategic plan is a multiple-year projection. The bottom line of the strategic plan is projected net returns. Companies need a strategic plan whether they are an established corporation or an entrepreneurial entrant. The strategic plan is a document that is used to operate the business.

For an established firm, the strategic plan sets out the results of the company's strategy process: the specific goals, the results of the external analysis and internal analysis, and the broad strategies to achieve the goals. The strategic plan makes revenue projections for the business. In addition, it defines the functional tasks that help to execute strategies and specifies their expected costs. The strategic plan also specifies what inputs the company needs to purchase, including activities that are outsourced to the company's suppliers. Although the company's strategies tend to be long-term plans covering several years, executing the strategy requires operating plans. The company's operating plans tend to be shorter term plans covering periods of a year or less. Some of this information is presented in the form of budgets. Multi-year budgets were the precursors of modern strategy making (see Exhibit 2.4).

For entrepreneurial start-ups, the strategic plan is used as a basis for a *business plan*, which is a vehicle for fund-raising. The entrepreneur uses the business plan to attract investment by venture capitalists or to secure bank financing. The business plan of an entrepreneurial start-up reflects the strategy process of the company's founders but is addressed to investors. Thus, the start-up's business plan also includes a summary of the company's goals, external and internal analysis, and broad strategies to achieve the goals. In addition, the start-up's business plan spells out the particular goods and services to be provided, projections of earnings, and analysis of the expected costs of the company's purchases, marketing and sales costs, and costs of operations. The business plan concludes with detailed financial information including profit and loss projections and cash flow forecasts.

Exhibit 2.4   Budgets and Management Strategy

Modern management strategy has its roots in accounting and corporate planning. Managers in the textile industry in the early 1800s employed cost accounting as a substitute for market prices when multiple stages in production were combined within the firm.[24] The railroads made advances in accounting as they solved the problems associated with constructing and operating large rail systems. The great business historian Alfred D. Chandler stated

(Continued)

Exhibit 2.4 (*Continued*)

that the "managers of large American railroads during the 1850s and 1860s invented nearly all of the basic techniques of modern accounting", including financial, capital, and cost accounting.[25] Andrew Carnegie used financial information at the Carnegie Steel Company as part of a competitive strategy of lowering cost below those of rivals and undercutting prices during periods of falling demand.[26] In the 1880s and 1890s, the railroads used cost data in carrying out a competitive strategy of system expansion.[27]

More advanced managerial accounting systems were developed by large diversified firms such as DuPont at the beginning of the 20th century.[28] The firms' managers began to allocate capital between competing activities within the firm based on the rate of return.[29] Chandler explains that the management techniques developed at General Electric, DuPont, and General Motors were widely adopted during the 1920s, particularly the multi-divisional organization structure, and "accounting, budgeting, and forecasting methods".[30]

The annual budget of a firm and associated financial controls can be viewed as a one-year plan. In the 1950s, *long-range planning* emerged, extending the annual budget by projecting the firm's cost and revenue growth over a multi-year period based on historical trends. Hax and Majluf observe that "long-range planning makes sense under the conditions that prevailed after World War II, that is, high market growth, fairly predictable trends, firms with essentially a single dominant business, and relatively low degree of rivalry among competitors".[31] Long-range planning reflects the emphasis placed on the growth of the firm in the 1950s and 1960s.[32]

In the late 1960s, faced with intense competition and slower economic growth, firms developed strategic business planning. This reflected a shift in emphasis toward marketing in the early 1970s, with a corresponding reduction of emphasis on production. Strategic business planning required careful identification and projection of anticipated changes in market demand, actions of competitors, technology, and government policy, rather than simple projections of company growth. Moreover, multi-year budgets and financial projections were modified to reflect expected market developments for each line of business.[33]

By the start of the 1990s, with the end of the Cold War with the Soviet Union, business shed the notion of "planning" corporate growth and development. Because of their inward focus, budgets

(Continued)

Exhibit 2.4 (*Continued*)

were no longer the way that companies engaged in strategy making. Rather, budgets became the means for implementing strategy within the organization by allowing management to target funding and expenditures for activities that support the company's strategy.

Strategy has moved beyond budgets and traditional planning to more comprehensive goals and strategic analysis. Companies have available sophisticated data processing methods and nearly instantaneous information about sales, production and purchasing. They are thus able to make real-time adjustments to prices, products, and purchases in a manner never before possible. Enterprise software and Internet communication technologies allow rapid responses to changing market conditions with updates to production schedules and instant contacts with principal suppliers. The ability to change course entails greater flexibility in strategic analysis.

2.3 *Strategy and Change*

Managers of established companies begin the strategy process with an existing organization and an existing market position. The manager must decide whether to remain on course or to initiate change. The decision to remain on course depends on how long the manager believes the company's competitive advantage can be sustained and on the attractiveness of other opportunities.[34] The company's goals and strategy must take advantage of emerging market opportunities. However, those goals and strategies must be feasible given the firm's organizational abilities. This section introduces a basic framework for thinking about change. The "4R" framework is summarized in Table 2.2.

At the corporate level, managers may want to change the collection of businesses the firm operates, diversifying the business or changing its direction. This would involve corporate restructuring. At the individual business level, managers may want to change the set of markets the business serves or change the structure of the business organization, or possibly both.

Managers change the company's strategy and its organization to maximize the value of the firm. The manager evaluates market opportunities and the organization's abilities relative to its competitors. Technological change and changes in customer demand also provide new market opportunities and ways to develop new resources and competencies. Table 2.3 shows how value-driven

Table 2.2: Matching organizational abilities with market opportunities.

Markets Organization	Serve Existing Markets	Enter New Markets
Use existing resources and competencies	Remain on course	Reposition the firm
Develop new resources and competencies	Redesign the firm	Restructure the firm

Table 2.3: Dynamics of value-driven strategy (changing the firm's abilities and its markets).

Markets Organization	Serve Existing Markets	Enter New Markets
Use existing resources and competencies	Remain on course Value of the firm 40 →	Reposition the firm Value of the firm 100
Develop new resources and competencies	Redesign the firm Value of the firm 65	Restructure the firm Value of the firm 150

strategy leads to change. In this example, the value of the firm in its current position is 40. The company might choose to enhance its resources and competencies and realize a value of 65. Alternatively, the company might pursue a new market opportunity and realize a value of 100. By restructuring the firm and changing both market position and organizational abilities, the firm realizes a value of 150.

Remain on Course

To *remain on course* is often the most attractive option. The company chooses to continue serving its existing markets with its existing organizational structure, resources, competencies, and incentives. For a successful company, remaining on course suggests that the company perceives that its competitive advantage can be sustained over the planning period. The company can vary its products through innovation and fine-tune its organizational structure.

The manager estimates the expected value of the firm if the same strategy is pursued over the planning period. The company may project continued profitability and growth with the same strategy. Then, the manager considers whether or not pursuing new market opportunities or changing the company's organization will increase the total value of the firm. If these changes are not expected to increase value, the company should remain on course.

Some companies can choose to remain on course because changing their markets and their organization simply appears too costly or speculative. Moreover, organizational inertia and natural conservative impulses can lead managers to choose the status quo. Often, highly successful companies are reluctant to make strategic changes because earnings will decline in the short run as the new business replaces the earlier one. For example, Thomas Watson Sr. of IBM was dead set against entering the computer business because he felt that computers posed a threat to the company's business of selling punch cards and tabulating machines. IBM entered into computers later under the urging of his son, Thomas Watson Jr.

Redesign the Firm

Companies cannot expect that their resources and competencies will retain their market value forever. Even if the company targets the same customers, the forces of competition, technological innovation, and changing customer preferences will lead the company to update its resources, competencies, and organizational structure. Companies *redesign* their organization to improve its effectiveness and performance. This includes redesign of business processes. The company continues to serve the same markets, but enhances its resources and competencies.

The company assembles a portfolio of both new and existing resources and competencies. Developing new resources and competencies need not require abandoning old ones; the company can expand its resource base and extend its competencies.

Companies should redesign their organization if competitors are able to deliver greater value to customers. Competitors may have greater production efficiency that lowers costs. Competitors may have better technology that allows them to enhance product quality. Also, competitors may offer greater convenience to customers, delivering products and services with greater speed or lower transaction costs.

To enhance the company's performance, it becomes necessary to enhance its resources and competencies in various ways. The company can develop them within the organization, create the resources in partnership with providers of complementary goods, obtain the resources from suppliers, or acquire the resources through merger with and acquisition of other companies. The company's resources include balance sheet assets, such as facilities and capital equipment, and information assets, such as brands, patents, and human capital.

Just as products can become obsolete, the firm's knowledge resources can become outdated.[35] Technological change allows competitors to surpass the company's internal processes. Moreover, competitors can learn new skills or copy others in the industry. Accordingly, knowledge-based resources require continual updating.

The company's distribution and production processes can easily become outdated. For example, a company in the newspaper business might perfect its paper printing and distribution process by improving print quality, speeding production runs, producing multiple editions, and maintaining a fleet of trucks for rapid delivery to homes and news-stands. However, the paper may face competition from online publications. An organization focused on the traditional paper printing and transportation process might miss the larger problem of delivering information, even if that requires a new medium of communication. Delivering information online can require the company to develop new competencies in website design and electronic commerce.

The company's competencies must be market-driven. For example, Microsoft launched one of the first exclusively electronic magazines, called *Slate*. They focused on the properties of the new electronic medium that allowed delivery over the Internet, providing the publication free of charge electronically. However, they soon discovered a demand for the magazine in print form. Reversing a trend, they produced a "hard copy" print version of their online magazine for delivery to subscribers in the traditional way.

Business process re-engineering has been a widely applied form of company redesign. Beginning in the 1990s, companies applied information technology to enhance the efficiency of their manufacturing and service operations. According to the General Accounting Office, business process re-engineering is "a systematic, disciplined approach for achieving dramatic, measurable performance improvements by fundamentally re-examining, re-thinking, and redesigning the processes that an organization uses to carry out its mission". Such re-thinking requires fundamental change in the company's businesses processes. There are three basic types of processes: Processes that deliver final goods and services are sometimes identified as mission-critical; processes that provide support services within the company often operate across functional areas, involving say engineering, operations, purchasing and other functions; and management processes include co-ordination and decision-making procedures.[36]

Reposition the Firm

When the company's resources and competencies have greater market value in new markets than they do in existing markets, it is time to *reposition* the firm. This requires developing new customers and new products and services based on the company's resources and competencies.

Table 2.4: Choosing strategies based on market position and relative strengths of competitors.

| | Relative Strength | |
Type of Firm	Stronger than Competitor	Weaker than Competitor
Entrant	Offensive strategy	Indirect strategy
Incumbent	Defensive strategy	Repositioning strategy

To achieve competitive advantage, it is critical to compare the abilities of the firm with those of its competitors. Matching the organization's abilities with market opportunities requires an understanding of the company's competitive advantage. Four basic possibilities are summarized in Table 2.4.

Entry into new markets involves repositioning the firm. When an entrant believes it is stronger than an incumbent firm, it will contest pivotal market segments served by the incumbent. Entrants facing established competitors pursue *offensive strategies*, which are strategies designed to take on other companies in head-to-head competition. Companies should follow offensive strategies only when they have clear competitive advantages in terms or costs, products or transactions.

The principle of concentration of strength against weakness means focusing marketing, product introduction, customer services, R&D, management and other resources on the segment of the market where competitive advantage is greatest. Finding the best point of attack is intrinsic to the art of competition. The market segment need not be the one least well served or least well defended, since determining the point of weakness is not an easy calculation. It should be the rival's critical vulnerability, what Clausewitz calls the opponent's "center of gravity".[37]

One way to find a competitor's Achilles' heel is to identify high-margin market segments. The challenger can offer better deals to customers and suppliers. Enhancing value added can take away the incumbent firm's high-margin business. For example, MCI initially targeted selected AT&T business customers. These customers were a major source of revenue for the incumbent — not a position of strength for the incumbent firm AT&T. MCI extended its entry to the entire long-distance market. Then, having built a long-distance network, MCI repeated similar strategic maneuvers to begin its entry into local exchange markets.

The United States Postal Service's productive inefficiencies made it a sitting duck for entrants. Although the Postal Service created overnight delivery service, Federal Express perfected the service with reliable delivery and modern tracking systems and captured most of the market. United Parcel Service and other private carriers snatched package delivery away from the Postal Service by offering reliability and convenience, which were weak points of the incumbent.

The incumbent's weak point can be a flagship product or cherished market segment. If the established market leader neglects its flagship brand as it pursues new projects, the entrant has an opportunity to make significant inroads. The incumbent firm's center of gravity can be on the supply side. Wooing away critical parts suppliers or research partners can unbalance the competitor's plans and create a competitive advantage.

When competitors are relatively stronger, an entrant often chooses *indirect strategies*, that is, strategies that avoid direct competition. The managers of the firm will avoid entering markets served by established firms whose abilities and resources are perceived as being superior. They will seek to enter different markets that are not well served by established firms or that are not viewed as critical to stronger competitors. This might mean serving distinct market niches, say business or residential customers, or different geographic regions.

Established companies often face challenges that require *defensive* strategies.[38] Managers of a company defending its markets from entrants or from expansion by competitors should expect rivals to concentrate their strengths against your weaknesses. Entrants will take advantage of their strengths and the incumbent's perceived weaknesses. Since an incumbent cannot anticipate every challenge, defensive strategies require flexibility of response. An incumbent who is overextended creates easy targets for entrants. To sustain a winning position requires being prepared for expansion of existing competitors and challenges from entrants. An incumbent facing entrants that have significant competitive advantages may prefer to *reposition* the company by exiting its current market and serving new markets.

Exhibit 2.5 Nokia

The history of Nokia, one of the leading mobile phone suppliers in the world, illustrates *restructuring*. The company began in 1865 as a wood-pulp mill by the river Emäkoski in southern Finland. The company manufactured paper and cardboard and exported to Russia, China, the UK, and France. The Nokia Group, formed in 1967, included the Finnish Rubber Works, established in 1898, and the Finnish Cable Works, opened in 1912. At the time, electronics generated only three percent of the group's net sales and employed 460 people. Cable Works' electronics department began

(Continued)

Exhibit 2.5 (*Continued*)

research into semiconductors that would lead to its development of a digital switch and a platform that formed the basis for Nokia's network infrastructure.

Jorma Ollila, who became head of the Nokia Group in 1992, chose to focus on telecommunications. The company divested non-core operations in the cable industry and television business, leaving it organized around three business units: mobile phones, networks, and communications products (terminals and monitors). In addition, the company has a venture capital unit that invests in a diverse set of information technology start-ups. Nokia's evolution involved both entering completely new markets and acquiring substantially different resources and competencies.

Nokia pursued an indirect strategy by not taking on the cellular phone giant Motorola in the United States. Nokia did not compete directly with Motorola because Motorola made analog phones while Nokia concentrated on digital phones.

Nokia focused on Europe in the early 1990s, where the company saw the coming standard for wireless phones as a market opportunity. Europe's standard, the Global System for Mobile Communications (GSM), was based on digital transmission and used narrowband time division multiple access (TDMA), which allows eight simultaneous calls on the same radio frequency. Nokia's 6050 wireless car phone and network equipment was used to make the first GSM phone call.

Nokia also turned its sights to Eastern Europe and Asia where the GSM standard became widely adopted. Nokia was an early entrant in Eastern Europe. In 1993, Nokia was the first and only supplier of GSM network equipment to Pannon, a consortium composed of Dutch PTT, Telecom Finland, Tele Denmark, Telenor of Norway, Antenna Hungária, and Wallis of Hungary. In 1994 Nokia Telecommunications and the Beijing Telecommunications Administration signed a contract for supply of a GSM digital cellular network to Beijing.

Nokia built its resources and competencies in the digital phone business. As the company's experience and abilities expanded, it was ready to take on competitors head-to-head on their home turf. In 1997, Nokia's wireless digital phones were adopted by AT&T. By building low-cost phones with less power consumption and extended battery life, and by emphasizing product design and stylish colors, the company was able to directly challenge other phone makers. Nokia became the largest wireless phone maker in the world, surpassing Motorola and Ericsson of Sweden.[39]

Restructure the Firm

The observation that the only thing that remains constant is change applies particularly well to the business environment. When a company discerns opportunities to serve new markets and finds that new abilities are needed for the tasks at hand, it is time to *restructure* the firm. Companies that *restructure* need to *both reposition and redesign* the firm. The company should restructure when the increase in the company's value from pursuing new opportunities is sufficient to justify the costs of developing new organizational abilities. The firm must change its goals and strategy and consequently its organization.

Restructuring also comes in response to significant competitive challenges. As a result of a downturn in telecommunications, Lucent Technologies had to restructure its company. Chairman and CEO Henry B. Schacht stated that "restructuring is about far more than cost reduction and financial discipline. It is about better serving our customers with efficiency, speed, quality and responsiveness". Schacht outlined the company's difficult restructuring efforts as a method of competitive advantage: the most complete portfolio of products for telecom service providers, the most extensive network management systems, and the largest in-house R&D program focused on service providers. The company made substantial changes in its organization and product offerings.[40]

Changes in markets imply that the company's goods and services must be upgraded or fundamentally altered. Customer tastes and needs change continually. Competitors offer improved products and new types of goods that are substitutes for the company's products. As a competitive response, companies need to offer innovative products and enter new markets. To serve markets with goods and services that differ significantly from the firm's current offerings is likely to require acquisition of additional assets, learning new skills, introduction of new brands, establishment of new production facilities, and extension of the company's research and engineering capabilities.

To serve new markets, organizations must establish new routines despite the costs of adjustment.[41] Managers setting new goals must take into account the costs of organizational learning. The existing skills of the organization can facilitate or constrain the acquisition of additional knowledge. The company may have a limited capacity to absorb new knowledge.[42] Investment in research and development activities is likely to enhance the company's awareness of scientific and technical developments.

Companies that become skilled at learning develop dynamic capabilities. Manufacturing companies in the automobile and computer industries needed to integrate learning into their organizations.[43] Organizational learning stems from investment in training employees both formally and through informal learning processes. Moreover, an organization can acquire additional skills through hiring skilled employees and merging with companies that have skilled employees or that have the desired organizational capabilities.

The restructuring company assembles a group of businesses that serve some combination of new and existing markets. The company can increase or decrease the number of markets it serves in several ways:

- **Specialize**: Serve some existing markets and exit some existing markets.
- **Diversify**: Serve existing markets as well as enter new markets.
- **Shift**: Exit existing markets and enter new markets.

When the firm's repositioning is substantial, there will likely be a need to redesign the organization to carry out the company's strategies.

By diversifying, the company uses its existing resources and competencies to develop and offer new products and services. This potentially generates enhanced revenues and cost economies. Sara Lee's competence in marketing allows it to diversify both internationally and across a wide range of products, including not only baked goods but also clothing and personal care items.[44] These new types of offerings are likely to be reflected in the business units established by the company and will result in changes in the company's organizational structure.

One method of repositioning the company is to extend its customer base by selling its existing products and services to new customers. Companies serving a niche market gain by extending their products and services to a wider audience, resulting in enhanced revenues and potential cost economies. Corona beer was a low-price staple in Mexico but became a high-selling luxury import when marketed in the United States and Europe as being associated with tropical vacations. As the company expands, the organization must adapt to serve new markets by expanding the product scope of its divisions or by establishing new divisions.

Companies that serve a specific geographic area can extend their reach to national, regional, or global markets. General Electric's (GE) international operations experienced more than double of its domestic operations. GE's overseas sales revenues were more than half of the company's total revenues, and the global share was expected to continue to grow. CEO Jeffrey Immelt noted that the business outside the U.S. was growing at 15 percent a year: "Now I always say we're an American company. I'm proud to be an American company. I like being an exporter. I like being a good citizen of the United States. I want Congress to root for GE as a company that can win on a global stage. I don't want people to say, 'Gosh, there's something wrong with GE because they want to be a global company'. I think that's where we have to grow". For Immelt, the global challenge meant: "Can we be in enough places?"[45]

Technological resources and core competencies often can be applied to multiple products. For example, Honda's expertise in engine design formed the basis for diverse product offerings including motorcycles, automobiles, lawn mowers, and outboard motors.[46] Honda's product development skills speeded the creation of its Acura division and associated product lines.[47] Canon's core competencies in precision mechanics, fine optics and microelectronics were applied in such diverse areas as cameras, laser printers, copiers, fax machines, video systems, and excimer laser aligners.[48] As the company develops its core competencies, applying them to new markets will entail changes in the company's product design capabilities as well as the new types of marketing and sales efforts. These will be reflected in changes in the company's organizational structure.

Some manufacturing companies have added services to their product offerings. By observing that customers of manufactured products require complementary services including training and repairs, companies identify new markets that can often be addressed with existing resources and competencies. General Electric extended service of customer equipment to enhance technological features of their customers' installed base of machines. By "reengineering the installed base", GE "will have the capability of returning [customers] to operation not just 'overhauled' but with better fuel burn rates in engines, higher efficiency in turbines, better resolutions in CT scanners, and the like".[49] GE experienced rapid growth in its services businesses, with services increases from 15 percent of sales to 70 percent (http://www.ge.com/investor/).

Another way for companies to use their in-house resources and competencies is to *sell functional processes as services*. Companies often develop significant skills in the provision of functional services such as operations management, logistics, distribution, marketing and sales, and customer services. These internal activities can be converted to services for other companies. As the company expands its service offerings outside the company, it will be necessary to develop new types of skills and competencies. Ultimately, the organization of the company will change as divisions focus on supplying services to external customers. If successful, such service businesses may ultimately be divested by the organization. Some companies should spin off their functional units if they have sufficient market value and the company would be better off outsourcing the functional task.

By offering functional processes as services, the company repositions its resources and competencies and realizes their market value. For example, Royal Dutch/Shell Group created Shell Services International from internal accounting, information technology, and other support functions. The unit operates as a profit center, serving both the parent company and outside customers. In turn, Shell departments can seek outside suppliers rather than dealing with the company services unit. Xerox created a quality services unit that provides consulting services to other companies on improving product quality. Scandinavian Airlines trains flight crews and maintains planes for other airlines. Walt Disney extended its internal training services by setting up the Disney Institute to hold leadership conferences and train other companies' executives.[50]

IBM built up a competence in managing large-scale computing systems under government contracts and in operating mainframe computers. Moreover, IBM had significant corporate resources in its labs and the human capital of its engineering staff. IBM used its resources and competencies to create a new business called IBM Global Services that became the fastest growing unit of the company.[51] Half of the company's 240,000 employees were involved in providing services. IBM used its own computers and command centers to operate data services for major clients such as Monsanto and Merrill Lynch. The advent of electronic commerce created increased demand for IBM services in operating websites and tracking customer data. IBM extended its services that were developed for large corporations to help small and medium-sized businesses with e-commerce. In addition, IBM partnered with AT&T, Sun Microsystems, Cisco Systems, and Hewlett-Packard to provide software management services to corporate customers, known as application services. Thus, IBM matched its existing software and hardware abilities with new market opportunities in computer services.

2.4 *Ethics and Stakeholders*

Companies have many stakeholders, that is, groups of people with a great financial stake in the company's outcomes. In addition to the company's owners, stakeholders include customers, employees, creditors, suppliers, and social groups affected by the company's activities. Choosing goals that maximize the value of the company to its owners does not mean that managers neglect the interests of other stakeholders. Rather, understanding and satisfying some types of stakeholder needs can be consistent with shareholder value. The manager must take into account many types of stakeholder interests in order to comply with standards of ethical behavior, social norms, laws, and government regulations.

Maximizing the value of the company is itself part of the manager's ethical duties. The CEO and other senior managers of the company, whether it is publicly-traded or privately-held, are in a position of *trust* for the owners of the company. To be in a position of trust, that is, to be a *fiduciary*, means that the manager must act in the interests of the company's owners. The company is a financial asset entrusted by its owners to the company's managers. Acting in the interest of the company's owners, the manager should seek to maximize the total value of the company.[52]

Managers must report information accurately and clearly to investors. Moreover, managers must act in the interests of the company's owners by not seeking to enrich themselves personally at the expense of shareholders. Managers should be motivated not only by the possibility of job loss and the threat of legal actions, but also by personal integrity and honesty. The energy firm Enron became the largest bankruptcy in United States history because executives allegedly hid information about the company's debt from investors and benefited personally from partnerships. Problems stemming from the Enron debacle led to the demise of the major accounting firm Arthur Andersen, which was found guilty of obstruction of justice. Managers in a number of other companies allegedly engaged in false accounting to cover up financial difficulties. WorldCom experienced serious financial difficulties after being charged with securities fraud stemming from accounting irregularities. A common issue at these and a host of other companies was the ethics of managers. The unethical actions of managers shook investor trust in the management of publicly-traded companies.

Although managers should choose goals that maximize the value of the firm, unfortunately not all managers perform their fiduciary duties. Managers that pursue market share or company sales growth without attention to the value of the company are not serving the interest of the company's owners. Some top executives seek corporate growth to satisfy the need for empire building or the urge to merge. Some managers pursue perks such as fancy offices and travel that are not in the interests of the company's owners. Managers may seek a quiet life by avoiding risks that might pay off for the business. Other managers take excessive risks in pursuit of unlikely market prospects. However, managers whose actions depart significantly from the interests of owners face dismissal and diminished job prospects. Managers also can be replaced through corporate takeovers.

Creditors are one type of stakeholder interest that can conflict with that of shareholders. By taking on too much risk, the company increases the risk of bankruptcy which affects banks and others who have lent money to the company. Shareholders might prefer more risk since they benefit from any earnings above the company's debt obligations. However, companies that create too much risk for creditors will face a greater cost of borrowing in the future, which could reduce the value of the company.

Although the company seeks to serve the customer better than its competitors, the company's goals are not the same as those of its customers. Some customers of a theater might prefer free admission. However, the theater then would not be able to pay its employees or earn returns for its investors. If the theater is able to operate profitably, it can serve many customers over a long period of time. The company's goal is to continue serving customers profitably.

Similarly, a company's employees may prefer higher pay and better working conditions, but paying too much to employees will not be profitable. For example, United Airlines increased the pay of its mechanics to avoid a strike, but the carrier faced losses and the possibility of bankruptcy. Moreover, the interests of different groups of employees will conflict. For example, United Airlines faced wage demands not just from mechanics but also from pilots and from baggage handlers, customer service representatives, and reservation agents. The airline must pay its employees enough to attract them in comparison with other occupations and to make sure they are motivated to perform well. United Airlines must balance these different labor interests while still earning enough money to stay in business. The company will serve its employees best by maximizing the value of the company.

The stakeholder perspective suggests that managers take into account not only multiple types of business performance measures, but also the objectives of others with a stake in the organization's activities, including customers, employees, political interest groups, and local, state, or national governments. These groups often have different and conflicting perspectives on the actions of the firm. For example, the consumer desire for lower prices is likely to conflict with environmental concerns that might raise the operating costs of the company. Although managers would do well to carefully consider the interests of others in their decision-making, it is not possible to satisfy conflicting objectives simultaneously. Managers still require a single criterion for resolving conflicting demands.[53] Managers should choose goals and strategies that maximize the value of the company while respecting ethical standards and social norms and complying with legal rules and government regulations.

Behaving in an ethical manner towards customers is part of an effective strategy. A customer buying a product is concerned about the quality of the good, including in some cases the risks of an accident or the long-term health effects of using the product. If the customer can be reassured about those risks, the customer will be willing to pay more for the product. A customer that is concerned about product risks will be forced to invest in additional learning about the product, purchasing consumer guides or seeking professional opinions. These costs inevitably decrease demand for the product since the customer evaluates the total costs of using the product, which is the purchase price plus the cost of information.

By providing consumers with the necessary information, the company increases its value added. By building a reputation for quality and safety, the company further increases customer value. For example, after tampering with their product resulted in tragic poisonings, the makers of Tylenol responded quickly by taking their products off the shelves, creating new tamperproof containers with multiple safe-guards, and increasing customer information. Any delay in the costly recall of the product would not only have posed a short-term danger but would also have hurt the company's reputation.

In contrast, Intel was slow to acknowledge seriously the flaw in their Pentium chip, emphasizing instead that customers were unlikely to encounter errors in their floating-point decimal operations. Intel management devised a replacement policy only after being stunned by an unexpected public outcry. Someone joked that the Pentium was not named the 586 since the chip could only calculate the number 585.9999999. Intel's reputation for uncompromisingly high quality was damaged by the delay in making restitution to its customers. Eventually, Intel spent millions of dollars settling class action suits.

Companies can secure a competitive advantage by quality leadership, certifying quality and safety through product information, warranties, and reputation. After fighting against passive restraints such as airbags, auto manufacturers eventually realized that safety sells cars, and they now advertise safety features including anti-lock breaks and side-impact protection. The movement toward total quality management (TQM) is a reflection of the competitive advantages from product differentiation through quality.

By earning a reputation for a healthy and safe working environment, companies also bid away skilled employees from companies with less attractive working conditions and reduce the need to pay compensation in wages for the risks of a dangerous workplace. Again, the managers need to provide employees with information and specialized training to enhance workplace safety and lower insurance costs. Also, by considering the effects of their activities on resource usage and the natural environment, companies potentially reduce the costs of environmental regulation and enhance their reputation. For example, companies voluntarily reduced some types of toxic chemical emissions through pollution abatement efforts.[54]

2.5 Overview

This chapter has presented the basic framework of value-driven strategy. The manager chooses goals and strategies that maximize the total value of the firm. The manager factors in the external and internal analyses when choosing the company's goals and strategies. The external analysis identifies market opportunities by considering the company's current and prospective customers, suppliers, competitors, and partners. The internal analysis identifies the company's abilities by evaluating the organization's structure, performance, competencies, resources, and incentives.

The company's market position must be continually defended against existing competitors, new entrants, and substitute products and services. Sustaining advantage requires focus on critical market segments and flexibility of competitive response. Steady improvement of one's own products and services and operations is the antidote to complacency. To obtain a competitive advantage, companies should take careful account of the relative strengths of their organization and those of rivals. Companies facing stronger rivals should pursue indirect strategies by entering underserved markets and targeting market segments. Companies should follow offensive strategies only when they are significantly stronger than competitors. When going head to head, a company should concentrate its strength against the opponent's weakness, improving customer service, pricing, and customer convenience in critical market segments.

Competitive advantage depends on making the best match between the company's organizational abilities and market opportunities. The organization's resources and competencies should be suited to its market entry goals. Companies should remain on course when they have achieved the best fit between abilities and opportunities. They should redesign by acquiring new resources and competencies when product improvements and lower costs are needed to continue serving existing customers. Companies should reposition by diversifying their product and service offerings when their resources and competencies can earn returns in new markets. Finally, companies should restructure when achieving a competitive advantage requires both acquiring new resources and competencies and serving new markets. Technological progress and changes in market demand ultimately force companies to restructure.

Questions for Discussion

1. What problems might managers encounter in pursuing multiple objectives simultaneously? If the company pursues more than one objective, how should different goals be compared?
2. Suppose that a company has a unique expertise in a particular manufacturing technique. Some examples might include LCD displays, small engines, plastic molding, fermentation processes, and wooden musical instruments. How should the company's management take this information into account in selecting the company's goals and formulating the strategies to achieve them?
3. Suppose that the company's managers discover an opportunity to provide the market with an innovative service but the company has no experience providing that service. How should the company's management take this information into account in selecting the company's goals and formulating the strategies to achieve them?

4. Select a particular industry and consider some major changes that have affected the industry, including technology, consumer tastes, and government regulation. How might firms in the industry respond to those changes?
5. What types of market changes would lead a company to seek out new markets for its existing products?
6. What types of market changes would lead a company to change its organizational structure?
7. What types of market changes would lead a company to change both its target markets and organizational structure?
8. How might improvements in communications and information technology cause a company to change its organizational structure?

Endnotes

1. The notion that both external analysis and internal analysis are vital for strategy making draws upon Kenneth R. Andrews, *The Concept of Corporate Strategy* (Homewood, IL: Irwin, 1971), p. 48, who wrote that "Economic strategy will be seen as the best match between qualification and opportunity that positions a firm in its product/market environment." Andrews stated that "Determination of a suitable strategy for a company begins in identifying the opportunities and risks in its environment" (p. 48). He observed that "opportunism without competence is a path to fairyland" (p. 70). Bourgeois, citing Andrews, describes "strategic fit" as follows: "The central tenet in strategic management is that a match between environmental conditions and organizational capabilities and resources is critical to performance, and that a strategist's job is to find and create this match"; L. J. Bourgeois, III, "Strategic Goals, Perceived Uncertainty, and Economic Performance in Volatile Environments," *The Academy of Management Journal* 28, no. 3. (September 1985), pp. 548–573. Our analysis builds on and extends that of Andrews by introducing a method of valuing alternatives and by considering the process of changing the company's strategy and organization.
2. "The Company's strategy centers on one primary objective — to grow the business profitably. The core of Atlas Air's strategy is unique, yet simple: It maintains a low-cost operating structure that enables the Company to operate as an air cargo carrier on a long-term contract basis on behalf of other international airlines at extremely competitive rates." See http://www.atlasair.com/about/strategy.asp.
3. William George, "Medtronic's Chairman William George on How Mission-Driven Companies Create Long-term Shareholder Value," *Academy of Management Executive* 15 (2001), pp. 39–47.
4. Michael W. Miller and Laurie Hayes, "Gerstner's Nonvision for IBM Raises a Management Issue," *The Wall Street Journal*, July 29, 1993, p. B1.

5. Ibid.

6. Ibid.

7. Ibid.

8. Steve Lohr, "At Last, a Sneak Preview of Big Blue's Strategy," *New York Times*, July 8, 1993.

9. Steve Lohr, "On the Road with Chairman Lou," *New York Times*, June 26, 1994, section 3, p. 1.

10. See Ira Sager with Amy Cortese, "Lou Gerstner Unveils his Battle Plan," *Business Week*, April 4, 1994. In addition to these points, Gerstner discussed reengineering the sales force to cut costs, but this is an organizational change. Also, the company had the goal of seeking new geographic markets, particularly in Asia.

11. The quotations in this paragraph are from Steve Lohr, "Broad Reorganization at I.B.M. Hints at Successor to Gerstner," *New York Times*, July 25, 2000, section C, p. C1.

12. Michael C. Jensen emphasizes the need for a single value measure in the corporate objective function. See Michael C. Jensen, "Value Maximization, Stakeholder Theory, and the Corporate Objective Function," Working paper, Harvard Business School, February 2001.

13. For example, the balanced scorecard approach suggests more than a dozen performance measures to supplement financial measures but does not give the manager a means for combining these measures into a coherent criterion for ranking strategic alternatives. See Robert S. Kaplan and David P. Norton, *The Balanced Scorecard* (Boston, MA: Harvard University Press, 1996) and Robert S. Kaplan and David P. Norton "Using the Balanced Scorecard as a Strategic Management System," *Harvard Business Review*, January-February 1996, pp. 75–85. For a critical analysis of the balanced scorecard approach, see Michael C. Jensen, "Value Maximization, Stakeholder Theory, and the Corporate Objective Function."

14. The account of Intel's new chips is based on Jay Wrolstad, "Intel Unveils Chips for Next-Gen Mobile Devices," www.wirelessnewsfactor.com, February 12, 2002, http://www.techextreme.com/perl/story/16290.html.

15. See Berger Wernerfelt, "A Resource-Based View of the Firm," *Strategic Management Journal* 5 (1984), pp. 171–180; I. Dierickx and Karel Cool, "Asset Stock Accumulation and Sustainability of Competitive Advantage," *Management Science* 35 (1989), pp. 1504–1511; Jay B. Barney, "Firm Resources and Sustained Competitive Advantage," *Journal of Management* 17 (1991), pp. 99–120; Katherine R. Conner, "A Historical Comparison of Resource-Based Theory and Five Schools of Thought Within Industrial Organization Economics: Do We Have a New Theory of the Firm?" *Journal of Management* 17 (1991), pp. 121–154; Raphel Amit and Paul J. H. Schoemaker, "Strategic Assets and Organizational Rent," *Strategic Management Journal* 14 (1993), pp. 33–46; and Berger Wernerfelt, "The Resource-Based View of the Firm: Ten Years After," *Strategic Management Journal* 16 (1995), pp. 171–174.

For a critical survey and discussion see Richard L. Priem and John E. Butler, "Is the Resources-Based 'View' A Useful Perspective for Strategic Management Research?" *Academy of Management Review* 26 (2001), pp. 22–40, and the reply by Jay Barney in the same issue. According to Priem and Butler, "the elemental concept of 'value' remains outside the RBV [resource-based view]."

16. Henry Sender, "Global Crossing's Many Creditors Quarrel Over Best Method to Recoup Their Money," *The Wall Street Journal*, March 6, 2002, p. C1.

17. "Procter & Gamble: Jager's Gamble," *The Economist*, October 30, 1999, p. 75.

18. Harold Demsetz observes that the complexity of organizations such as General Motors or IBM "defies easy analysis" so that the reasons for success are not readily understood. See Harold Demsetz, "Industrial Structure, Market Rivalry and Public Policy," *Journal of Law & Economics* 16 (1973), pp. 1–10. A related phenomenon has been called uncertain imitability. See Steven A. Lippman and Richard P. Rumelt, "Uncertain Imitability: An Analysis of Interfirm Differences in Efficiency Under Competition," *Bell Journal of Economics* 13 (Autumn 1982), pp. 418–438.

19. Arik Hesseldahl. "The Return of Iridium," Forbes.com, November 30, 2001, http://www.forbes.com/2001/11/30/1130tentech.html; David Barboza, "Iridium, Bankrupt, Is Planning a Fiery Ending for its 88 Satellites," *New York Times*, April 11, 2000. As of January 2002, the service is offered by the Canadian company Infosat Telecommunications, a subsidiary of Telesat Canada. See http://www.infosat.com/.

20. The account is based closely on Gregory L. White, "How Ford's Big Batch of Rare Metal Led to $1 Billion Write-Off," *The Wall Street Journal*, February 6, 2002, p. A1.

21. "Face Value: A Busy Bee in the Hamburger Hive," *The Economist*, March 2, 2002, p. 62.

22. http://www.jollibee.com.ph/default.htm.

23. Richard Rothstein, "'Goals 2000' Scorecard: Failure Pitches a Shutout," *New York Times*, December 22, 1999.

24. See H. T. Johnson and R. S. Kaplan, *Relevance Lost: The Rise and Fall of Managerial Accounting* (Cambridge, MA: Harvard Business School Press, 1986), p. 23. They observe that the textile factories provide the earliest examples of cost accounting (p. 44).

25. Alfred D. Chandler, *The Visible Hand: The Managerial Revolution in American Business* (Cambridge, MA: Harvard University Press, 1977), p. 109.

26. See Johnson and Kaplan, *Relevance Lost*, pp. 33–34.

27. The strategy was to achieve "self-sufficiency," which led to bankruptcy and consolidation (Chandler, *The Visible Hand*, pp. 170–187). After the consolidation of the railroads in the early part of the 20th century, strategic planning was no longer a major concern for the railroads (ibid., p. 186).

28. After 1900, with the growth of mass production, came the development of modern factory cost accounting (Chandler, *The Visible Hand*, p. 278). Chandler observed that the management innovations at DuPont beginning in 1903 combined financial, capital, and cost accounting and thereby "helped lay the base for modern asset accounting" (p. 447).

29. See Johnson and Kaplan, *Relevance Lost*, p. 86.

30. Chandler, *The Visible Hand*, p. 464.

31. A. C. Hax and N. S. Majluf, *Strategic Management: An Integrative Perspective* (Englewood Cliffs, NJ: Prentice Hall, 1984), p. 11.

32. Peter F. Drucker states that "the 1950s and especially the 1960s indulged in a veritable 'growth craze' in economy as well as in business" and "growth — on the order of '10 percent growth in sales and 10 percent growth in profit each year' — was what management promised"; Peter Drucker, *Management: Tasks, Responsibilities, Practices* (New York: Harper & Row, 1973), p. 770. For a historical perspective on planning systems, see H. I. Ansoff, "The State of Practice in Planning Systems," *Sloan Management Review* (Winter 1972), pp. 61–69, and H. I. Ansoff, "Strategic Issue Management," *Strategic Management* (April-June 1980), pp. 131–148.

33. In the 1970s, General Electric initiated a project titled "Profitability Impact on Marketing Strategies" (PIMS) that gathers market data at the line-of-business level, such as financial data, market share, advertising, R&D, and investment. See, for example, R. D. Buzzell and F. D. Wiersema, "Successful Share-Building Strategies," *Harvard Business Review* 60 (1981), p. 135.

34. Andrews notes that "In each company, the way in which distinctive competence, organizational resources and organizational values are combined is or should be unique"; Andrews, *The Concept of Corporate Strategy*, p. 70.

35. Dierickx and Cool, "Asset Stock Accumulation" pp. 1504–1511.

36. General Accounting Office, "Business Process Reengineering Assessment Guide," Washington, DC, 1995. See www.dtic.mil/c3i/bprcd/.

37. As military strategist Robert Leonhard points out, in the game of chess, the opponent's source of strength is his queen, but the critical vulnerability is the king. Although the king is a weak piece, its checkmate wins the game. See Robert R. Leonhard, *The Art of Maneuver* (Novato, CA: Presidio Press, 1991), pp. 20–21.

38. The discussion of indirect, offensive, and defensive strategies draws from Daniel F. Spulber, *The Market Makers: How Leading Companies Create and Win Markets* (McGraw-Hill/Business Week Books, 1998).

39. Information on Nokia is drawn from www.nokia.com/inbrief/history/early3.html.

40. Henry B. Schacht, "Chairman's Message: Creating the Foundation for the Future," Lucent Technologies, Annual Report, 2001.

41. On the rigidities of organizational routines, see Richard R. Nelson and Sidney G. Winter, *An Evolutionary Theory of Economic Change* (Cambridge, MA: Harvard University Press, 1982).

42. W. M Cohen and D. A. Levinthal, "Absorptive Capacity: A New Perspective on Learning and Innovation," *Administrative Science Quarterly* 35 (March 1990), pp. 128–152.

43. See M. Iansiti and Kim B. Clark, "Integration and Dynamic Capability: Evidence from Product Development in Automobiles and Mainframe Computers," *Industrial and Corporate Change* 3 (1994), pp. 557–605.

44. Robert E. Hoskisson and Michael A. Hitt, *Downscoping: How to Tame the Diversified Firm* (New York: Oxford University Press, 1994), pp. 146–147.

45. Richard McCormack, "General Electric CEO Jeffrey Immelt: The U.S. No Longer Drives Global Economic Growth and Must Decide If It Wants To Be a Competitive Nation," *Manufacturing and Technology News*, November 30, 2007, Volume 14, No. 21, http://www.manufacturingnews. com/news/07/1130/art1.html.

46. C.K. Prahalad and Gary Hamel, "The Core Competence of the Corporation," *Harvard Business Review*, May-June 1990, pp. 79–91.

47. George Stalk, Philip Evans and Lawrence E. Shulman, "Competing on Capabilities: The New Rules of Corporate Strategy," *Harvard Business Review*, March-April 1992, pp. 57–69.

48. Ibid.

49. General Electric, Annual Report to Shareholders, 1998.

50. These examples are drawn from Claudia H. Deutsch, "Letting Side Business in the Side Door," *New York Times*, January 22, 1998, p. C1.

51. Saul Hansell, "Now, Big Blue Is at Your Service," *New York Times*, January 18, 1998, Section 3, p. 1.

52. The company's owners can adjust their consumption patterns and earnings risk by borrowing and lending and adjusting their personal portfolios. Accordingly, the interest of each shareholder is that the company maximizes the market value of their shares of the firm. This means that the company should maximize its market value. For the classic reading on this subject, see Irving Fisher, *The Theory of Interest* (New York: Augustus M. Kelly, 1930, 1965).

53. On the stakeholder approach see, for example, R. Edward Freeman, *Strategic Management: A Stakeholder Approach* (Boston: Pittman Books Ltd., 1984). For a critique of the stakeholder approach, see Michael C. Jensen, "Value Maximization, Stakeholder Theory, and the Corporate Objective Function."

54. John Maxwell, Thomas P. Lyon, and Steven C. Hackett, "Self-Regulation and Social Welfare: The Political Economy of Corporate Environmentalism," *Journal of Law & Economics* 43 (October 2000), pp. 583–617.

PART II

THE MARKET COMPASS

Customers

Partners

Competitors

Suppliers

CHAPTER 3

CUSTOMERS AND SUPPLIERS

Contents

The manager performs an *external analysis* that examines the company's prospective customers, suppliers, competitors and partners. The manager is guided by the "market compass" (see Figure 3.1). The manager should understand the characteristics of the company's customers, suppliers, competitors and partners. The firm both competes and cooperates with its customers, competitors and partners. The market compass helps the manager identify sources of competitive advantage and the potential for value creation. Information gathered through external analysis must be updated as the manager revises the firm's strategy.

Connections between the firm and its customers and suppliers are referred to as *vertical* market relationships. Competitive and cooperative interactions with firms in the industry and with firms that supply complementary products are referred to as *horizontal* relationships. After you have completed this chapter you will have some of the tools that you need for analyzing potential customers and suppliers. The next chapter addresses the firm's competitors and partners.

Customers are the main source of the firm's value because they generate the company's revenue. Customer characteristics determine the firm's market opportunities. The external analysis seeks to identify the firm's potential customers. This will guide the manager in choosing the firm's goals. Customer characteristics, preferences, and willingness to pay are summarized by the description of overall market demand schedules. By understanding market demand, managers can make decisions about whether to enter or exit markets or even to create new markets. Customer analysis also guides strategy because the firm's

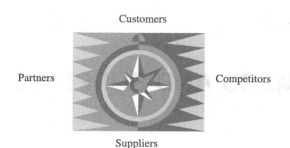

Figure 3.1: The market compass.

products, services, prices, production and contract terms should be responsive to changes in market demand.

Suppliers also are a source of value because most of the firm's costs are payments to its suppliers. More efficient suppliers reduce costs and increase the value of the firm. Suppliers provide goods and services whose quality can affect the firm's productive efficiency and the features of the firm's final product. Retailers depend on the products obtained by manufacturers and wholesalers. Manufacturers depend on the critical parts and components obtained from input suppliers. The characteristics of suppliers, including their technology and costs, are summarized by supply schedules in the markets for inputs. Supplier analysis guides the firm's choice of goals and its strategy since the quality and prices of input supplies fundamentally affect the firm's products and costs.

The manager's external analysis takes account of the firm's current market position and determines how market forces will affect the value of the firm. The manager attempts to determine which markets present the most attractive business opportunities and makes entry and exit decisions. The combination of market demand for the firm's products and market supply of the firm's inputs affect the firm's ability to create value. Table 3.1 sets out some of the important questions that managers need to ask in carrying out their external analysis.

Chapter 3: Take-Away Points

The manager's external analysis starts with customers and suppliers to obtain a picture of the potential demand for the firm's products and the potential supply of critical inputs:

- Customers are the beginning of external analysis because successful strategy requires understanding and satisfying customer preferences.
- Customer willingness to pay is the potential source of the firm's value.

Table 3.1: Conducting an external analysis.

	Evaluate current market	Evaluate prospective market
Prospective customers	Who are current customers and how are their preferences and income changing?	Which customers do we want to retain? What new customers should we try to attract? What are the characteristics of customers in markets we seek to enter?
Prospective suppliers	Who are our current suppliers and how are their technologies and opportunities changing?	Which suppliers do we want to retain? What new suppliers should we try to attract? What are the characteristics of suppliers that will be needed to carry out our entry plans?
Prospective competitors	Who are our current competitors and how are their technologies and opportunities changing?	Which competitors will we face as our firm enters new markets? What competitors will we face as companies develop substitute products, new production processes, new types of transactions, and new combinations of products, services and transactions?
Prospective partners	Who are our current partners and how are their technologies and opportunities changing?	Which partners do we want to retain? What new partners should we seek among customers, suppliers, and makers of complementary and substitute products? What new partners should we attract as we enter new markets and develop new products?

- The market demand schedule defines the firm's market opportunities.
- Suppliers are integral to external analysis because the firm's ability to realize its market opportunities depends critically on the types of services, parts, equipment, and technology that the firm receives.

- Supplier costs are important determinants of the firm's ability to create value.
- The market supply schedule affects the firm's ability to realize its market opportunities.

3.1 *Customers*

In the manager's external analysis, customers necessarily come first because customer preferences drive the business. The business exists to satisfy a customer need. The company creates products and services that target critical customer needs. Accordingly, the business defines itself in terms of the characteristics of its customers. Customers provide the company's revenue, its motivation and its direction. Customer willingness to pay for the firm's products is the source of the firm's value.

Companies generally describe their business in terms of how close they are to the customer. Wal-Mart is a retailer because its stores serve customers, even though the company's extensive warehouse and transportation system make it a significant wholesaler as well. A company can serve at the retail, wholesale, manufacturing, primary manufactured input, or resource level. Allied Metal Industries supplies manufacturers with steel, aluminum, and other metals. Within each vertical level, companies define themselves by the customer segment they serve, not just the products and services that they offer. Blockbuster, for example, serves the home entertainment market. Gapkids sells clothing for children while Gap's Old Navy sells value-priced clothing to adults and children.

Strategic analysis is necessarily forward-looking so that external analysis looks at the company's *prospective customers*, those customers it wishes to retain and new customers it wishes to attract. The company's customers are described by demand schedules that detail their willingness to pay for the company's goods and services. Market changes are described by shifts in the demand schedules due to changing customer preferences, income and information, changes in the company's product and service characteristics, and the offerings of competitors and partners.

The manager's inquiry begins with an examination of customer characteristics. This is potentially useful for evaluating how these characteristics affect customer *willingness to pay* for the firm's products. Customer willingness to pay is based on the benefit the customer receives from the firm's product in relation to the best alternative. By considering customer willingness to pay, it may be possible to sketch out a demand schedule that associates the relationship between the firm's price and the total quantity purchased. With the demand schedule in hand, the company can start to explore the effects of raising or lowering prices on sales and revenues based on an overall evaluation of what is the customer's responsiveness to changes in the firm's price. By examining the effects of price changes on revenue, the company can make projections of incremental revenue at different price points.

Who Are Our Customers?

The external analysis must begin with understanding the company's customers. Strategic action necessarily involves change. The company might gradually change its products and services or drastically change by exiting and entering markets. These changes imply that the firm's customer base is likely to change in consequence. The company needs to have a handle on which of its existing customers it is trying to retain and what new customers it hopes to attract. Demographic changes in the population in such areas as age and family size are crucial factors.

Entering new markets requires learning about new customers. Companies planning international expansion will need to devote resources to learning about the different characteristics of customers in the countries they plan to serve. Changing distribution channels, for example, adding online distribution in addition to retail outlets, also entails new sets of customers.

Modern methods of data collection, storage, and analysis allow the creating of vast databases on purchasing patterns and customer information. Customer information has become an essential strategic instrument. What are the key characteristics of prospective customers?

Customer Characteristics

Get close to the customer! This enduring strategic advice means that companies should get to know their customers' characteristics so as to tailor their products and pricing to meet customer needs.[1] Information about customers can be summarized by their *willingness to pay* for the firm's goods and services. This key piece of information identifies the maximum amount that a customer is willing to pay for the firm's product and reflects the benefits received relative to the best alternatives.

What are the underlying determinants of willingness to pay? Willingness to pay is the outcome of individual consumer choices, whether conscious or emergent. Consumer choices can result from a rational process of observing market prices, knowing the budget, and assembling a market basket of products that yields the greatest benefit for the consumer. Think of a customer preparing a shopping list before going to the grocery store that reflects careful planning about upcoming meals and detailed knowledge of prices, promotions and product availability. Alternatively, the consumer's choices can be less a product of careful planning and more a response to unexpected daily needs. A customer who shops on impulse reacts to retail sales personnel, product displays, and in-store promotions. Most customers engage in some combination of planning and impulse buying, but in either case some choices are being made.

The consumer's willingness to pay is the outcome of four main factors: the consumer's budget, the consumer's preferences, the set of substitute and complementary products that the consumer might purchase, and transaction costs. The consumer's budget depends on wealth, employment income, and decisions about borrowing and savings. The consumer's budget constraint is affected by the relative prices of the many products the consumer buys.

The consumer's preferences depend on many underlying factors, including the consumer's experience, cultural background, social interaction, and personal traits. A consumer who is accustomed to listening to a particular type of music, such as country, rock, jazz or classical, is likely to prefer that type of music. Tastes in music are often affected by cultural background, including religion and nationality. Enjoyment of music often is a social experience and the consumer's choices are likely to depend on the consumer's social environment.

Transaction costs include the costs of gathering information about available products and the indirect costs of dealing with retailers. Consider, for example, a consumer that wants to have a fence around his yard. The consumer may spend considerable effort searching for a fence company, gathering estimates, and negotiating with sales personnel and may be willing to pay $1,500 for the fence. Alternatively, the consumer may be willing to pay $2,000 for a fence if it is readily available with a minimum of additional effort. The consumer pays the additional $500 as a way of avoiding transaction costs.

As a first pass, the company can describe its customers by observations of characteristics such as income, level of expenditure or purchasing patterns over time. The company can design promotions or contracts to induce customers to reveal their characteristics. The company can discover the distribution of these features in its customer population through demographic surveys. Knowing the distribution of customer characteristics allows the company to define and target market segments with pricing and promotions.

3.2 Demand Characteristics

It is useful to understand how the market segments for a given product fit together so the company can combine or separate segments to adjust its product features and price offers. Representative market segments are building blocks for analyzing the firm's total market demand. Information about customer willingness to pay for the firm's good can be summarized in a useful manner as a demand schedule.

Willingness to Pay and the Demand Schedule

Each customer has a benefit level equal to $V that determines the customer's willingness to pay for the firm's product. The willingness to pay tends to differ across individual customers depending on their preferences, income, attractiveness of alternatives, travel costs, product knowledge, and other personal characteristics. The customer's willingness to pay is an upper limit. The customer will be willing to buy a unit of the firm's product as long as the product price is less than or equal to $V. One way to think about market segments is that the customers in each segment have similar willingness to pay levels.

The total demand that the firm experiences is the sum of demands by individual consumers in each of the firm's market segments. To calculate total demand, use the *adding down* method. Suppose that each customer wishes to buy at most one unit of the firm's product. Then the total demand at a uniform price can be calculated by adding the number of customers at each benefit level. Adding customers shows that the demand curve slopes downward. The lower the price the more people will buy the product simply because more customers have benefits greater than the price. Notice that it does not matter how many customers are in each segment. Demand increases as the price decreases simply because customers are added together.

The adding down method works as follows. First, estimate how many customers are in each market segment at the various benefit levels, as in the first and second columns of Table 3.2. Then, add down the middle column to see how much demand there is at each price. In the example, the three prices are 14, 9, and 4, each of which is $1 less than the benefit level for each market segment. The method can be easily adapted for larger numbers of customers and more gradations of benefit levels.

The total demand experienced by the firm depends on how many consumers derive benefits from the firm's product and how much they are willing to pay. Willingness to pay for the good depends not only on customer preferences but also on customer income levels because

Table 3.2:   The adding down method of calculating demand.

Benefit V	Number of customers	Demand "Add down" column 2
15	6	6 (at price 14)
10	12	18 (at price 9)
5	4	22 (at price 4)

customers must allocate their income across many alternatives. Willingness to pay for the good is also affected by the transaction cost, including the costs of shopping to obtain the good, the time and effort required to use the good, and additional costs such as transportation and taxes that the customer must pay. The benefits that potential customers derive from the firm's product depend on the customer preferences and the features of the product, as well as the availability of substitute products.

Two goods are said to be *substitutes* for a consumer if an increase in the price of one good causes the consumer to switch some consumption to the other good at any given price. For example, an increase in the price of orange juice will switch some demand to lemonade or pineapple juice. Other examples of substitutes are desktop and notebook computers, regular and premium gasoline, and railroad and airline transportation.

Goods can be substitutes even if they have different physical features. The only thing that counts however is the customer's benefits from the two goods. Thus, a glass jar and a metal can may be substitutes as containers even though they look different. The way in which the two goods are manufactured or provided also does not matter. A cellular telephone is a substitute for a pay phone as a communication device even though the cellular phone is wireless and the pay phone uses wireline transmission. Delivery channels can differ as well — news delivered on the Internet competes with television news and newspapers. The two goods can be substitutes even if one good has a much higher price than the other good. The consumer compares the relative benefits from the two goods with the difference in prices.

Because the firm's products compete against substitute products, it follows that the firm's market opportunities and threats depend on the quality and prices of those substitutes. The manager should evaluate competing products in terms of the benefits customers ultimately receive. This means that managers should not just base decisions on a narrow view of competing products. The manager should anticipate that technological change can create all kinds of substitutes whose features are substantially different from those of existing products.

The strategic implication of substitutes can be subtle. For example, companies have applied nanotechnology to develop stain-resistant fabrics. By purchasing clothing made with the new fabric, customers may be able to reduce their use of special detergents that remove stains. Thus, customers may substitute the use of clothing made with stain-resistant fabric for purchases of stain-removing detergents. Accordingly, the new fabric poses a competitive threat to companies that make certain types of detergents while offering opportunities for some companies to develop new types of detergents that work with the new types of cloth. Companies that manufacture detergents should not expect that the only competitive threats come from other makers of detergents. In this example, nanotechnology applied to clothing poses a threat.

Elasticity of Demand

Selecting the price points for the company's goods and services is one of the most important strategic activities. The key number for pricing is the price elasticity of demand. That number measures the price responsiveness of the firm's customers. In markets where the firm faces significant competition from firms offering similar products, the elasticity of the firm's demand will be very large, indicating that customers are very price-responsive. In markets where products are significantly differentiated or the firm does not face competition from firms offering similar products, the price elasticity of the firm's demand will tend to be lower. The elasticity number provides a convenient summary of the price responsiveness of customer demand, and gives an indication of the attractiveness of the firm's products.

Increasing the price charged to consumers reduces the total amount that they demand. How sensitive are consumers to a price increase or to a price reduction? Knowing price sensitivity is essential to adjusting the firm's pricing strategy. A numerical indicator of customer price sensitivity is called the price elasticity of demand. This number is very important because it measures the effect of price changes on the firm's sales revenues and therefore serves as a useful guide for pricing.

Example 3.1   Price Elasticity

To calculate the price elasticity of demand in practice requires determining the effects of price changes on the quantity that customers purchase. The following information is available. The initial price is $10 and the firm increases the price to $11. The firm sold 200 units at the price of $10 in a given period of time. Raising the price causes demand to fall to 160 units for a comparable period of time. The elasticity of demand is the ratio of the percentage fall in demand to the percentage rise in price. Given the information in this example, the price elasticity of demand equals 2, as Table 3.3. shows.

What happens when a firm raises its price? Does revenue go up or down? It all depends on elasticity of demand because price increases have two results: the firm *sells fewer units* and *earns more per unit*

Table 3.3: Calculating the elasticity of demand.

Percentage rise in price	Percentage fall in quantity	Price elasticity of demand
Initial price $10 New price $11	Initial quantity sold 200 New quantity sold 160	Divide column 2 by column 1
$\dfrac{\$11 - \$10}{\$10} = 10\%$	$\dfrac{200 - 160}{200} = 20\%$	$\dfrac{20}{10} = 2$

sold. The relative size of these two effects is measured by the price elasticity of demand, which gives the company a numerical measure of the customer price responsiveness, at least in the short run. The elasticity of demand is critical for determining how changing the company's price point affects sales and revenue.

If the elasticity of demand is greater than one, as in the example, demand is "elastic". This means that customers are very price-responsive. A small change in price leads to a greater response in demand. If the firm increases the price, demand falls by a greater percentage as customers reduce their purchases or switch to other goods. If the firm reduces the price, demand rises by a greater percentage as customers increase their purchases or switch away from other goods toward the firm's product.

On the other hand, if elasticity of demand is less than one, demand is "inelastic", meaning that customers are not very price-sensitive. A small change in price leads to an even smaller response in demand. If the firm increases the price, demand falls by a smaller percentage as customers reduce their purchases or switch to other goods. If the firm reduces the price, demand rises by a smaller percentage as customers increase their purchases or switch away from other goods toward the firm's product.

The elasticity of demand has important implications for the firm's pricing policy. Should the firm increase or lower its price? The elasticity number predicts the effects of the firm's price on its revenue. If the elasticity of demand is greater than one, a price increase lowers revenue. If the price elasticity of demand is less than one, a price increase raises revenue.

Example 3.2   Does Revenue Rise or Fall?

At the initial price of $10 the firm sold 200 units and at the higher price of $11 the firm sells 160 units. As predicted by an elasticity of demand greater than one, revenue falls (see Table 3.4).

Table 3.4: **Revenue decreases with an increase in price when elasticity is greater than one.**

Price P	Quantity sold Q	Revenue P × Q	Change in revenue
10	200	$2,000	
11	160	$1,760	$(240)

Exhibit 3.1   Wal-Mart's Private Labels

Wal-Mart stepped up its use of private-label products.[2] The private-label or store brand products are almost always sold for less than nationally advertised products. Because private-label products are not advertised, Wal-Mart substantially lowers its costs. Also, because Wal-Mart buys directly from the manufacturer and handles its own wholesaling, the company avoids additional costs of distribution.

Among Wal-Mart's private labels are *No Boundaries* cosmetics, *Mary-Kate and Ashley* clothing for preteen and teenage girls, *Apple Express* cereal, *Ever Active* batteries, and *Great Value* food and beverages. Wal-Mart's *Sam's Choice* brand applies to a wide range of items including detergents and food products, such as soft drinks and juices. In addition, Wal-Mart purchase nationally recognized brands such as *White Stag* in women's apparel and *White Cloud* toilet tissue.

Wal-Mart offered 1,259 private-label products in 2001, up substantially from the previous year. The discounts were substantial compared to national brands. Wal-Mart's *No Boundaries* "Paint Me" nail polish sold for over 72 cents below the national brand (*Cover Girl*) while its *Sam's Choice* detergent sold for over $2 below the national brand (*Tide*). Why were these price differences so large? Why were Wal-Mart's discounts seemingly calculated to the nearest penny rather than simply rounded up or down?

Wal-Mart devoted considerable amount of attention to the price differences between its private-label products and the price of the comparable nationally advertised brand. If the discount for the private-label product was too small, customers would be likely to favor the brand-name product. If the discount for the private-label product was too big, Wal-Mart would forgo revenue. Consider the case of *Great Value* cookies. Wal-Mart originally priced the cookies at $1 a bag. The company then cut the price from $1 to 88 cents, a cut of exactly 12%. The number of bags sold then soared

(Continued)

Exhibit 3.1 *(Continued)*

by 54%. What was the elasticity of demand for *Great Value* cookies at Wal-Mart? The elasticity was equal to 54/12 = 4.5.

The price cut caused Wal-Mart's revenue from *Great Value* cookies to increase. Although the company sold each bag for 12 cents less, revenues went up because the company sold 54% more bags at 88 cents each. The price responsiveness of customers buying the *Great Value* brand of cookies demonstrated that for some customers the discount outweighed any perceived advantages of national brands. The high degree of customer responsiveness to price showed that managers did well to carefully evaluate the effects of small price changes on customer demand. Moreover, the revenue boost from the discounts helped explain why Wal-Mart might discount substantially below the competing brand-name product. All things considered, the customer's high degree of price responsiveness demonstrated the value created by Wal-Mart's private-label strategy.[3]

The demand elasticity itself changes in response to the firm's price. The firm generally faces different demand elasticities at different price points along the demand curve. For substantial price changes, it is necessary for the manager to reevaluate estimates of the demand elasticity.

Elasticities of demand are useful in evaluating demand differences across the firm's market segments. In fact, elasticity differences are useful in defining market segments. Customers of different income levels are likely to have different price sensitivities. Elasticity of demand can differ across marketing channels. Customers purchasing a particular good in supermarkets may be more price sensitive than customers in convenience stores.

The manager can also use elasticity of demand to identify geographic segments because elasticities often differ across countries or regions. Table 3.5 shows elasticities of demand for six different market segments and the whole industry for automobiles for five countries: Belgium, France, Germany, Italy, and the United Kingdom. The size of the new car market for these countries in 1990 was 10.2 million cars. The elasticity estimates are based on the demand by dealers for manufacturers' cars and thus indirectly reflect final customer demand. The differences in demand elasticities across countries reflect different trade barriers and country differences in income, preferences, and the strength of domestic and international competition. Notice, for example, that Italy has consistently lower price elasticities than the other countries, primarily reflecting barriers to trade. Significant price differences have been observed between countries, in some cases as high

Table 3.5:   Elasticities of demand in the European car market.

Segment	Belgium	France	Germany	Italy	United Kingdom
Mini and small	12.35	9.62	10.53	6.1	9.8
Medium	10.87	8.70	8.85	6.21	9.09
Large	10.87	8.26	8.20	6.41	8.47
Executive	9.62	7.58	8.77	6.85	8
Luxury	7.46	7.46	6.29	5.32	7.04
Sports	6.94	6.54	6.58	5.71	5.03
Whole Industry	10.64	8.77	8.4	6.13	8.77

Source: Adapted from Frank Verboven, "International Price Discrimination in the European Car Market", *Rand Journal of Economics* 27 (1996), pp. 240–268.

Table 3.6:   Elasticities of demand for cars in the United States in 1990.

Model	Price	Price elasticity
Mazda323	$5,049	6.358
Sentra	$5,661	6.528
Escort	$5,663	6.031
Cavalier	$5,797	6.433
Accord	$9,292	4.798
Taurus	$9,671	4.220
Century	$10,138	6.755
Maxima	$13,695	4.845
Legend	$18,944	4.134
TownCar	$21,412	4.320
Seville	$24,353	3.973
LS400	$27,544	3.085
BMW 735i	$37,490	3.515

Source: Steven Berry, James Levinson, and Ariel Pakes, "Automobile Prices in Market Equilibrium," *Econometrica* 63 (1995), pp. 841–890.

as 90 percent for the same car models although relaxation of trade barriers and increased competition has narrowed price differences.[4]

The elasticities in the European car market run slightly higher than elasticities in the United States. Table 3.6 shows elasticities of demand for cars arranged in order based on the final purchase price. The elasticities are calculated for final customer demand. Elasticities of demand tend to decrease for more expensive cars. This may be due to the availability of substitutes in the sub-compact and compact segments as well as to customer value of individual brands in the luxury car segments. In general, elasticity of demand for any particular good varies with the price point. The manager should expect that if the price is increased or decreased substantially, the firm will encounter different price elasticities than for small price changes. The effect of the price level on the elasticity itself has important implications for pricing strategy.

A study of demand for automobiles shows that if the market demand is split between first-time buyers of a car model and repeat buyers of that model, the elasticity of demand tended to be substantially lower for the repeat buyers. Presumably, the buyers that purchased the car again were satisfied with the car's features and less price-responsive in their purchase decision (see Table 3.7). Moreover, repeat buyers were able to avoid the costs of searching for different cars and the effort required to learn about their different features. Notice also that imported cars tended to have more elastic demand than domestic cars.

Customer price elasticity changes over time. Customers often are insensitive to price changes in the short run due to inertia in buying patterns and the cost of searching for alternatives. This means that the firm is likely to retain customers in the short run after a price increase and to attract few new customers after a price decrease. After some time has passed, customers adjust to the change in relative prices so that the firm may lose more existing customers after a price increase or attract more new customers after a price decrease.

Monitoring changes in the elasticity of demand allows the company to consider changes in customer demand for its goods. As substitute

Table 3.7: Elasticities of demand in the U.S. car market for domestic and imported cars.

Class	Origin	Elasticity (first-time buyer)	Elasticity (repeat buyer)
Subcompact	Domestic	3.62	2.98
	Imports	5.25	2.95
Compacts	Domestic	4.87	3.15
	Imports	5.72	3.37
Intermediate	Domestic	5.32	2.84
	Imports	6.22	4.93
Luxury	Domestic	2.60	1.11
	Imports	3.13	2.00
Sports	Domestic	2.35	1.40
	Imports	3.02	1.14
Pickups	Domestic	5.14	3.16
	Imports	3.98	2.15
Vans	Domestic	5.50	3.98
	Imports	4.88	2.44

Source: Adapted from Pinelopi Koujianou Goldberg, "Product Differentiation and Oligopoly in International Markets: The Case of the U.S. Automobile Industry", *Econometrica* 63 (1995), pp. 891–952.

products become available they will impact the elasticity of demand for the firm's product. With a greater number of substitutes or with more attractive substitutes, the firm's demand will tend to become more elastic. This means that as consumer alternatives improve, their demand for the firm's goods becomes increasingly price-responsive. Then, a price increase will cause more consumers to shift to the alternatives. Thus, introduction of frozen yogurt by a competitor can be expected not only to reduce the demand for ice cream but to alter the price responsiveness of demand for ice cream. As customer attraction to yogurt develops, the elasticity of demand for ice cream is likely to increase.

As the company contemplates entry into new markets, demand elasticity again provides a useful guide to pricing and product introduction. By evaluating the elasticity of demand in market segments, the company can determine the responsiveness of customers to price increases and discounts. For what price ranges is demand elastic and in what price ranges is demand inelastic?

Evaluating Demand Elasticities

How are demand elasticities evaluated? Both formal and informal measures are available to managers. Formal methods of evaluating customer demand yield more precise figures but are necessarily based on past demand. Applying demand elasticity estimates based on past demand requires either that the market demand remain fairly stable over time or that the manager identify and estimate specific factors that will alter future elasticities.

Formal methods of measuring demand elasticities fall into four main categories that are listed in Table 3.8. The firm can collect data on actual purchases or on customer preferences and intentions. Data can be gathered in the normal course of business or it can be collected with experimental controls. After data are gathered, statistical analysis is used to estimate elasticities. The manager can rely on aggregate

Table 3.8:   Techniques for measuring price sensitivity.

	Market Data	**Experimental Data**
Customer Purchases	Sales data and consumer panel data	Laboratory and in-store experiments
Customer Preferences	Customer questionnaires, demand estimation	Controlled studies of purchasing and willingness to pay

Source: Thomas T. Nagle, *The Strategy and Tactics of Pricing* (Englewood Cliffs, N.J.: Prentice Hall, 1987), Exhibit 11.1, p. 266.

company sales data, store audit data, or consumer panel data. Company sales data and store audit data are readily available but may lack sufficient detail. Consumer panel data is obtained from marketing research companies that intensively study the purchasing patterns of a group of households. The company can learn about its customers' preferences and intentions through direct questioning and buy-response surveys. More precise data can be obtained by controlled pricing experiments, which can take place in selected stores or at a research facility. Finally, the company can conduct experimentally controlled studies of customers' preferences and intentions by purchase surveys or by estimating customer willingness to pay for product attributes.[5]

Managers often rely on informal measures because formal methods can be costly and time consuming. Informal methods have the disadvantage of yielding less precise estimates than formal methods. However, they have the advantage of being forward looking, allowing managers to identify the factors that will raise or lower customer price sensitivity in the future. Informal methods of evaluating customer demand are part of the firm's overall marketing and sales activities and support the firm's strategic analysis.

The manager needs to evaluate all aspects of the company's customers. She or he weighs the effects of customer characteristics on willingness to pay for the company's products. Customers can also respond to a price increase by curtailing their purchases of the product and saving their income for other purposes. The manager can make inferences about demand elasticities based on information about customer incomes and the share of income devoted to purchasing the company's product.

The main effect of a price increase is that it can induce customers to switch to a competitor's product that is similar or that acts as a substitute. The manager should determine what types of substitutes are available and how they are priced relative to the firm's product. When substitutes are available, the extent of customer price sensitivity will depend in part on brand loyalty.

Customers often encounter switching costs when they must learn how to use another product or adapt to a product with unaccustomed features. Customers may choose to upgrade to a new version of their existing software rather than purchase a competing product if it is costly to learn to use a different software program. Loyalty programs such as airline frequent-flyer miles or discounts for product upgrades are intended to increase customer switching costs and reduce their price sensitivity.

The manager also should determine whether the customer faces transaction costs associated with switching products. Such transaction costs can include the costs of searching for competing suppliers. The consumer can encounter costs of obtaining information about the prices and characteristics of competing products. Customers may remain loyal to a domestically produced product if they are unfamiliar with an international brand. Also, the customer can incur costs of communication and negotiation in establishing a relationship with another company.

A crucial aspect of switching costs is the value of the customer's time. A customer that spends time shopping or making a purchase could be using that time for other activities. The value to the customer of the best alternative activity is the customer's opportunity cost of time. By providing greater convenience for customers, a company reduces the customer's total purchase costs while at the same time lowering the customer's price sensitivity. Customers with a high opportunity cost of time will be more sensitive to convenience and less price-sensitive.

Informal methods of evaluating demand are important in monitoring prospective customers. In an industry experiencing rapid technological change, competitors are continually developing innovative substitute products. Moreover, competitors might offer customers lower transaction costs or lower costs of learning about their products. The expansion of Internet commerce is making it easier for customers to explore alternatives and to switch suppliers with the click of a computer mouse. The manager must continually make predictions about the identity and characteristics of the company's customers.

3.3 *Suppliers*

While customers define the business, the company's *suppliers* are its foundation. No company is an island; companies rely on others for financing, services, manufactured inputs, and technology. The manager's external analysis identifies the company's suppliers with almost as much care as it gives to understanding its customers. The quality of the company's suppliers often makes the difference between success and failure in delivering customer satisfaction.

The company's organization is defined by its choice of suppliers. Cisco Systems, a leading provider of networking hardware and software for the Internet, manages 37 plants around the globe but owns only two of them, relying heavily on its suppliers.[6] In contrast, a vertically integrated company such as BPAmoco is present at every level, from exploration, to crude oil mining, to refining, to wholesaling gasoline, to retailing gasoline at the pump, doing practically everything within its organization.

The company stands as an intermediary between its suppliers and its customers. A firm obtains products and services from its suppliers and adds value through its manufacturing, assembly, distribution, or sales activities. By carefully choosing what activities to leave to its suppliers and what activities to perform within the organization, the company concentrates on those activities that are the key to its value creation and the source of its profitability.

The military strategist von Clausewitz observed that when an army begins an operation, whether attacking, invading, or taking up positions along its own borders, "it necessarily remains dependent on its sources of supply and replenishment and must maintain communications with them. They constitute the basis of its existence and survival."[7] A company's suppliers are essential for its economic survival. Many companies have learned the importance of developing good working relationships with their suppliers and keeping the lines of communication open.

Forging relationships with suppliers is a crucial component of market-focused strategy. The company's profit is affected in large part by the interplay between the costs of goods and services offered by suppliers and revenues from the goods and services demanded by customers. The company defines its markets by connecting suppliers and customers in innovative ways. The quality and cost of goods and services obtained from suppliers are crucial to the company's ability to provide quality goods and services to its own customers or to keep its costs competitive.

By *prospective suppliers* we mean both those with whom the firm has continuing relationships and new suppliers of goods and services. Suppliers include investors who provide various forms of financing, prospective employees, providers of services and manufactured inputs, and developers of technology.

What are the key characteristics of prospective suppliers? It is useful to divide suppliers into four categories: goods and services, technology, labor services, and finance capital. Even though all types of markets have some fundamental similarities, the basis of the classification lies in differences in contracting for productive inputs. Contracting in the four types of markets has implications for ownership and control of the company. The intermediation activities of companies in these markets can differ as well.

The company must be customer-focused, but effective procurement behind the scenes is essential. The competitive ability of the firm depends in many ways on how well it purchases goods and services, selects productive technology, hires personnel, and obtains financing. Just as marketing and sales are the company's connection with its customers, the company's purchasing, engineering and design, human resources, and finance functions are the company's contacts with its suppliers. Companies must have effective procurement strategies in each type of input market.

Goods and Services

By *goods and services* we mean all products purchased by the firm, including manufactured components, capital equipment, and natural resources such as land and raw materials. This also includes all the services received by the firm, including advertising, accounting, legal, data processing, consulting, and R&D. Companies receive utility services such as water, electric power, natural gas, and telecommunications. Manufacturers depend on the reliability of critical suppliers of raw materials, manufactured components, and outsourced services. Wholesalers and retailers depend on access to manufactured goods.

The company's purchases of goods and services are carried out through spot and contractual exchange. The procurement of productive goods and services is a crucial part of any business whether the company is a manufacturer or distributor. A manufacturer must decide whether to make or buy each productive input. If the input will be produced by the company, the organization must expand through investment and hiring or through mergers and acquisitions. If the input is to be purchased, the company must choose between multiple sources and determine what types of contracts with input providers are required. These procurement decisions determine the company's degree of vertical integration.

Labor Services

The company employs many types of *labor services*, such as managers, engineers, accountants, production personnel, and clerical staff. Companies must have a coherent employment policy, hiring the most talented people and investing in training and employee development.

The company's purchase of labor services is not simply a set of spot transactions at a fixed wage, as in the traditional economic model of labor markets. The company's employees form the organization, managing and staffing the firm. Certainly, the company has a legal identity that is distinct from its employees. However, the company's knowledge and skills generally rest with its personnel. The employees are not just a flow of services that passes through the firm, although that may be the case for individual workers. Collectively, groups of employees have substantial bargaining power whether through unions or internal divisions of the firm. Because most employment contracts are of long duration, the purchase of labor services is a long-term agreement in which the provider of labor services has a stake and an influence on the decisions of the firm.

Technology

Companies utilize diverse types of *technology*, including basic scientific information, engineering specifications, operations management, employee training, licenses, computer software, and other intellectual property. This category broadly represents the marketplace of ideas, including the output of commercial, government, and university laboratories and research centers. Obtaining technological innovations is critical to keeping and holding a competitive lead.

Markets for technology and other informational inputs have unique characteristics. Information may be scarce and costly to produce, but particular know-how can be used repeatedly without being depleted, it can be reproduced at minimal cost, and in some instances it can be easily copied. Ownership rights to "intellectual property" are notoriously difficult to define, complicating market transfers and contractual arrangements. Companies can obtain technological knowledge through in-house research and development, association with universities and research laboratories, and technology transfers from other companies. Information has become the primary driver of the economy. The creation, processing, and transmission of information is a pervasive activity that is reshaping every type of business. The information economy does not simply mean an increased usage of computers and telecommunications. Rather, it represents fundamental changes in the design of organizations and the structure of market transactions.

Finance Capital

Finance capital refers to the firm's sources of capital investment, including securities, bonds, bank loans, and venture capital investments. Winning markets requires an effective financial strategy to obtain the funds for growth and cost-reducing investment. Companies compete through capital expenditures on facilities, equipment, and research and development.

The corporate financing process has unique qualitative features. To obtain financing, companies not only provide returns to investors but surrender various degrees of control to new debtors and shareholders. The financial sector is distinguished in part by its sheer size, the efficiency and rapidity of its markets, and the existence of major centralized exchanges. The securities markets allow transfers of assets that provide residual returns after the claims of the company's debtors have been satisfied. They are also markets for *corporate control* because securities confer control rights to shareholders. The debt that companies obtain from banks and other lenders also confers some rights of supervision and control, particularly when agreements are renegotiated or in the event of bankruptcy. This implies that the company's financing decisions are not simply purchases of capital in return for a payment of a share of profit or interest. Rather, they entail choices about the ownership structure and control of the firm by investors.

3.4 *Supply Characteristics*

The main strategic aspects of supplying goods and services include supplier costs, technology, and price elasticity of supply. Just as with customer demand, it is important to understand how the supplier market segments fit together. The company can combine its suppliers to adjust the types of inputs it uses to produce goods and services. As technology changes, the characteristics of the company's suppliers will change accordingly. As the company develops new products and enters new markets, it may need to develop new supplier relationships.

Who Are Our Suppliers?

Companies depend on a wide range of suppliers for their goods and services. Large wholesalers and retailers deal with thousands of manufacturers. Large manufacturers can have thousands of parts suppliers. Understanding the characteristic of suppliers is almost as important as understanding a company's customer characteristics. Accordingly, it is necessary for companies to keep track of the performance of their suppliers and to scan markets to determine prospective suppliers. Interaction with suppliers entails significant costs of transaction and communication, so that improving the efficiency of transaction with suppliers can have significant returns.

Exhibit 3.2   U.S. Auto Manufacturer Interaction with Suppliers

In the automobile industry, interaction with suppliers requires extensive coordination to assure that parts are available when they are needed and that excess inventories do not accumulate. Additional coordination efforts ensure that parts are properly designed and that quality standards are met and even exceeded. A key aspect of interaction with suppliers is the exchange of information. Manufacturers need to transmit orders and technological specifications to parts makers. In turn, parts makers need to send information about their technical abilities, bids to supply parts, and billing information.

General Motors (GM) has over 30,000 suppliers and operates a World Wide Purchasing (WWP) organization. Harold Kutner, GM vice-president for WWP, observed that "it is an awesome achievement to supply billions of parts to a corporation that produces

(Continued)

Exhibit 3.2 (Continued)

vehicles in 30 countries and sells them in over 170 countries in the world."

WWP's responsibilities for procurement are divided into four regions: Asia Pacific, Europe, Latin America, and North America. Commodity purchasing has four main areas: metallic, chemical, electrical, and machinery and equipment. Commodity executive directors within each area are responsible for worldwide purchasing. Metallic goods include axles, cables, castings, catalytic converters and exhaust systems, exterior decorative trim, fasteners (attaching devices — bolts, screws, nuts, plastic fasteners, plugs, rivets, wheel locks), forgings, jacks, camshafts, engine valves, steering systems, shock absorbers, silencers, stampings, steering systems, transmissions, and wheels. Chemical includes items such as air bags, blow-molded plastic parts, chrome-plated plastic parts, glass, fluids (anti-freeze, engine oil, and transmission fluid), injected molded plastics, carpets, floor mats, paints, resins, trunk trim, headliners, acoustics, door handles, cargo nets, mirrors, name plates, sealing strips, sealants, seats, seat belts, tires, and wheel covers. Electrical products include alternators; batteries; bulbs; horns; electronics; communication and entertainment systems; heating, ventilation, and air conditioning (HVAC) systems; wiper systems; lighting systems; lock systems; remote keyless entry systems; sensors; control modules; spark plugs; switches; thermostats and water pumps. Machinery and equipment products include industrial, building, janitorial and business supplies; spare parts; construction services; mobile equipment; machine presses; dies; corporate services; health care; and sales and marketing services and agencies.[8]

According to GM, the mission of WWP is "to globally supply required direct and indirect materials, machinery & equipment, and services that provide more value to our customer and improve customer enthusiasm through quality, service, technology and price". To achieve its mission, GM outlines a number of strategies. The company seeks to make purchases based on globally competitive quality, service, technology, and price and to coordinate purchasing across regions. GM directs suppliers to engineer products carefully, manufacture locally, and achieve results on a global basis. GM encourages suppliers to participate in its global sourcing network. GM seeks to "balance the sourcing footprint" by matching sales and purchases in each individual country, where possible, to reduce the impact of currency exchange fluctuations.

(Continued)

Exhibit 3.2 (*Continued*)

Also, GM coordinates its purchases of parts for production and those for service. GM's Process Control and Logistics handles the flow of materials, components, and vehicles within the organization. GM's Global Supplier Network (GSN) website provides worldwide access for its suppliers.

GM engages in a complex formal procedure for quality improvement with its suppliers. An eight-step process of planning and prevention begins with an on-site evaluation of the supplier's ability to meet requirements and continues with approval by GM's sourcing committee, definition of the steps necessary to monitor quality, pre-production meetings, prototype sample approval, production part approval, physical verification of the supplier's production process, and a review of the supplier's early production plans. GM also establishes formal procedures for problem resolution that include defining the supplier's responsibility for continuous quality improvement, problem reporting routines, quality workshops and meetings, and finding new sources if the original supplier does not meet quality and price targets. In addition to its sourcing activities, GM has invested in the development of over 5,000 suppliers in 43 countries, sharing information and techniques such as lean manufacturing.

GM spun off its internal parts manufacturing unit, Delphi Automotive Systems Corp., in 1999 to create the world's largest auto parts supplier with over 200,000 employees. Delphi's headquarters are in Troy, Michigan, with regional headquarters in Paris, Tokyo, and Sao Paulo. Its four business areas are Dynamics and Propulsion, Safety, Thermal and Electrical Architecture, and Electronics and Mobile Communications. Delphi has 168 manufacturing sites, 38 joint ventures, 51 customer and sales offices, and 27 technical centers in 36 countries. In addition to specific parts, the company offers integrated systems and modules to automobile manufacturers.[9]

Ford Motor Company's purchasing organization has seven units: Vehicle Center/Powertrain; Commodity Management; Facilities, Materials and Services; Supplier Technical Assistance; Total Cost Management; Purchasing Process Leadership; and the Purchasing Business Office. Total Cost Management works with suppliers to reduce costs and apply lean manufacturing techniques. Purchasing Process Leadership coordinates communication and the Purchasing Business Office develops the organization's long-term strategy and competitive analysis.

(*Continued*)

> ### Exhibit 3.2 *(Continued)*
>
> Ford created an Internet website that serves as a portal for its thousands of suppliers. The use of non-proprietary Internet protocols allows easier access to the network for many suppliers. Called the Ford Supplier Network, the portal reduces Ford's cycle time and inventory requirements by streamlining the process of data interchange with its suppliers. Suppliers can check on the status of a bid, track engineering changes, monitor assembly rejection rates, and handle transactions through the network. Ford launched its portal with 14,000 individual accounts at 1,500 companies, with a goal of reaching one million accounts at 16,000 companies.[10]
>
> A trade association of North American vehicle manufacturers and suppliers, the Automotive Industry Action Group, established an Internet-based network called the Automotive Network eXchange (ANX) to facilitate electronic communications. The service bills itself as "the world's premier business-to-business managed communications infrastructure."[11] The automotive industry also has a high-speed network called the Automotive Industry Exchange that connects large manufacturers with suppliers, allowing transmission of computer-aided design and computer-aided manufacturing (CAD/CAM) information.

Supplier Characteristics

Companies have learned to focus on the preferences of their customers. This essential knowledge about the characteristics of the demand side needs to be supplemented by understanding the characteristics of the supply side. Companies also acquire knowledge about their suppliers. As global competition has intensified, companies also realize that few can go it alone. Alliances, partnerships, and other long-term relationships with suppliers are necessary to assure performance. What are the critical features of suppliers?

For suppliers of manufactured parts and manufacturing equipment, the critical feature is their technology. The supplier's technology is a shorthand way of referring to many things, including technical manufacturing know-how, proficiency in product design, basic scientific knowledge, and engineering expertise. The supplier's technology can provide methods for transforming inputs into outputs, as in a refiner's production process for transforming crude oil into gasoline and other by-products. It can be artistic skills and artisan traditions such as making musical instruments.

For suppliers of services such as marketing, accounting, legal support, and management consulting, human capital is an essential component. Human capital refers to the training, education, and abilities of the firm's employees. Firms also possess organizational capital, which is the firm's accumulated information about the skills and abilities of its employees. The organizational capital of suppliers is important and not easily duplicated by raiding a few key employees. In addition, critical supplier characteristics include their intellectual property such as software and wisdom distilled from industry experience. The supplier's organizational design, management style, and corporate culture are also important determinants of performance.

Because companies increasingly rely on suppliers in alliances and long-term partnerships, the trustworthiness of suppliers is important as well. Before entering into such relationships, companies evaluate the reputation of their suppliers in the industry. Can the suppliers be relied upon to adhere to pricing terms, to provide on-time delivery, and to maintain quality targets? Will suppliers stay up-to-date with technological change? Will suppliers seek continual improvements in performance? Will suppliers adapt to changing market requirements?

Because relationships with suppliers are often complex and time-consuming, managers evaluate whether the supplier is easy to work with. The costs of transacting with a supplier includes not only the purchase price of goods and services but also the indirect costs of negotiating with the supplier, exchanging information, and monitoring their performance. The supplier's management, corporate routines, and information processing should be a good fit when extensive interaction is necessary.

Supplier costs also are crucial. The supplier's costs reflect many underlying determinants including the supplier's technology and operating efficiency. Costs also depend on how much the supplier pays for its inputs, including technology, parts, human resources, and capital. The supplier's manufactured input costs depend on the supplier's own procurement relationships. The supplier's human resource costs reflect its hiring and the productivity of its employees. The supplier's capital costs reflect the type of financing it obtains and the risks perceived by its investors. Accordingly, companies sometimes review the financial solvency of their suppliers. Companies evaluate the costs of their suppliers to evaluate the ability of the supplier to perform effectively, maintaining competitive prices and performance.

It is useful to highlight an important aspect of costs known as *opportunity cost*. Because suppliers tend to have limited capacity, their provision of a product or service generally requires them to forgo some other opportunity. They may miss the other opportunity because their production capacity or available time is limited. Suppliers obviously forgo other opportunities when they sign an exclusive supplier agreement. The return from the best opportunity forgone represents the supplier's opportunity cost. If a company must forgo multiple opportunities that could have been taken all at once, opportunity costs are the total of the opportunities forgone.

Example 3.3   Opportunity Cost

A supplier receives two orders but only has the capacity to fulfill one order. The first order will be worth $15 in net earnings to the supplier. This is the supplier's opportunity costs from taking on the second order, because taking the second requires forgoing $15. Suppose that the supplier has operating costs of $7 from taking the second order. How much must the second order pay? The revenue from the second order must cover the supplier's total costs, which equal the direct cost of $7 and the opportunity cost of $15, for a total of $22.

Supplier Costs and the Supply Schedule

A company chooses its suppliers on the basis of their characteristics. Recall that customer willingness to pay is an important underlying determinant of market demand. In a very similar way, supplier costs are an important underlying determinant of market supply. To bring forth additional supplies of an input, companies may have to provide price incentives to a particular supplier or to a set of suppliers. Accordingly, companies are highly interested in the price responsiveness of their suppliers. For a standardized manufactured input or a specific service, *supplier costs determine their price responsiveness.*

Consider a company that sources a particular input from many suppliers. For example, a company purchases a standardized component from many suppliers. Costs differ across individual suppliers depending on their technology and their own input costs. Suppliers would be willing to provide the component as long as the price offered is greater than or equal to their unit costs. Suppliers belong to the same market segments when their costs are similar.

The supply experienced by the firm is the total offered to the firm by individual suppliers. To calculate the total supply, use the *adding up* method. First estimate how many suppliers are in each market segment at the various cost levels, as in the first and second columns of Table 3.9. Then, *add up* the middle column to see how much supply there is at each offer price, which in this example is $1 more than the cost level. The last column in Table 3.9 summarizes the supply schedule. This method can be adapted for more suppliers and more gradations of costs. To verify that cost is an upward-sloping line, make a graph using the first and last columns. The supply curve resembles a

Table 3.9: The adding up method of calculating supply.

Cost Levels	Number of Suppliers at Each Cost Level	Supply "Add up" Column 2
30	7	32 (at price 31)
20	16	25 (at price 21)
10	9	9 (at price 11)

staircase. The supply schedule and the supply curve summarize information about the company's suppliers.

The higher the price offered by the firm for the input, the more suppliers will offer to provide the product because more suppliers have costs below the offer price. As in the case of demand considered earlier, it does not matter how many suppliers are in each segment. Supply increases as the offer price increases because suppliers are added together.

Elasticity of Supply

How are prices determined for purchased inputs? When suppliers have market power, they post prices for the inputs they supply to the firm. When both suppliers and the firm that is purchasing inputs have market power, prices for inputs are set through negotiation. A company that has market power in dealing with its suppliers can select the price to offer suppliers or require suppliers to bid competitively to supply the firm.

In markets where there are many potential buyers of the inputs, the elasticity of supply will be very high since suppliers have many potential customers. In markets where there are few potential buyers, or where the purchasing requirements of the firm are very large relative to other buyers, the firm will have market power as a buyer. For example, large retailers such as Wal-Mart and Target have significant buying power in relation to manufacturers and wholesalers. Large manufacturers such as General Electric have significant buying power in relation to their suppliers of parts and equipment.

In the case of a firm that has substantial buying power in dealing with suppliers, a key number is the price elasticity of supply. The price elasticity of supply summarizes the price responsiveness of the company's suppliers and gives an indication of their costs of serving the firm. The firm is concerned about its total expenditures on inputs. Strategic pricing in procurement recognizes that to purchase more inputs the company may have to pay more to its suppliers. If the payment to suppliers is in the form of a per-unit price, purchasing more of the inputs requires raising the price offered to suppliers. To evaluate the effect of purchasing more of the input on its total expenditure, the firm calculates how much expenditure rises when more of the input is purchased. The effect of increasing the amount purchased on total expenditure is called the marginal expenditure.

When the firm increases the price it offers to suppliers, two results occur:
More units are purchased.
More is paid per unit purchased.

The relative size of these two effects is measured by the size of the supply elasticity. The price elasticity of supply is the percentage change in the quantity purchased from suppliers divided by the percentage change in the price offered to suppliers. Example 3.4 illustrates how to calculate the elasticity of supply.

Example 3.4   The Elasticity of Supply

A company initially offers suppliers a price of $20 per unit of the input purchased from those suppliers. Then, the company increases the price offered to its suppliers by $1 to $21, that is, a 5 percent increase in the price offered to suppliers. Suppose that the firm purchased 500 units at the price of $20 each. Suppose that raising the price to $21 induces suppliers to provide 550 units. This is an increase in supply of 10 percent. This example provides enough information to calculate the elasticity of supply. Divide the percentage increase in supply by the percentage increase in the offer price to obtain an approximate measure of the supply elasticity (see Table 3.10).

Table 3.11 shows industry supply elasticities for selected U.S. industries.[12] The categories are rather broad and more targeted estimates could presumably be obtained for narrower groups of firms. The numbers in the table are useful for understanding the general range of supply elasticity. Because the elasticity numbers include the overall supply response, they mingle together short-run and long-run supply

Table 3.10:   Calculating the elasticity of supply.

Percentage rise in price	Percentage rise in quantity	Price elasticity of supply
Initial price $20 New price $21	Initial quantity sold 500 New quantity sold 550	Divide column 2 by column 1
$\dfrac{\$21 - \$20}{\$20} = 5\%$	$\dfrac{550 - 500}{500} = 10\%$	$\dfrac{10}{5} = 2$

Table 3.11: Some industry elasticities of supply in the United States.

Industry	Elasticity of Supply
Lumber	1.06
Sawmills	1.06
Millwork and plywood	8.85
Building paper and board	5.59
Drugs	2.14
Paints	3.72
Asphalt and paving materials	1.79
Tires	14.49
Stone, clay and glass products	5.35
Glass products except containers	9.62
Cement	3.80
Structural clay	5.00
Construction pottery	5.18
Concrete and plaster	8.85
Metal barrels	9.43
Electronic equipment	5.81

Source: Adapted from Table II in John Shea, "Do Supply Curves Slope Up?" *Quarterly Journal of Economics* 108 (1993), pp. 1–32.

effects. Presumably, the price elasticity of supply is greater in the long run as companies adapt to price increases by increasing production and as new entrants respond to incentives provided by higher prices.

Elasticities of supply are useful for evaluating supplier differences across the firm's supply markets. Elasticity can vary over time as existing suppliers and potential entrants respond to changes in the offer price. The supply response differs considerably across the industries shown in Table 3.11. Not only are the supply elasticities different, but the reaction times also are considerably different. For example, the reaction time between the impact and the peak effect of a price change is 12 months for lumber and for sawmills but 3 months for millwork and plywood. For any given industry, the longer-run supply elasticities are greater as companies are able to change their production plans.

Elasticities of supply can be estimated formally or informally, as with demand elasticities. Informal analysis of supply elasticities is useful as a way of evaluating changing market conditions. Managers must determine what alternative activities suppliers can pursue. How dependent are the firm's suppliers on its business? Do suppliers have information about alternative markets for their products and services? Suppliers can encounter search costs in finding new customers and they may encounter switching costs in tailoring their products and services to the requirements of other customers. Generally, search costs and switching costs make suppliers less responsive to price reductions.

The elasticity of supply also reflects the suppliers' own costs. Technological change that lowers the suppliers' operating costs should increase total supply and make suppliers more price-responsive. Similarly, lower costs of purchasing inputs by suppliers, such as lower labor cost for example, should also increase the price responsiveness of suppliers. New technology that allows suppliers to have greater flexibility to serve other customers also will increase their price elasticity of supply.

Quantity Discounts

The discussion of price elasticity of supply is based on offering suppliers higher per-unit prices to provide incentives to individual suppliers to produce more or to find high-cost suppliers when low-cost suppliers are already at full capacity. Large companies need not pay more per unit when procuring significant quantities of an input. Instead, they often can become eligible for quantity discounts from suppliers. Quantity discounts stem from two sources, supplier market power and supplier scale economies.

If suppliers have market power, they may choose to give special deals to their largest customers. Under these special deals, the larger customers pay more in total, but pay lower per-unit prices than do smaller customers. This is a form of price discrimination practised by suppliers. Suppliers earn more money by charging high per-unit prices to smaller customers and lower per-unit prices to their larger customers. For example, manufacturers of food products typically charge less per unit to supermarket chains than to smaller independent grocery stores. By charging uniformly high prices to both types of stores, the supplier would sell less to the large stores. By charging uniformly low prices, the supplier would forgo revenue that could be earned from the small stores.

In addition to supplier market power, quantity discounts result from supplier economies of scale in serving customers. Manufacturers of food products or their wholesalers typically incur lower costs of serving a supermarket chain than do smaller independent grocery stores because invoicing and ordering costs are spread over a larger volume. Also, large deliveries can be made to a central warehouse rather than incurring the transport costs of making many deliveries to each store location.

3.5 Overview

The external analysis always begins with understanding the characteristics of the firm's customers. The purpose of understanding customers is that they are the foundation of the business. The company earns its revenues by providing products that satisfy customer needs. The ability of the firm to earn revenues depends on the extent of customer willingness to pay for the firm's products. By understanding customer preferences, the firm can improve the combination of features of its products and adjust the variety of products that are offered. Understanding customer preferences is also essential to determining the firm's pricing strategies. By understanding customer preferences better than competitors and tailoring prices and products accordingly, the firm obtains a competitive advantage.

In the external analysis, the manager considers prospective customers through informal methods as well as formal evaluation of total demand and price elasticity of demand. The manager segments markets based on specific characteristics of groups of consumers. The manager also takes into account the price elasticity of demand from particular demographic groups in choosing the firm's pricing policies and product offerings.

The external analysis then embraces the company's prospective suppliers. The purpose of understanding suppliers is that the company's expenditures are the costs of the business. The company's value depends on what portion of revenues is taken by costs. By understanding market supply, the manager can tailor the company's purchases and the prices it offers to suppliers. Moreover, the efficiency of suppliers and the quality of its products are critical to the company's operation. A manufacturer depends on the quality of parts and the cooperation of its suppliers. A retailer or wholesaler depends on the products it can resell. Firms that understand their suppliers better than their competitors do, and adjust their purchases accordingly, can gain a competitive advantage.

In the external analysis, the manager chooses prospective suppliers based on their technology and costs, their reputation, and the transaction costs of working with the supplier. The price elasticity of supply usefully describes the supplier market when many suppliers provide a standard product or service. The price elasticity of supply is also helpful in determining price incentives to offer suppliers. Companies must pay more per unit to elicit additional supplies. In contrast, when suppliers have market power or experience cost economies in serving larger customers, larger customers can obtain quantity discounts on larger orders.

Value-driven strategy requires continual monitoring of changing customer requirements and changing supplier capabilities. The effectiveness of the firm's strategy depends in large part on the quality of the information generated by the firm's external analysis. Companies integrate the information about prospective customers and suppliers to maximize the value of the firm. Whether the firm is a retailer, wholesaler, manufacturer, or service company, the firm creates value by bringing together buyers and sellers in innovative ways.

Questions for Discussion

1. What is the difference between vertical market relationships and horizontal market relationships?
2. Why are customers the most important part of external analysis?
3. How can a firm's customer base change drastically within a short period?
4. Consider a recent purchase that you made (clothing, travel, textbooks, entertainment). What are some transaction costs that you encountered in making your purchase? What types of transaction did the company from whom you made a purchase encounter?
5. The consumer's willingness to pay is the outcome of four main factors: budget, preferences, the set of products that the consumer might purchase, and transaction costs. Describe in detail what types of things might determine these four components of the consumer's willingness to pay.
6. Give a few examples of transaction costs that might affect consumers' net willingness to pay.
7. Why should managers keep track of the price elasticity of customer demand?
8. Describe the formal techniques for measuring demand elasticity and give examples for each technique.
9. How do the firm's relationships with its suppliers affect its ability to provide products to customers?
10. What are some types of transaction costs that a firm encounters in obtaining supplies of parts and equipment?
11. What are some types of transaction costs that a firm encounters in hiring personnel and obtaining financing?

Endnotes

1. Getting close to the customer can also mean relying less on intermediaries and more on direct contact with customers, not only to gather information but also to capture margins from providing services.
2. The information in this exhibit is drawn from Constance L. Hays, "The Heavyweight Goes In-House," *New York Times*, July 8, 2001, Section 3, p. 1.
3. The calculations in the exhibit are based on information in Hays, ibid.
4. See Frank Verboven, "International Price Discrimination in the European Car Market," *Rand Journal of Economics* 27 (1996), pp. 240–268.
5. The discussion in this paragraph is based on Thomas T. Nagle, *The Strategy and Tactics of Pricing* (Englewood Cliffs, N.J.: Prentice Hall, 1987), chapter 11, pp. 265–298.
6. John T. Chambers, CEO of Cisco Systems, http://www.cisco.com/warp/public/750/johnchambers/.

7. Carl von Clausewitz, *On War* (Princeton: Princeton University Press, 1976), p. 341.

8. http://www.gmsupplier.com/apps/gsnhome/public/purchasing/docs/selltogm/.

9. See http://www.delphiauto.com

10. David Joachim, "Ford Rebuilds Extranet as Supplier Portal," *Internet week*, May 17, 1999.

11. On the Automotive Industry Action Group (AIAG), see http://www.aiag.org/scriptcontent.index.cfm

    The quotation was taken from the former website of the Automotive Network eXchange (ANX). Science Application International Corp. (SAIC) bought the ANX Network from AIAG in 1999, see http://www.anx.com/?page=Company

12. The data are based on Table II in John Shea, "Do Supply Curves Slope Up?" *Quarterly Journal of Economics* 108 (1993), pp. 1–32. The table reports price effects associated with output increases. Accordingly I calculated the reciprocal to show the price elasticity of supply and rounded to two decimal places for ease of presentation. The article uses various different estimation methods. I report the estimates using one of these methods.

CHAPTER 4

COMPETITORS AND PARTNERS

Contents

We now turn to the other two directions of the market compass. In addition to determining who the firm's prospective customers and suppliers are, the manager's external analysis examines the firm's prospective competitors and partners. The firm's prospective *competitors* are those the firm will face as it attempts to maintain its existing market position or tries to enter new markets. The firm's prospective *partners* are makers of related products and services with whom the company can form long-term contractual relationships, strategic alliances, and joint ventures.

The firm's strategic success requires outperforming competitors. The firm's *competitive advantage* depends on identifying ways to create greater value than competitors. This requires understanding the products, technology, and organizational abilities of current and prospective competitors. The manager will then compare these benchmarks with the firm's own characteristics to find ways of doing things better. By understanding prospective competitors, managers identify potential threats to the company's ability to create value and opportunities to serve customers better than competitors.

In addition, analysis of competitors is useful in devising competitive strategies. The firm's *competitive strategies* are a best response to what managers believe the firm's competitors will do. So, to compete effectively the firm must try to determine who its competitors are and to *anticipate* what strategies they are pursuing. The problem of identifying competitors appears easy at first blush since most companies are intimately familiar with their closest rivals. What makes competitor analysis difficult in practice is that the firm's products and services are likely to face competition from innovative entrants, sometimes from seemingly different industries.

Industry analysis summarizes information about the company's multiple competitors. The manager examines the *market structure* of the industry, measured in terms of the number and size of competitors. The manager must consider how *attractive* the industry is for new entrants. Most companies face both close and distant competitors as determined by their impact on the firm's economic performance. Within industries, the firm's competitors can sometimes be classified into *strategic groups* whose activities are primarily confined to industry segments.

The firm's success often is tied to cooperation with partners. Finding the right mix of competition and cooperation is a critical part of strategy and is enhanced by external analysis. The firm's prospective partners are producers of complementary goods that enhance customer demand for the firm's goods and services or those goods that are useful to the company's suppliers. The firm's prospective partners also can include suppliers and customers. A firm can partner with a manufacturer of a critical input. In addition, firms often partner with key distributors. Finally, partners can be similar companies operating in different geographic areas.

Companies join with partners to form strategic alliances and joint ventures. Partnerships can augment the firm's ability to serve customers and deal with suppliers since the firm may not have sufficient resources to handle strong competitors. Partnerships are very useful for firms forming networks in communications, transportation, and other industries. Companies can coordinate strategy with partners that produce products and services that are purchased by the firm's customers along with its own products. Alliances with partners, even including competitors, set technological standards and share the costs of research and development.

Chapter 4: Take-Away Points

The manager's external analysis examines prospective competitors to anticipate their competitive strategies and examines prospective partners to evaluate opportunities for cooperation:

- The manager should examine prospective competitors to determine possible sources of the firm's competitive advantage.
- The manager should take into account competition from substitute products even if the products are supplied by different industries and involve product innovations.
- The manager should anticipate competition from potential entrants.
- The manager's industry analysis should emphasize the competitive conduct of potential competitors, not just the existing market structure.

- The external analysis should examine prospective partners that offer opportunities to provide complementary goods to consumers.
- The manager should evaluate potential partners for alliances and joint ventures to reduce costs of production and innovation, to form market networks, and to set product standards.

4.1 *Competitors*

Identifying competitors begins with a clear understanding of the firm's customers and suppliers. What services is the firm providing to its current and prospective customers? What benefits is the firm deriving from its current and prospective suppliers? How do the firm's activities coordinate its relationships with customers and suppliers? The firm's competitors are companies that provide similar or substitute services to the firm's customers and companies that provide related market coordination services. Defining the firm's markets and identifying prospective competitors require management insight and creativity.

Competitors can look different in every way; they might provide different types of products and services, produce products and services in different ways, transact in different ways, and put together different combinations of products and services. Ultimately, customer choices determine the nature and extent of competition.

Porter's Five Forces Framework

Identifying the new competitors that enter its markets and established competitors the firm encounters as it enters new markets is critical for any firm. In his classic 1980 book *Competitive Strategy*, Michael Porter defined the "five forces" of business:

Customers
Suppliers
Established rivals
Substitutes
Potential entrants

This highly valuable checklist emphasizes that managers should not restrict attention to obvious competitors but also consider producers of substitute goods that compete for customer attention and expenditures. Moreover, the firm must be aware of opportunities for entry of potential competitors who will pose future challenges. The firm's established rivals, suppliers of substitutes, and potential entrants are included in the list of prospective competitors.

The characteristics of prospective competitors can be qualitative. Porter outlines four questions to ask: What drives the competitor? What is the competitor doing and what can it do? What assumptions does the competitor hold about itself and the industry? What are the competitor's strengths and weaknesses? By answering these questions the manager forms a *competitor response profile* consisting of the competitor's goals, assumptions, strategies, and capabilities. The profile attempts to infer whether the competitor is satisfied with its current position, what likely moves or strategy shifts the competitors will make, where the competitor is vulnerable, and what will provoke the greatest and most effective retaliation by the competitor.[1]

Porter's analysis of competitor strengths and weaknesses includes many factors, including the competitor's product quality, distribution, marketing and sales, operations, research and engineering, overall costs, financial strength, organization, general managerial ability, and corporate portfolio. In addition, an external analysis examines the competitor's core capabilities, ability to grow, quick response capability, ability to adapt to change, and staying power.[2]

Porter also emphasizes that the bargaining power of the firm's customers and that of its suppliers are also competitive forces. The firm competes with its customers and suppliers for a share of the total value created by the firm's transactions. The firm's earnings are lowered by the extent to which it shares value created with customers and suppliers. On the other hand, the firm must share value to attract customers and suppliers away from competing opportunities. The firm should evaluate the market alternatives available to customers and suppliers, the information available about those alternatives, and their costs of switching.

Industry Analysis

Industry analysis provides the manager with summary descriptions of competition that are useful for strategic decision-making. The manager attempts to define the relevant markets that the company is serving. Then, the manager attempts to characterize the extent of competition.

Market definition is a difficult process that is closely related to defining the nature of the company's business. Managers should try to identify the relevant market from the perspective of customers. The manager must ask what problem or need customers are trying to address and what benefits customers receive from the product. Recall the earlier discussion of substitute products. The relevant market is not defined simply by the physical characteristics of the company's product or service. Rather, the relevant market includes products and services that are substitutes for the company's offerings.

Managers must continually reevaluate their product definition to determine whether it is too narrow or too broad. If managers use a description of the product that is too narrow, the market definition will be narrow as well and the company will miss the threats posed by many potential competitors. If managers use a definition of the product that is too broad, they may be distracted by perceived threats from many competitors. The manager's product definition should be based on the extent to which price increases in the company's product induce the consumer to switch to competing products or services. If air fares rise, do some travelers switch to other modes of transportation, such as driving, or to other modes of communication such as teleconferencing?

A company's market may have geographical boundaries, that is, it may be local, national, regional, or global. For example, Legend focused on the domestic computer market in China. Exxon-Mobil operates in the worldwide market for crude oil. *Corona Beer* went from being a domestic beer in Mexico to an international brand. A company's market may be defined by customer groups, such as households or business customers. For example, Taiwan's Quanta Computer produces notebook computers under contract for manufacturers such as Dell, Gateway, or Hewlett-Packard. Dell Computer supplies equipment to both households and businesses.

The manager's evaluation of the *extent of the market* should broaden as the planning horizon increases. Managers should adjust the market definition to reflect those competitors they expect to challenge. In the short run, say within the coming year, the company's immediate competitors are more easily identified. In the medium term, say one to three years, producers of substitute products and producers introducing low-cost production methods enter into the mix. In the longer term, say over three years, innovative competitors and entrepreneurial start-ups will increasingly affect profitability.

Having defined the relevant market, managers evaluate *market structure*, that is, the number and size of firms in an industry. A market is said to have a concentrated market structure if there are only a few large firms. If there is only one firm in the market, the industry is said to be a *monopoly*. If there are only two firms, the industry is a *duopoly*. With a few firms, the industry is an *oligopoly*. With many small firms, the industry is said to be *fragmented*. There may be other industry structures, with one or more *dominant* firms wielding market power and a *competitive fringe* with many smaller firms acting as price-takers.

It bears emphasis that these measures of concentration depend on market definition. Simply put, tallying the number and size of companies in a market depends on what one thinks is the relevant market. The U.S. market for cable television included AOL-Time Warner and AT&T. Yet, the U.S. market for subscription television included not only AOL-Time Warner and AT&T, but also the direct broadcast satellite television duopoly DirecTV and EchoStar. Defined as the market for television, the market also includes all of the other forms of television broadcasting, including local television stations. What would be the competitive impact of a merger between DirecTV and EchoStar? The answer goes beyond market definition.

Although market structure provides a quick and useful snapshot of the industry, what matters more is the nature of competition. This requires understanding *conduct*, which refers to the behavior and competitive strategies of companies in the industry. Do firms compete on price, product features, service, choice of distribution channels, or technological innovation? Do firms respond slowly or quickly to competitor moves? Is competition primarily from established companies or is the primary challenge from potential entrants? The competitive strategies and entry plans of competitors affect how many companies are currently in the industry and how many competitors will enter later on.

Because competitive conduct is what counts, managers should avoid relying on market structure as a simple predictor of industry attractiveness. In some industries, a few large firms may compete intensely for customers; whereas in others, firms might tacitly coordinate strategies to reduce competition. In some industries, a single firm may wield monopoly power; while in others, the market power of a single firm with a large market share is reduced by potential competition. In some industries, many small firms compete vigorously; whereas in others, many small firms may have substantial local market power due to market frictions such as customer travel costs and imperfect information.

Competitive strategies and entry decisions are crucial determinants of the number and size of firms in an industry. The relative success of established companies determines which firms stay in the market and which firms exit the market. The effectiveness of the companies' strategic plans affects their relative market shares and their rates of growth and decline. The number of new firms entering a market depends on the entry strategies of start-ups and of established companies expanding through entry. Therefore, market structure depends on the conduct of firms.

At the same time, market structure generates feedback effects on the conduct of firms. Established companies choose strategies in anticipation of the actions of their competitors and the potential entry of new firms into the market. Also, companies contemplating entry into a market examine the strategies of firms already operating in the industry. In this manner, the number and size of established firms in an industry affect the competitive strategies of incumbent firms and decisions about entry by those outside the industry. Figure 4.1 illustrates these feedback effects.

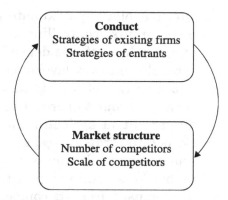

Figure 4.1: Interdependence of conduct and market structure.

Industry Attractiveness

What makes an industry attractive to a company? The practical answer is profitability. However, it is helpful to remember that it is not what incumbents are earning that counts — it is what the company itself will earn after entering the market. Accordingly, industry attractiveness is a measurement of opportunity tempered by an evaluation of the relative strength of the firm's competitors.

Forecasting the firm's profitability requires identifying those economic forces that impact all of the firms within the industry, even if the impacts are distributed unevenly. Measures of industry growth begin with analysis of total market demand, as detailed in the previous chapter. This allows estimates of industry revenues. In addition, measures of costs of productive inputs such as capital, labor services, natural resources, and manufactured inputs are important. For critical inputs, it is necessary to measure the supply of inputs into the market, as discussed in the preceding chapter.

Industry profit estimates based on total revenue and cost forecasts generally provide an upper bound on the company's earnings after entry. The company's projected market share provides a rough guide to that share of industry profit, although such a prediction must be adjusted using more detailed firm-specific cost and revenue forecasts. The company's profits can be expanded not only by competition for market share but in some cases by efforts that expand the market overall.

An important issue for managers is determining how much of the firm's profit is due to the industry that the firm operates in and how much of the firm's profit is due to the strategy of its managers.[3] The profits of an industry will differ from those of other industries as a result of demand and input supply shifts that impact the industry disproportionately relative to the economy as a whole. A study by Anita McGahan and Michael Porter shows that on average industry-level effects explain approximately one fifth of the overall variation in company profits, controlling for firm-level differences. The strength of industry-level effects depends a great deal on the economic sector. For example, industry-level effects account for about 64 percent of variance in company profitability in lodging/entertainment, about 47 percent in services, about 40 percent in wholesale/retail trade and transportation, 30 percent in agriculture/mining, and 10 percent in manufacturing. Corporate strategy seeks out industries that the firm will find profitable to enter. Business strategy aims to position the business within its markets and to employ the most effective strategies to distinguish the company from its competitors.

Industry Analysis and Strategic Groups

Just as managers identify the boundaries of markets, they often find it useful to identify market segments. As with market definition, segmentation analysis begins with characteristics of customers. Are there natural groupings of customers? For example, electricity companies often divide the market into households, commercial establishments, and industrial users. Such distinctions generally are summaries of a number of features including the size of purchases and the end use of the product or service. There are many other bases for classification of customers, including geographic location, income, frequency of purchase, distribution channel used, and extent of information. When customers are businesses, it can be useful to classify them by the industries they serve. IBM's marketing force is divided by industries served, focusing on such specialties as banking, retail, or telecommunications. Based on this information, companies may choose to target segments with special prices and product promotions. Market segments are defined based on how products and services address the needs of customers.

Dividing a market into segments has strategic value. Companies can concentrate their efforts on those segments in which their competitive advantage is greatest. Companies identify attractive segments in a manner that is analogous to industry attractiveness. Companies target segments where their expected profits are highest. In addition to companies already serving that market segment, potential competitors are those companies that enter the industry and serve that segment or companies already within the industry that expand or reposition their product offerings.

Companies may choose to serve only a few market segments depending on their attractiveness. The analysis of segment attractiveness is basically a refinement of analysis of industry attractiveness. The effect of market segments on company profitability is difficult to separate from industry-level and firm-level factors.[4] For example, there are differences in distribution methods and customer demand for national and regional companies in the beer industry.[5] The number and size of the firms serving a market segment do not necessarily determine the effect of competition on profits, just as industry structure is not a good predictor of profits within the industry. Instead, it is the competitive conduct of firms that matters and the underlying cost and demand factors. The entry of firms into market segments results from their competitive decisions.

Companies may not compete directly with all the firms within their broadly defined industry. Rather, managers may wish to focus their attention on selected rivals. In some cases, companies within an industry may be segmented into *strategic groups*. Companies compete with other firms in their group as well as with firms outside their group. The definition of strategic groups depends on the intensity of competitive interaction and similarities in the competitors' customers, suppliers, and technology. For example, strategic groups might be defined on the basis of size, with larger companies deriving benefits from economies of scale.[6] Group stability results from costs of entering market segments or costs of imitating the features of competitors within the industry. Costs of changing the company's position within the industry are referred to as *mobility barriers*.[7]

4.2 Industry Dynamics

The manager's evaluation of the current state of competition in the market must be supplemented with some projections about how competition will evolve over time. Evaluating industry attractiveness depends on *entry* and *innovation*. The manager should consider what types of companies are expected to enter the market. He or she should also evaluate how the innovation will change industry technology with the introduction of new types of production processes, product designs, and transactions.

Entry and Market Structure

Industries evolve through the exit of unsuccessful firms and the entry of new contenders. In evaluating opportunities in a given market, managers should attempt to project what types of entry and exit will occur. The manager should ask whether the industry's market structure will remain stable, that is, whether the number and size of firms in the industry will remain the same or whether the number of firms will contract through shakeouts and consolidation or expand through additional entry. The manager is not simply interested in market structure, but also in the characteristics of its future competitors.

The manager should examine whether industry membership will remain stable or whether new entrants will displace existing firms. Among other factors, the manager considers whether prospective competitors are single-product firms or are multi-product firms with economies of scope; domestic firms with loyal customers or international businesses with global brand recognition.

A key to understanding how markets evolve over time is the presence of entry barriers. Entry barriers result from competitive advantages of established firms. When entry barriers are low, companies can cross industry boundaries with relative ease. Moreover, entrepreneurial start-ups can establish beachheads and challenge incumbents. Markets with low entry barriers tend to be much more competitive. The effect of entry barriers on industry attractiveness depends on the trade-off between the cost of overcoming the barrier and the returns of operating in the market after the barrier has been surmounted. The effects of entry barriers should be measured relative to the returns of operating in the market. Industry attractiveness thus depends on the cost of entry as compared to the returns of being in the market. The impact of barriers to entry on competitive conduct and market structure is illustrated in Figure 4.2.

Because entry barriers have the potential to shield established firms from competition, managers should be aware of specific barriers. If entry barriers are effective, they can be important determinants of industry attractiveness. The greater the size of the barriers, the lower

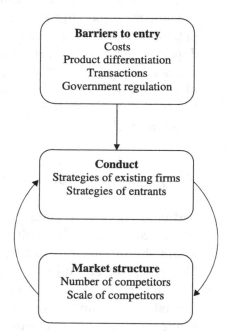

Figure 4.2: Effects of barriers to entry on industry conduct and market structure.

Table 4.1: Barriers to entry.

Type of Barrier to Entry	Source
Cost advantage	Efficiencies in plant and equipment, sunk costs
Differentiation advantage	Product development and design, marketing, and brand equity
Transaction advantage	Procurement and contracting, sales, market information
Government	Lobbying, regulations that grandfather incumbents, licensing and government franchises, trade protections such as tariff and nontariff barriers

the return to entrants and the higher the return to incumbents. However, even with high entry barriers, companies in the industry may compete vigorously, thus reducing profitability. Moreover, other factors besides entry barriers affect company profits, such as demand and cost shifts and the effectiveness of management. Thus, even if entry barriers exist, they do not guarantee incumbents' profitability.

Entry barriers reflect competitive advantages of incumbents relative to entrants. There are three main sources of competitive advantages: production cost, product differentiation, and transactions. These entry barriers are listed in Table 4.1. Each type of entry barrier can be seen as some type of cost entrants face that established firms do not.[8]

Entry barriers are likely to be *temporary* because changes in customer preferences, manufacturing technology, market organization, and government regulations can easily create opportunities for entry. Moreover, creative entrants can devise strategies that counteract entry barriers. If the established firm does not operate efficiently and its management is complacent, a well-run entrant can overcome possible entry barriers. Chapter 12 looks at strategies for surmounting entry barriers.

In evaluating industry attractiveness, the manager should determine the costs of entering the market. Are the set-up costs of establishing production and distribution facilities high relative to earnings after entry? Are the set-up costs of marketing and of research and development substantial relative to expected earnings? The manager should evaluate how the entering company's expected production cost, product features, and transactions compare with those of established companies and determine whether any differences are expected to persist after entry has occurred.

Production costs are a source of possible differences between the entrant and established firms. The manager of a prospective entrant should evaluate whether the company will be able to achieve costs that are at or below those of incumbent firms. The entrant has yet to incur set-up costs and is concerned about the need to make irreversible investments, also known as *sunk costs*. The manager contemplating entry weighs the risks inherent in sunk costs against the attractiveness of the industry.

Product characteristics are another source of differences between the entrant and established firms. An established firm may benefit from brand loyalty, such as customer preference for *Campbell's soup* or *Oreo cookies*. Alternatively, customer loyalty can be the result of high costs of switching, such as the time involved in learning the features of a new type of computer software. Customer loyalty can simply result from habit and inertia and difficulty in adapting to new products. An entrant attempting to establish a new brand or introduce a new type of product faces costs of marketing and sales that the incumbent has already incurred. The manager of a potential entrant compares the costs of developing and introducing new brands with the attractiveness of the industry.

Transaction costs are another possible difference between the entrant and incumbent firms. Established firms may derive advantages from existing transactions that create loyal suppliers. Entrants will not be able to deal easily with suppliers of key services, manufactured components, and raw materials who have contractual relationships with established companies and limited capacity to supply new customers. Alternatively, suppliers who provide customized parts will incur switching costs to supply entrants with different parts. The established firm has already invested time and effort in its supplier relationships. An entrant attempting to establish operations faces costs of procurement, including search and negotiation, that the incumbent has already incurred. The manager of the potential entrant should evaluate the transaction costs of entry in comparison with industry attractiveness.

Although competitive advantages related to cost, product differentiation and transactions tend to be temporary, governments can create more durable barriers to entry. Licensing and other entry restrictions control entry into regulated industries. Hospitals must obtain certificates of public convenience and necessity from state regulators. Established firms may have access to subsidies and tax breaks not given to entrants. Environmental, product safety, and workplace health and safety regulations often apply unevenly to entrants and incumbents, with established firms being grandfathered in with exemptions.

National governments create all sorts of barriers to the entry of international businesses in favor of domestic companies. The primary set of barriers are the tariffs that most countries impose on imported goods and services. Many developed countries including the United States, Canada, and members of the European Union, have placed tariffs on imported steel from developing countries. Many countries erect various types of non-tariff barriers such as quotas, licensing requirements, health and safety rules for imports, exchange and financial controls, and domestic content requirements. The United States imposed voluntary export restraints on Japanese automobile manufacturers.

Government entry barriers often are much more difficult to overcome than demand, supply, and technological entry barriers because they are legally enforced. Regulatory constraints can be effective entry barriers when compliance is sufficiently costly. Administrative red-tape can delay the entry of new businesses. Established firms may have invested heavily in political and regulatory lobbying and in legal actions needed to obtain favorable treatment. The cost to a prospective entrant of establishing a business includes the anticipated costs of these non-market activities. Managers should carefully evaluate government entry barriers and the costs of non-market activities when considering industry attractiveness.

Innovation and Industry Dynamics

Business history is replete with examples of firms who are blindsided by new competitors. Despite its pioneering retail experience, Sears did not accurately foresee the new type of retailing that Wal-Mart represented. Despite its technology leadership, IBM did not fully understand the implications of the personal computer. Many traditional retailers and wholesalers took some time to understand the impact of electronic commerce and incorporate the technology into their sales strategies.

New competitors arise in many ways. Some competitors emerge through growth and diversification. Established companies enter into related markets. Entrepreneurs start up new firms. Start-ups and local companies expand and become regional, national, or global competitors. Mergers and acquisitions change the direction and focus of established businesses. Managers can make some predictions about growth and diversification by monitoring existing rivals and by understanding the changing needs of their customers and the evolving abilities of their suppliers.

Competitors also arise or change fundamentally as a consequence of innovation. Managers exercise foresight by understanding the effects of innovations. Managers are sometimes tempted to ignore innovations because the firm's products and services are currently successful. Andrew Grove of Intel observes, however, that "only the paranoid survive," meaning that successful companies are those that are aware of competitive threats and react to them in time. Grove

Table 4.2:   Innovation generates prospective competitors.

Type of Innovation		Generates Prospective Competitors
Process innovation	$\Rightarrow$	Cost-leaders
Product innovation	$\Rightarrow$	Substitutes
Transaction innovation	$\Rightarrow$	Intermediaries and entrepreneurs

refers to the times of fundamental change as *inflection points*.[9] The difficulty lies in identifying those critical inflection points and distinguishing them from the normal course of technological change.

What makes identification of prospective competitors so difficult is that it requires anticipating the competitive impact of innovation. Three different types of innovation are associated with the three types of competitive advantage: process innovation, product innovation, and transaction innovation. Each of these types of innovations generates different types of competitors, as listed in Table 4.2.

Process innovations improve manufacturing and distribution technology, giving rise to lower-cost competitors or higher quality products at competitive prices. Product innovations create unique product characteristics that allow competitors and entrants to offer substitute products that customers may find more appealing. Transaction innovations allow competitors to secure suppliers and attract customers through greater convenience and to intermediate between buyers and suppliers in novel ways. Intermediaries, market makers, and entrepreneurs often create new businesses through transaction innovations. By understanding the effects of innovation on competition, the manager is able to understand how new competitors develop.

Potential innovation has important implications for competitive advantage. As already noted, no competitive advantage is sustainable for very long — advantage is temporary by its very nature. Competitors can surpass incumbent firms by taking advantage of progress in research and development efforts. Attempts to maintain the firm's competitive advantage thus require continual innovation to discover new sources of advantage.

Entrepreneurs and other market entrants consider the extent to which they can copy and surpass the incumbents' production processes, product features, and transactions. Some types of advantage are sustainable at least for some time because imitation is costly. Even if another company's personnel seem more talented, identifying key employees is difficult for outsiders, and raiding another company's work force can be a costly exercise. Copying innovations is difficult as well because it is costly to reverse engineer an innovation or to invent around patent protections. Emulating a competitor's distributor or supplier networks entails transaction costs of search and negotiation. These costs suggest that imitation only occurs if the entrant's economic rents outweigh the transaction costs of imitation. Where imitation is costly, independent innovation may offer greater returns.

In evaluating industry attractiveness, managers must consider the effects of technological change. Although, by its very nature, technological change is unpredictable, managers can attempt to examine how the industry will be affected by technological developments that have already occurred but have not yet been brought to market. Managers should anticipate that by the time the company is able to enter the industry, the products and production processes that competitors offer are likely to have changed. Opportunities for entry depend upon the rate of technological change in the industry, the costs of R&D, and the company's technical capabilities.

4.3 *Partners*

The manager concludes the external analysis by identifying prospective partners. Companies need not recognize all of their potential partners, but awareness of partners has strategic value. Companies that produce products that are complements for customers can form marketing and sales partnerships. Companies can form production partnerships by coordinating manufacturing and R&D. The company's customers, suppliers and competitors are also potential partners. In evaluating industry attractiveness, it is useful for managers to consider its potential partners.

Partnerships

Depending on the extent of commitment of the partners, partnerships can take many forms. Companies can enter into informal strategic alliances, explicit long-term contracts, or joint ventures and mergers and acquisitions (see Figure 4.3). These alternatives entail different strategic advantages and involve different types of transaction costs.

Informal purchasing agreements allow firms to form and dissolve partnerships quickly in response to changing market conditions. The partners avoid some of the costly negotiations required to form formal contracts and the potential costs of unwinding the agreement. Informal agreements have the disadvantage that the partners often need to renegotiate as market conditions vary. Moreover, the companies may not build up sufficient trust to allow sharing of information or taking complex strategic actions. Informal agreements can be stepping stones to longer-term relationships between companies. The classic example of a successful informal partnership is that between Hewlett-Packard and the Japanese company Canon, which invented the laser printer. Canon supplied the printer engine and Hewlett-Packard provided the

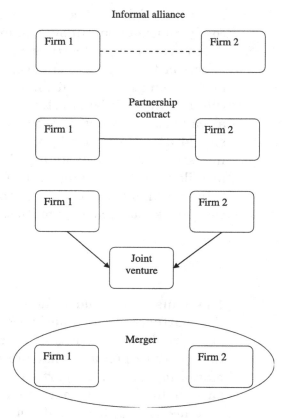

Figure 4.3:   Varieties of partnerships between firms.

software, control technology branding, marketing, and sales. The relationship between Canon and Hewlett-Packard was built on trust rather than on a long-term contract.[10]

Long-term contracts require costly negotiation and investment of management time and effort. Moreover, long-term partnerships entail commitment of resources by the parties involved and some loss of control. Although partnerships are more flexible than expansion of the organization through vertical integration, long-term contracts may fail to adjust to changing market conditions. The contract may no longer provide sufficient benefits to the companies involved or the partners may find they are forgoing other opportunities. Accordingly, one or more of the partners may wish to reevaluate the terms of the relationship. Companies may encounter significant costs associated with changing the terms of partnership agreements. Accordingly, all formal partnerships should include mechanisms for reevaluating the partnership and procedures for amicable dissolution when it is mutually beneficial.

Another potential problem with partnerships is that they can create incentive problems. There is always the possibility of free riding on the other members or pursuing objectives that are inconsistent with the partnership. A partner can use the alliance to gain valuable information, and then act opportunistically by using the information to compete against alliance members in the future; that is, the partnership can act as a Trojan horse. Alliances should be carefully structured to maintain incentives for partners to share information and resources.

Complementary Products

Partnerships between companies that produce complementary products are designed to increase the revenues of the two companies. Two goods are *complements* in demand when the use of either one of the goods enhances the benefits the consumer obtains from using the other good. Examples of products that are complements include cameras and film, computers and software, and automobiles and gasoline. When two goods are complements, reducing the price of one good increases the customer's demand for the other good.[11] A reduction in the price of peanut butter is likely to increase the demand for jelly as consumers make more sandwiches. See Figure 4.4(a) for a comparison of complements (cameras and

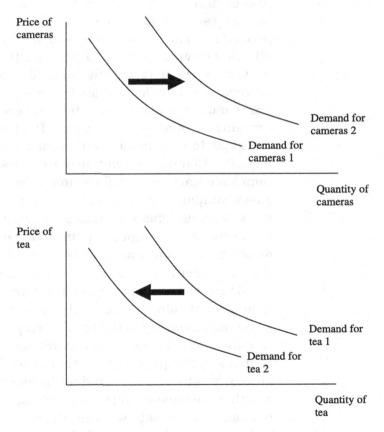

Figure 4.4: (a) Complements: Effect of a reduction in the price of film on the demand for cameras. (b) Substitutes: Effect of a reduction in the price of coffee on the demand for tea.

film) and Figure 4.4(b) for a comparison of substitutes (tea and coffee). Two companies that offer complementary goods can enhance their revenues through a partnership.

Improving the quality of a complementary product while holding prices constant works like a price reduction because it stimulates customer demand for the other product or service. Thus, better peanut butter increases demand for jelly. When complementary products are used together and the benefits of using the products jointly are affected by the compatibility of the products, any increase in the compatibility of the two products acts as a quality enhancement, stimulating the demand for both products.[12] To capture some of these benefits, companies form partnerships to jointly promote complementary goods, such as chips and salsa.

Producers of complements are potential partners because coordinated action offers benefits. Because consumers obtain additional benefits from combining complementary goods, joint promotions can increase sales for two companies offering complements. Companies producing complements can capture these benefits by coordinating pricing and product introductions. For example, a camera maker and a film manufacturer can introduce a new camera and film designed for that camera and benefit from the lower cost of joint marketing. A new camera can include coupons for the purchase of film. Complementary products, such as software and a computer, may be sold together, allowing companies to set a price for the bundle.

Customers may benefit from the convenience of product bundles. Prearranged bundles reduce the costs of searching for complementary goods and assembling compatible products into a bundle and allow for the convenience of one-stop shopping. Product bundles are a way of suggesting ideas to customers about what complementary products work best together. Companies may increase sales by partnering with makers of complementary goods to form unique bundles of goods and services. Video game equipment companies such as Sony, Nintendo, or Microsoft partnered with designers of video game software. Internet service providers partnered with computer manufacturers to offer bundles of Internet access service and computer equipment.[13] These new product combinations can offer advantages over competitors selling separate products.

When examining prospective partners for marketing and sales, managers should determine the extent to which the goods they supply are complementary to the firm's own products. The manager should try to determine the extent to which demand for one good responds to changes in the price of the other good. Moreover, the manager should assess whether joint product promotions would be effective and whether customers prefer to purchase products as a bundle. The manager should evaluate whether the company should focus on cultivating a few partnerships or seek to develop many industry relationships. A company whose product is complementary for many types of goods may need to form a wide range of partnerships.

Partners who sell complementary products also gain from sharing market demand information. United Airlines offers passengers discounts on customized product bundles that include airfare, hotel, car rental, entertainment, and meals at a restaurant. United Airlines and its partners benefit from sharing information about travelers. Companies whose products are complementary may gain from sharing technical information. For example, Microsoft lets developers of applications software know about changes in its Windows operating system software (see Exhibit 4.1).

Exhibit 4.1   Microsoft and Silicon Valley

Silicon Valley, which includes Santa Clara county and parts of Alameda, San Mateo, and Santa Cruz counties,[14] measures approximately 1,500 square miles and has a population of 2.3 million people. Silicon Valley employment is approximately 1.2 million, and the software and computers/communications industries account for about 150,000 jobs.

Microsoft's Windows operating system became the most popular personal computer operating system in the world by the end of the 1990s, when it made over 80 percent of operating system license shipments and at least 177 million computers around the world used the Windows operating system.

A number of companies in Silicon Valley produced goods and services that were economic complements to Windows. Complementarity is a two-way street because producers of complementary goods and services provide additional demand and thus incremental revenues for each other. Moreover, consumers derive additional benefits from the consumption of goods and services whose complements experience price reductions and quality enhancements. The primary function of Windows is to serve as a common user interface and operating system for the computer. Users are typically interested in working with software applications or Web browsers that are based on Windows, rather than working with Windows itself. Because of this fact, Windows has a unique linkage with the companies that produce goods and services based on Windows. In economic terms, they can be described as complements since they are typically used together and the demand for one is linked to the demand for the other.

One of the features of economic complements is that the success of one good is linked to the success of the other. The more Windows-related machines there are in the world, the more demand there will be for providers of complementary goods and

(Continued)

Exhibit 4.1 (*Continued*)

services. Likewise, the more goods and services based on Windows there are, the more demand there will be for Windows. Similarly, a fall in the price of Windows or an improvement in its product features will increase the demand for complementary goods and services, just as quality improvements or price decreases of complementary goods and services will increase the demand for Windows. Six types of companies in the Valley produce products that are complementary to Windows.

First, independent software vendors (ISVs) and application developers include companies that develop or design software applications to be used in the Windows environment. A reduction in the price of Windows or an increase in the quality of Windows would be expected to increase demand for the software applications that ISVs and application developers provide since more widespread usage of Windows will create more potential ISV customers. A reduction in the prices of software applications or an increase in their quality would be expected to increase demand for Windows as more customers acquire Windows in order to use the enhanced software applications ISVs develop.

Second, system integrators (SIs), Microsoft Certified Solution Providers (MCSPs), and consultants develop customized, Windows-related computer solutions for businesses that incorporate the latest technologies, including updating old systems, interconnecting existing, non-compatible systems, and installing new systems. These firms not only provide turnkey computer systems, but also specialize in data processing services, hardware and peripherals manufacture, and software development. They focus on developing customized business applications such as interconnection and installation of computer systems and data processing services. Computer consultants assist businesses in establishing and operating their computer systems.

Third, Windows-based computer training and education includes any companies or consulting firms that provide training, education and/or certification for applications that run on the Windows operating system. An improvement in the quality of Windows should stimulate demand for Windows-related computer training and education as more users employ Windows and some of those users seek training and education. Similarly, as Windows-related computer training becomes more widely available at lower prices and higher quality, demand for Windows should increase.

(*Continued*)

Exhibit 4.1 *(Continued)*

Fourth, Web-based companies are those whose customers interact with them through Internet browsers. As more users obtain PCs using the Windows operating system, and as Windows undergoes quality improvements that make Web-based commerce even easier, this will stimulate demand for the services of Web-based companies. Conversely, enhanced service by Web-based companies stimulates demand for complementary products used to access the Internet, including PCs and Windows.

Fifth, original equipment manufacturers (OEMs) are those companies that build PCs, servers, laptops, palm-tops, and related hardware and bundle Windows with their finished product. A reduction in the price of Windows or an improvement in the quality of Windows should stimulate demand for the products of OEMs, while falling prices for computers and related equipment clearly stimulate demand for the Windows operating environment. The fall in PC prices below $1,000, generated additional sales of PCs, and in turn greater demand for Windows operating systems.

Sixth, Microsoft has made a significant investment in some companies and formed strategic alliances/partnerships with others that develop software and other services for the Windows platform. As Windows' quality improves or price decreases, sales of other goods and services based on Windows, including those from Microsoft itself, increase.

In addition to these categories, several other types of companies are indirectly connected to Windows. The largest such category includes the suppliers of components such as memory chips, processors, and peripherals that are used in PCs and PC networks, including such companies as Intel and Cisco Systems. Other companies provide online information services, which can be accessed via computer using a variety of programs, including many based on Windows or accessed through Windows-based Web browsers.[15]

Competition among providers of complementary goods and services, such as Windows-related applications and PC manufacturers, tends to result in improved quality and competitive pricing of these products. This, in turn, is beneficial to Windows, since it increases demand for Windows. Similarly, competition and potential competition between the Windows operating environment and other operating environments (such as Unix, OS/2, or Java), is good for users and suppliers of complementary goods and services since it results in continual innovation and quality enhancements and serves as a check on prices.

(Continued)

Exhibit 4.1 *(Continued)*

Technological advancement in computer software and equipment, such as advancements in microchip technology, also enhance demand for the Windows operating system and other complementary goods and services. Moreover, expectations of further technological enhancements serve to stimulate demand as well to the extent that goods and services remain compatible with each other. Thus, Microsoft research and development investment that enhances the expected performance of the Windows operating environment should serve to stimulate current demand for applications and equipment that are compatible with that operating environment.

The existence of a Windows operating environment standard offers benefits to providers of complementary goods and services. Increases in the market share of a leading standard simplifies the task of providers of complementary goods and services by allowing them to achieve scale economies in research and development, production, and distribution that would not otherwise be present with wide diversity of operating environments or with diverse versions of the same operating environment that varied in terms of compatibility standards.

The benefits of complements are transmitted through the market and do not require external subsidies or policy interventions to be achieved. These benefits, although divided among suppliers of complementary products and customers of complementary products, are internalized through market transactions. When compatibility issues are present, they can typically be resolved through informal information sharing and formal contractual arrangements. Microsoft established formal and informal relationships with producers of complements. These relationships involved sharing information about the details of future Windows upgrades with programmer and value-added resellers. In addition, Microsoft gathered valuable information about the concerns of programmers and resellers, and feedback regarding customer experiences using Windows and related applications software.

Joint Ventures

Joint ventures are separate organizational units formed by two or more firms. The partners to a joint venture can be any type of firm, from small entrepreneurial businesses to multinational corporations. The joint venture can be directed at practically any business activity, including procurement, R&D, manufacturing, assembly, marketing, and distribution. By creating a joint venture, the partners separate the activity from the parent companies while sharing the costs and the benefits of the enterprise.

Companies may choose to pool capital investment in a joint venture to manufacture products. Distribution joint ventures share marketing and sales knowledge. Research and development joint ventures share scientific knowledge and technical personnel. Such ventures are particularly common in industries experiencing accelerating technological change. The largest 20 pharmaceutical firms are involved in hundreds of joint ventures with biotech firms.[16]

What can make a joint venture an attractive alternative to full vertical integration is that the partners share the total costs of distribution, manufacturing, procurement, or other activities. Cost sharing through joint ventures helps to avoid duplicating investment in production capacity. Partners reduce the risks associated with large-scale investment. For example, in a 50/50 partnership to manufacture a component, the two firms divide the cost of investment in production facilities. Moreover, the joint venture avoids some of the management costs of operating a vertically integrated organization. Partnerships allow large companies to avoid the antitrust and regulatory scrutiny that can come with large-scale mergers.

Many companies form cost-reducing joint ventures by supplying complementary inputs. Two inputs are said to be *complementary* if increased use of one input raises the productivity of the other input. Companies with complementary skills, such as production and engineering, can benefit from a joint venture without the cost of acquiring the other competency, particularly when the joint venture partners are different types of firms such as parts manufacturer and an original equipment manufacturer. Partners can contribute complementary inputs such as manufacturing facilities and engineering personnel.

The joint venture can sometimes speed up market entry if the partners already possess the necessary personnel, investment, and facilities. The joint venture benefits from economies of scale and scope that would not necessarily have been attained had the partners taken on the activity separately. The partners share the risks of making capital commitments associated with entering a market. The partners to the joint venture acquire and share knowledge with each other, possibly lowering the transaction costs of information transfer in comparison with other contractual alternatives. These types of partnerships may be necessary for smaller firms to catch up and overtake larger companies or other partnerships.

The joint venture can reduce transaction costs for the partners because the enterprise formed by the joint venture serves as an intermediary between the two partners, with specialized personnel and facilities that help to coordinate interaction between the parent companies. If the partners to a joint venture form a durable relationship, transaction costs are reduced by the specification of roles of the parent company and greater flexibility than might be achieved by contracts made directly between the parent companies.

When joint ventures involve complex business processes, the transaction costs of forming the joint venture can exceed the costs of contracting between the parties. joint ventures can create difficulties for the parent company if free riding occurs — when one partner provides a greater share of resources or derives a disproportionate share of the benefits. Assignment of intellectual property rights from the joint venture poses some difficulties. With unsuccessful joint ventures, one partner may use the joint venture to gain information about the partner's business plans, especially when the partner is acquired by another company. Managers establishing a joint venture should form an exit strategy, formally specifying how the joint venture will be dissolved if there are changes in market conditions or in the objectives of the partners.

R&D joint ventures allow companies to share the costs and benefits of basic research without duplicating their efforts. The inherent uncertainties of R&D heighten the need for cost sharing. Moreover, companies can share the development cost for new products or production processes. R&D joint ventures are mechanisms for companies to cooperate in setting standards for new products, while sharing the costs of product development.

International joint ventures fulfill additional purposes, allowing companies to formalize their relationship with a local partner. Joint ownership of the ventures allows international companies to comply with domestic content regulations and ownership requirements imposed by the host country, while providing the company in the host country with access to international capital investment and technology. The international manufacturing joint venture, Cablecom International, was formed to produce connector cable assemblies for computer and mobile communications manufacturers. Based in Hong Kong, with production facilities in Shenzhen, China, Cablecom International is a partnership between the U.S. company, ITT Industries, whose Cannon division provides technology of connector manufacturing and design, and EDA Inc., of Taiwan, which provides knowledge of low-cost supply channels and experience operating in China.

International partnerships take advantage of the market knowledge and relationships of local partners, that is, companies in the host country. International partners provide access to new suppliers and customers and technological information. Transaction costs are reduced in comparison to going it alone, and a partnership is easier to form and maintain than a full merger. Partners can leverage relatively small investments to create international networks, as with airline alliances. The members maintain their corporate identity, management, and ownership structure intact, but take advantage of production efficiencies. The strength of partnerships is that companies can rapidly expand their scope without the costs and loss of flexibility that would result from a merger or internal growth. The partnerships can be dissolved or refocused as market conditions change. Additional partners can be added to expand the market reach.

Exhibit 4.2   Concert

AT&T and British Telecom formed a 50/50 joint venture to provide international telecommunications services, called Concert. AT&T contributed $2 billion in assets while British Telecom put in $1.4 billion, for a total of $3.4 billion. The venture provided a full spectrum of global voice and data services and had 7,000 employees worldwide. Prior to the venture, AT&T's international unit earned $3.1 billion and British Telecom's international unit earned $2.1 billion, for a total of $5.2 billion.[17] The joint venture earned about $7 billion per year.

British Telecom sought international partners after the collapse of its attempt to acquire MCI. The company lacked the resources to be a major international service provider without significantly expanding its customer base and network facilities. Other telecommunications providers were forming international partnerships. As a primarily domestic carrier, British Telecom would have been at a disadvantage in serving international business customers, particularly without access to the U.S. telecommunications market. As a formerly regulated utility, British Telecom also sought to increase its technological and marketing knowledge through partnerships with other carriers.

By putting together their two units, the companies avoided the costs of duplicating each other's international long distance networks. By sharing assets, the joint venture achieved economies of scale and scope. The joint venture offered the benefits of a global brand and allowed the two partners to share marketing and sales costs. Business customers obtained the benefits of one-stop shopping for international telecommunications services.

According to British Telecom, "Concert serves the global communications needs of multinational companies, traditional and emerging carriers, wholesalers and Internet service providers." Concert combined the international network assets of the two parent companies including software and network infrastructure. The joint venture shared the international traffic of the parent companies and jointly offered international products for business

(Continued)

Exhibit 4.2 (*Continued*)

customers. Concert's international voice service extended to 230 countries or territories and to nearly 1,000 cities, with over 200,000 international private line circuits. The Concert network handled 28 billion minutes of international voice traffic, and it included 75,000 km. of fiber and an Internet backbone serving 21 cities in 17 countries. The joint network had over 800 points of presence, covering more than 90 percent of the world's leading markets. Concert encompassed over 21 joint ventures covering 85 percent of the European Union (EU) market as well as Asia Pacific and the Americas. Moreover, Concert had network operations centers in Bermuda, Europe, Asia Pacific, and the United States and planned to establish more in Asia and Latin America. Finally, Concert maintained relationships with 250 other telecom companies.[18]

According to AT&T, "Concert support means a lot more than problem resolution. We bridge time zones, languages, currencies and cultures. We employ systems to integrate billing and reporting for all our services, providing you with uniform worldwide reporting and analysis with a choice of language and currency." AT&T's marketing offered to "take your business to the next level in worldwide communications ... with a single global network".[19]

A downturn in telecommunications markets, however, meant that Concert had lower earnings than expected. Despite the advantages of the joint venture, the two companies reevaluated Concert in the wake of corporate restructuring at both AT&T and British Telecom. Managers of the two parent companies examined various options, including acquisition by one of the parent companies, reorganization, divestiture, or closing the joint venture. The possibility of AT&T acquiring Concert raised concerns that European regulators might object.[20] AT&T and British Telecom decided to close the Concert joint venture, dividing the assets, and retaining some of the venture's employees. The establishment and dissolution of Concert shows how the operation of international joint ventures depends on market conditions and the changing needs of their parent companies.

4.4 Alliances

In a competitive environment, companies sometimes form large-scale alliances with prospective partners. Companies form alliances to construct networks, sharing the costs of providing network services and assuring interconnection. Network alliances exist in telecommunications,

Internet transmission, airlines, rail, and trucking. Companies also form partnerships to negotiate and promote technology standards in such industries as computer hardware and software, product labeling, and broadcast communications.

Building Networks

Companies form partnerships with suppliers and distributors to build networks. Supplier partnerships offer benefits from supplier knowledge about technology and markets for primary inputs, and distributor partnerships offer benefits from distributor marketing experience and knowledge about customer demand. By forming networks of independent suppliers and distributors, the firm avoids having a large vertically integrated organization. With external networks of suppliers and distributors, the firm has a leaner, more focused organization. This allows the firm's managers to concentrate on providing specific goods and services rather than on administering input manufacturing and output distribution.

Supplier partnerships have been used effectively by some of the major manufacturing companies in Japan. Ozawa describes the auto industry as 11 core companies (Toyota, Nissan, Honda, Mazda, Mitsubishi, Fuji, Daihatsu, Isuzu, Suzuki, Hino, and Nissan Diesel) on the top of the pyramid that assemble the vehicles.[21] Next are 168 primary subassemblers and subprocessors, then 4,700 secondary subcontractors and 31,600 tertiary producers.[22]

The major U.S. automakers have relied heavily on vertical integration with in-house production of auto parts. Henry Ford's slogan "From Mine to Finished Car, One Organization" illustrates preference for vertical integration in the automobile industry in the early part of the 20th century. Alfred P. Sloan's 1921 organization plan for General Motors featured the Accessories Group for units that sold more than 60 percent of their output outside the company and a Parts Group for units that sold 60 percent of their output within the company.[23] This system continued for over 75 years but could not be sustained in the face of global competition. Jack Smith, GM's chairman, told *The Economist* on October 10, 1998, "As the world opened up to free trade, Sloan's system was not competitive." In 1999, GM spun off its parts manufacturing unit, Delphi Automotive Systems Corp., to create the world's largest auto parts supplier with over 200,000 employees. Ford's parts manufacturing unit Visteon has 82,000 employees in 21 countries around the world with 120 manufacturing, engineering, sales, and technical centers. Visteon operates with some autonomy and has sought to grow its sales outside of Ford.

About 6,000 suppliers serve General Motors, Ford, DaimlerChrysler, and Japanese companies that manufacture in the United States.[24]

Traditionally automakers procured auto parts by providing suppliers with detailed specifications and even designs of parts to their subcontractors, who then competed on the basis of price. Automakers began to change their relationships with parts producers by involving them in the design process and sharing engineering and design knowledge.[25]

In network industries such as transportation, telecommunications, electric power transmission, and the Internet, companies form partnerships to improve interconnections. When companies interconnect, customers benefit by being part of a more extensive network.

Airlines create networks through strategic alliances — there are over 500 airline alliances.[26] For example, the Star Alliance includes United Airlines, Lufthansa, Scandinavian Airlines, Air Canada, Asett Australia, Air New Zealand, Thai Airways International, VARIG, and All Nippon Airways. Passengers benefit because they can put together a trip on multiple airlines with greater ease of making connections and take advantage of shared airline amenities such as lounges in airports for business travelers. The airlines benefit because they can feed passengers to each other's routes, thus creating a global hub-and-spoke system. The Star Alliance advertises that it is "the airline network for Earth" with over 210,000 employees and flights to 720 destinations in 110 countries.[27] It seeks worldwide recognition for the Star Alliance brands of air travel. Customers flying with the Star Alliance earn miles with any member airline and can redeem the miles on any member airline.

British Airways and American Airlines, which together provide over 60 percent of transatlantic service, formed the competing OneWorld, a global marketing alliance that also includes Canadian Airlines, Cathay Pacific Airways, Qantas, Linea Aerea Nacional de Chile, Finnair, and Iberia. The OneWorld Alliance goes to 630 destinations in 138 countries. As a promotion, the alliance announced a 42,000-mile trip from and back to London in which a journalist and a photographer visited the seven wonders of the world in a record-setting eight days: the Taj Mahal, the Great Wall of China, the Sydney Opera House, the Golden Gate Bridge (viewed in flight), the Empire State Building, the Leaning Tower of Pisa, and the Pyramids.

Setting Standards

Setting technological standards is closely related to forming networks. Companies form alliances with customers, suppliers, and competitors to set a common standard. Standards enhance convenience for customers and manufacturers alike, allowing companies to compete on prices and product features. By enhancing convenience and eliminating multiple formats, technological standards boost overall demand for compatible products and services. When entire industries compete with each other, standard-setting alliances have major strategic impacts.

Research alliances provide ways to reduce the costs of R&D duplication. These alliances yield results that can diffuse immediately without the need to copy or "reverse engineer" rival innovations. Research consortiums can achieve economies of scale and make efficient use of scarce scientific and engineering talent. They also have the advantage of being isolated from short-term pressures within the company to develop greater basic research. By focusing on a specific set of industry problems, they can be better positioned to develop useful solutions than a general research lab. Consortia also can benefit from public subsidies.

Of course, R&D consortia are no panacea; they are as subject to free riding and disagreements as any partnership. Moreover, the partners must each receive a gain in information and cost reductions to outweigh the competitive advantages of homegrown innovation. Each of the partners also must be assured that cooperation with rivals does not diminish their own relative market position in comparison to independent R&D.

IBM worked on a network of technology alliances in the early 1990s. For example, it collaborated with Toshiba to make large flat-panel color computer screens. The company designed microprocessors in Austin, Texas, with Motorola, Apple, and Groupe Bull to create chips based on reduced instruction set computing. It worked with Siemens and Toshiba to develop 256-megabit memory chips. IBM played a key role in forming the Sematech research consortium.

Research consortia can be particularly important as a competitive response to the formation of rival consortia, as is the case in the high-definition television (HDTV) battle between three alliances: Zenith Electronics and AT&T; General Instruments and M.I.T; and Thomson Consumer Electronics, Phillips Electronics, Sarnoff Research Center, and NBC. The first two groups entered into cross-licensing arrangements to mitigate the effects of racing. General Motors, Ford, and Chrysler formed 12 research consortia on electric vehicle batteries, wiring, auto safety, and parts recycling.[28]

Informal industry alliances are a mechanism for increasing innovative speed. Eric von Hippel observes that trading informal proprietary know-how can occur even among competitors. He notes that, among U.S. steel minimills, such know-how trading occurs through "an informal trading network that develops between engineers having common professional interests."[29] Minimills produce steel from scrap that is melted in an electric arc furnace, cast in continuous casters and rolled into the desired shapes. Considering the four largest firms, Chaparral, Florida Steel, Northstar, and Nucor, and another seven smaller companies, von Hippel observes that technical know-how about the steel-making process is of value to the firms and that technical abilities vary across them. He characterizes informal trading as a means of reducing transaction costs because it is an "inexpensive, flexible form of cross-licensing."[30]

R&D can be enhanced by close collaboration with suppliers and customers. Thus von Hippel questions the common assumption that product innovations always are made by product manufacturers. He shows instead that the functional source of innovations can often be customers or suppliers. For example, most innovations were created by users of the innovation in the cases of specialized scientific instruments such as the gas chromatograph, nuclear magnetic resonance spectrometer, ultraviolet spectrometer, and the transmission electron microscope. Most of the innovations in machines used to attach connectors to electric wires and cables were not due to the manufacturers of those machines, but rather to the suppliers of connectors. These observations suggest that customers or suppliers can be important sources of accelerated innovation.

Exhibit 4.3   The Symbian Alliance

The Symbian alliance was formed to set standards for the third generation of wireless phones.[31] Psion, a maker of palm-top computers and designer of the EPOC software for the phones, led the alliance.[32] Alliance partners included the leading wireless phone manufacturers Nokia, Motorola, Ericsson, Matsushita, and later the Sony-Ericsson joint venture, which would incorporate the technology into their phones. Colly Myers, CEO of Symbian, stated that "our strategic relationships have demonstrated to the industry that Symbian is the focal point for delivering a new networked economy, fueled by ubiquitous wireless information devices."[33]

In addition to the alliance members, Symbian had a wide array of technology partners, including Oracle, Sybase, Texas Instruments, Cirrus Logic, NEC, Lotus Development, and Compuserve. Palm Computing formed a partnership with the Symbian alliance, bringing its expertise in handheld devices, operating system software, and applications software. The Symbian alliance was a competitive challenge to Microsoft and its partners, which promoted an alternative standard based on Microsoft's Windows CE operating system for wireless devices. Whether EPOC, Windows CE, or some other standard wins the marketplace will have a significant effect on the success of individual companies in wireless communications. Google and members of the Open Handset alliance offered an open source operating system platform known as Android.

(Continued)

Exhibit 4.3   (*Continued*)

With rapid technological change in communications and computers, such standard-setting alliances were an important part of competitive strategy. Intensifying competition among cell phone manufacturers increased interest in standard setting. Nokia, which owned just less than half of Symbian, purchased the rest of the company. Nokia announced plans to make the Symbian operating system available without royalties to other members of the Alliance: Sony Ericsson, Motorola, NTT DoCoMo, AT&T, LG, Samsung, STMicroelectronics, Texas Instruments, and Vodaphone.[34]

4.5 *Overview*

The manager's external analysis encompasses the company's customers, suppliers, competitors, and partners. Managers begin their evaluation of competition by attempting to define the relevant market in which their company will operate. Managers should focus on the conduct of competitors; that is, on their behavior and competitive strategies, as well as demand and cost characteristics. The number and size of firms within the industry provide useful summary information about the *identity* of competitors, but managers should understand that market structure is affected by the *conduct* of competitors.

The manager tries to anticipate competition by examining the effects on prospective competition of structural factors, particularly barriers to entry and the direction of innovation. The main types of barriers to entry are irreversible investments required for marketing, purchasing, productive technology, and overcoming legal and regulatory constraints. Established companies have written off these entry costs whereas entrants must assess the risk that they will not recover their investments. In later chapters, we consider how to anticipate competitor strategies and how to surmount barriers to entry.

The evolving nature of competition means that information about industry structure and competition can be of limited value in strategy making. Future profitability is a function of industry dynamics. Barriers to entry can affect the extent of potential competition. In addition, managers anticipate prospective competitors by understanding the impact of innovation. Three main types of innovation are useful for competitor analysis: product innovation, process innovation, and transaction innovation. Product innovation creates new substitutes that will compete with the firm's product. Process innovation improves production processes, creating lower cost competitors. Transaction innovation creates more convenient ways to transact with the company as well as new combinations of products and services that fundamentally change the nature of the business. By examining the effects of innovation, the manager is better able to define the types of competitors the firm faces.

Potential company partners include customers and suppliers. They also include manufacturers of complementary goods with whom the company can coordinate product features, promotion, and pricing. The company forms strategic alliances to share costs, to set up sourcing and distribution networks, and to establish technological standards.

Accelerating technological change and increasing global competition rule out complacency. Strategy fundamentally entails change, as companies alter their products and services and enter new markets. With strategic change, the company is likely to encounter changing market conditions. This accentuates the need for managers to conduct an external analysis that makes projections about the company's prospective customers, suppliers, competitors, and partners. Ultimately, industry attractiveness is a measure of the company's competitive advantage. Companies should focus attention not where they have the greatest strength or where rivals are at their weakest. Rather, companies should enter where their strength is greatest in relation to their competitors. Managers need to combine information obtained from the external analysis with the firm's internal analysis, which is the subject of the next chapter.

Questions for Discussion

1. Your company plans to launch a new product. The product consists of software that allows users of personal computers to organize files (including text documents, images, and sound) and to use them more easily. How would you define the relevant market?

2. Select an industry (automobiles, steel, soft drinks, newspaper publishing). Explain how market structure might affect the strategic conduct of firms and how, in turn, the strategic conduct of firms affects market structure.

3. Select a specific company and perform a five-forces analysis by listing the company's customers, suppliers, established rivals, substitutes, and potential entrants.

4. Suppose that a certain market is served by three firms, each having exactly one third of the sales in the market. Another company considers entry to that market. Why might the profit earned by the entrant differ from that of the incumbent firms? How will entry affect the conduct of firms in the industry?

5. For a particular industry (computers, banking, retail, fast food) suggest a particular cost advantage, differentiation advantage, and transaction advantage that an incumbent firm might have in comparison with potential entrants.

6. Give an example of a government regulation that might create a barrier to entry. What arguments would an established firm use to defend the regulation? What arguments might an entrant use to criticize the regulation?

7. Give an example of competition between different product standards (say in electronics, computer software, or video game formats). How should a company go about establishing its product specifications as the industry standard? How should a company maintain its position as the leading standard? How might a company create a new standard that replaces the leading standard?

8. Give an example of two complementary products.

9. What are some advantages for two companies in forming an alliance rather than merging with each other?

10. What are the advantages and disadvantages of research and development joint ventures?

Endnotes

1. These questions are drawn directly from Figure 2-1 in Michael E. Porter, *Competitive Strategy: Techniques for Analyzing Industries and Competitors* (New York: Free Press, 1980), p. 49.

2. Ibid., pp. 64–67.

3. See especially Richard Schmalensee, "Do Markets Differ Much?" *American Economic Review* 75 (1985), pp. 341–351; Richard P. Rumelt, "How Much Does Industry Matter?" *Strategic Management Journal* 12 (1991), pp. 167–185; Anita M. McGahan and Michael E. Porter, "How Much Does Industry Matter, Really?" *Strategic Management Journal* 18 (1997), pp. 15–30.

4. For additional discussion see J. McGee and H. Thomas, "Strategic Groups: Theory, Research and Taxonomy," *Strategic Management Journal* 7 (1986), pp. 141–160; K. Hatten and M. L. Hatten, "Strategic Groups, Asymmetrical Mobility Barriers and Contestability," *Strategic Management Journal* 8 (1987), pp. 329–342; J. B. Barney and R. E. Hoskinson, "Strategic Groups: Untested Assertions and Research Proposals," *Managerial and Decision Economics* 11 (1990), pp. 187–198; R. Wiggins and T. Ruelfli, "Necessary Conditions for the Predictive Validity of Strategic Groups: Analysis without Reliance on Clustering Techniques," *Academy of Management Journal* 38 (1995), pp. 1635–1655.

5. V. J. Tremblay, "Strategic Groups and the Demand for Beer," *Journal of Industrial Economics* 34 (1985), pp. 182–198.

6. See Michael E. Porter, "The Structure within Industries and Companies' Performance," *Review of Economics and Statistics* (1979), pp. 214–227.

7. See Michael E. Porter, *Interbrand Choice: Strategic and Bilateral Market Power* (Cambridge, MA: Harvard University Press, 1976), and Richard

Caves and Michael E. Porter, "From Entry Barriers to Mobility Barriers," *Quarterly Journal of Economics*, 1977, pp. 241–261.

8. This definition of entry barriers is due to George J. Stigler, *The Organization of Industry* (Homewood, IL: Irwin, 1968), p. 67.

9. Andrew S. Grove, *Only the Paranoid Survive* (New York: Doubleday, 1996).

10. Paul Klebnikov and Benjamin Fulford, "Canon on the Loose," *Forbes*, July 23, 2001.

11. Adam M. Brandenburger and Barry J. Nalebuff refer to companies that produce complementary goods as "complementors" and emphasize the importance of cooperation. See Adam M. Brandenburger and Barry J. Nalebuff, *Co-opetition* (New York: Doubleday, 1997).

12. Suppose that the price of peanut butter falls or the quality improves, resulting in an increase in the demand for jelly. When the demand for jelly increases, there are two effects. First, the buyers of jelly experience an increase in consumer surplus (measured as the difference between what each consumer would be willing to pay for a unit of jelly and what they actually have to pay). Second, earnings rise for the jelly suppliers, to the extent that the price of jelly exceeds the marginal cost of producing the jelly. Therefore, a fall in the price of peanut butter or an increase in the quality of peanut butter increases both consumer surplus for buyers of jelly and earnings for jelly suppliers. In this sense, some of the benefits of a lower price or increased quality of a good are shared by consumers of complementary goods and producers of complementary goods.

13. Peter H. Lewis, "It's a Land of the Free (Computer)," *New York Times*, July 8, 1999, p. D1.

14. While there is no single definition of Silicon Valley, this definition is consistent with that used by Joint Venture Silicon Valley, a consortium of businesses and other organizations in the Valley. Silicon Valley encompasses the cities of Atherton, Belmont, Campbell, Cupertino, East Palo Alto, Foster City, Fremont, Gilroy, Los Altos, Los Altos Hills, Los Gatos, Menlo Park, Milpitas, Monte Sereno, Morgan Hill, Mountain View, Newark, Palo Alto, Redwood City, San Carlos, San Jose, San Mateo, Santa Clara, Saratoga, Scotts Valley, Sunnyvale, and Union City.

15. Daniel F. Spulber, "Economic Activities of Silicon Valley Firms Developing Goods and Services Complementary to the Microsoft Windows Operating System," Report, August 1998.

16. "Hold My Hand," *The Economist*, May 15, 1999, pp. 72–74.

17. Seth Schiesel, "AT&T and British Telecom Merge Overseas Operations," *New York Times*, July 27, 1999, p. A1.

18. All of the information in this paragraph is drawn from http://www.concert.com/initial.asp.

19. http://www.att.com/global/concert/.

20. Simon Romero, "AT&T and British Telecom Study Possible Closing of Joint Venture," *New York Times*, July 2, 2001, p. C10.

21. See T. Ozawa, "Japanese Multinationals and 1992," in B. Burgenmeier and J. L. Mucchielli, eds., *Multinationals and Europe 1992: Strategies for the Future* (London: Routledge, 1991), pp. 135–154.
22. Ibid., p. 148.
23. Alfred D. Chandler and Stephen Salsbury, *Pierre S. du Pont and the Making of the Modern Corporation* (Harper & Row: New York, 1971), p. 495.
24. James Bennet, "Detroit Struggles to Learn Another Lesson From Japan," *New York Times*, June 19, 1994, section 3, p. 5.
25. Ibid.
26. "Hold My Hand," *The Economist*, May 15, 1999, pp. 72–74.
27. See http://www.star-alliance.com.
28. "What's the Word in the Lab? Collaborate," *Business Week*, June 27, 1994, pp. 78–80.
29. See Eric von Hippel, *The Sources of Innovation* (New York: Oxford University Press, 1988).
30. Ibid., p. 89.
31. The third-generation mobile services platform called wideband code division multiple access (W-CDMA) is "based on modern, layered network-protocol structure, similar to the protocol structure used in GSM networks." See http://www.symbian.com/corporate/news/1999/pr990525.html.
32. According to Symbian: "EPOC is an operating system, application framework and application suite optimized for the needs of wireless information devices such as Smartphones and Communicators, and for handheld, battery-powered computers. EPOC also includes connectivity software for synchronization with data on PCs and servers." http://www.symbian.com/corporate/news/1999/pr990525.html.
33. http://www.symbian.com/corporate/news/1999/pr990525.html.
34. Mikael Ricknäs, IDG News Service, "Nokia Buys Rest of Symbian, Will Make Code Open Source," *PC World*, June 24, 2008, http://www.pcworld.com/businesscenter/article/147470/nokia_buys_rest_of_symbian_will_make_code_open_source.html.

PART III

THE ORGANIZATIONAL GRID

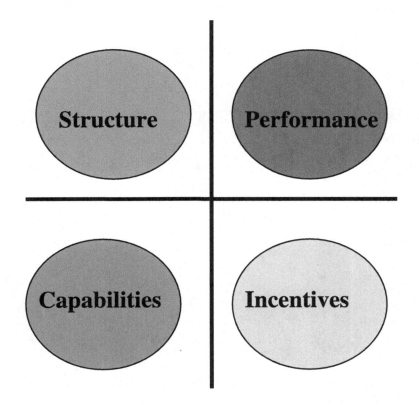

CHAPTER 5

ORGANIZATIONAL STRUCTURE
AND PERFORMANCE

Contents

Having examined the external landscape, the manager shifts attention to an internal analysis. Because strategic success requires finding the best match between market opportunities and the characteristics of the company, the manager attempts to determine whether the organization can implement the company's strategies and achieve its goals. By observing the company's strengths and weaknesses, the manager considers what goals and strategies will be feasible for the company and whether there is a need to change the company's organization in response to market forces.[1]

The manager's internal analysis considers four aspects of the company's organization: (1) the organization's *structure*, including the boundaries of the firm and the allocation of tasks across the divisions of the firm; (2) the organization's *performance*, in terms of profitability and efficiency both in the company's current activities and in planned activities; (3) the organization's *capabilities*, including its resources and competencies; and (4) the *incentives* for employees to execute the company's strategies and attain its goals. Given this information, the manager is better able to evaluate whether the company's market opportunities and its organization are well matched. The manager's internal analysis is summarized by the "organizational grid" (see Figure 5.1).

Table 5.1 presents some of the questions that the manager should ask in preparing the internal analysis. The first two aspects of internal analysis, organizational structure and performance, are covered in the current chapter. The next chapter examines the other two aspects of internal analysis, organizational abilities and incentives.

155

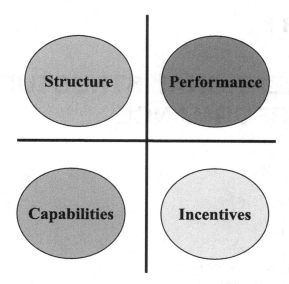

Figure 5.1:   The organizational grid.

Table 5.1:   Conducting an internal analysis.

Aspects of the Organization	Analysis of Current Conditions	Analysis of Prospective Conditions
Structure	What is the company's current organizational structure, including activities, divisions, decentralization, efficiency, and delegation of authority?	How does the company's structure match up with the company's goals and strategy? How should the company's structure change to carry out the company's strategies and achieve its goals?
Performance	How is the company performing relative to industry and financial benchmarks?	How can the company's performance be maintained or improved? What new measures of performance should be applied?
Abilities	What are the company's current tangible and intangible assets? What are the company's unique skills in productive technologies and operating processes?	How do the company's abilities match up with market opportunities and constraints? Can the company's abilities be applied to realize market opportunities? What new abilities should be developed to carry out the company's strategies and achieve its goals?

(*Continued*)

Table 5.1: (*Continued*)

Aspects of the Organization	Analysis of Current Conditions	Analysis of Prospective Conditions
Incentives	What are the company's incentive mechanisms? What types of performance are rewarded by the company's incentives?	What types of performance measures are used as the basis for the company's incentives? How should incentives be changed for employees to carry out the company's strategies and achieve its goals?

Chapter 5: Take-Away Points

The manager's internal analysis examines organizational structure and performance:

- The manager examines the company's structure, including the firm's boundaries and the allocation of tasks across divisions.
- The manager determines whether the company's structure is suited to serving the company's target markets and carrying out the company's strategy.
- The manager observes the organization's performance in terms of profitability and efficiency both in the company's current activities and in planned activities.
- The manager determines whether the company's performance measures reflect its strategic objectives.
- The manager evaluates how well the organization implements the company's strategies and how effective the company's strategies are in attaining its goals.

5.1 *Boundaries of the Firm*

The manager's internal analysis begins with an examination of the company's organizational structure. This analysis is more difficult than drawing the traditional organization chart. The manager seeks to determine what advantages the organization might offer in the marketplace and what constraints the organization places on the choice of goals. In addition, the manager tries to evaluate whether there are benefits to be achieved from changing the organization.

The manager begins by asking what products the business provides and how it provides them. The *boundaries of the firm* depend on the

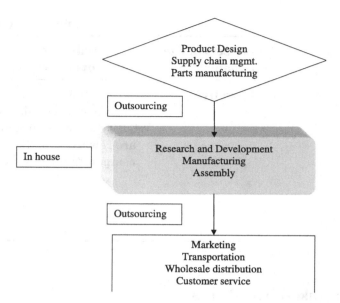

Figure 5.2: The boundaries of the firm.

range of *functional activities* that transform primary inputs into final products. The *boundaries of the firm* also depend on the variety of final products the firm provides and the types of customers the firm serves.

Crucial management decisions about what tasks to undertake and what tasks to leave to others determine the boundaries of the firm. Consider the boundaries of the firm, as in Figure 5.2. Some activities are handled within the firm, such as research and development, manufacturing, and assembly. The manager chooses to outsource some activities such as product design, supply chain management, parts manufacturing, and marketing, transportation, wholesale distribution, and customer service. The manager compares the value created for the company if it carries out an activity with the implications for the value if the activity is left to others — customers, suppliers, competitors, or partners. *The manager selects the set of tasks that will maximize the value of the firm.*

The manager reviews the final products currently provided to customers. What are the features of these products? Are they manufactured goods, customer services, or some combination? Have product features changed frequently or have they been stable? Do they use cutting edge technology? Are the products targeted at market segments? Does the company offer full product lines or only a few varieties? Are the company's products customized or one-size-fits-all?

An important question is whether the company's products resemble those of competitors or whether they are *differentiated* in some manner. Is the quality of the company's products higher or lower than that of competitors? In particular, are the company's products more durable, more effective, or easier to use than competing products? Are product differences with competitors simply a matter of taste, as in *Coke* versus *Pepsi*?

The manager determines whether the company should provide customers with bundles of complementary products and services. For example, auto manufacturers provide cars as well as repair service during the warranty period. The manager decides whether the company should provide narrow product lines or extensive variety and determines the extent of diversification — the categories of products the firm should provide. The manager considers what form products should take — whether they should be manufactured goods or customer services. Moreover, she or he chooses the set of products and the target customers that will maximize the value of the firm.

The decision about what products the organization will provide to final customers is perhaps the most important decision the manager makes because it is closely tied to the nature of the business. Recall that the goals of the business are what markets the company tries to serve. Serving specific markets will require maintaining or changing the company's product offerings. Choosing to supply some products but not others creates opportunities for other companies. The types of services that are provided with the company's products also affect what tasks are shared with customers. The company's decisions about its products can involve deferring to competitors, depending on partners to supply complementary goods or relying on customers, and these decisions in turn affect the boundaries of the firm. See Table 5.2 for some examples.

Table 5.2: **Examples of different choices that affect the boundaries of firms.**

Relationship to Others	Outside the Firm	Inside the Firm
Customers	IKEA relies on customers to take home and assemble furniture	Ethan Allen delivers fully assembled furniture to its customers
Suppliers	Kmart outsources shipping to its warehouses and stores to transportation suppliers	Wal-Mart owns and operates its own trucks to provide shipping to its warehouses and stores
Competitors	Southwest Airlines serves a limited set of routes leaving others to competitors	United Airlines maintains an extensive set of routes
Partners	Dell sells computers with Microsoft's Windows operating system installed	Apple Computers provides the operating system for its personal computers

Next, the boundaries of the firm are determined by *how* goods are produced, that is, what functional activities the firm keeps in-house and what functions are outsourced. For example, Nike designs sneakers and markets them to retailers, but practically all of Nike's manufacturing is outsourced to independent suppliers abroad. The Gap also designs its products and contracts for manufacturing but sells its brands through its own retail stores, including Banana Republic, Gap, and Old Navy. Spanish clothing chain Zara operates stores in over 30 countries and manufactures much of its clothing in-house.

Any functional activities are candidates for outsourcing. Functional activities include input manufacturing, purchasing, financial management, R&D, product design, human resource management, information processing, final output manufacturing and assembly operations, transportation, accounting, marketing, and sales. The manager decides what functions to keep in-house and what functions to outsource so as to maximize the value of the firm.

Some vertically integrated companies supply manufactured inputs both to themselves and to final customers. For example, before spinning off its Bell Labs and Western Electric divisions to form Lucent, AT&T generated R&D internally through Bell Labs and obtained many types of switching equipment from Western Electric that it used to operate its own network. Also, AT&T sold telecommunications products made by its Western Electric division to many other companies. AT&T's competitors were reluctant to purchase switching equipment from AT&T, but as a free-standing business, Lucent became a major telecommunications equipment provider. Some activities qualify as functional processes within the firm and also represent final goods and services supplied by the firm to others.

The company's collection of functional activities is determined by a number of important factors. What is the relative cost of internal production in comparison with market alternatives? What are the less tangible organizational costs associated with managing internal functional processes? What are the transaction costs of procuring services and monitoring outside suppliers? What special competencies does the firm have in its functional areas? What advantages and skills do outside suppliers possess?

The manager's assessment of the company's final goods and services and functional processes identify current boundaries. These are useful in determining what markets the company is currently serving and what markets the company could service in the future. Internal functional processes are sometimes businesses in their own right. In evaluating whether or not to carry out such activities internally, managers should consider the potential returns to operating the business as an independent company.

Choosing new goals and the strategies to achieve them are likely to entail changes in the boundaries of the firm. Entering and exiting markets will change the configuration of goods and services provided to final customers, and the company will need to alter its collection of functional support services to correspond. The company relies on outside suppliers to reduce the organizational costs of its market entry strategy. Chapter 8 examines some of the main factors that should be weighed in determining whether or not to undertake functions within the organization.

5.2 *The Value Chain*

The activities of the business organization, from sales to operations to purchasing, are known as *functions*. The functions of the firm include both market trans ctions and internal processes. For each business operated by the firm, the sequence of functional tasks within the organization constitutes its *value chain*. The value chain is a set of activities that transform purchased inputs and the firm's assets into final products for customers, thereby creating value.

Organizational Functions

All business organizations have to carry out certain basic functions. First, the business interacts with customers to provide *outputs*. The business informs customers about its products through its *marketing* function. The business carriers out customer transactions through its *sales* function. The business may also perform an additional *customer service* function, such as product repair and maintenance.

Second, the business must procure productive *inputs*. Through its *finance* function, the business interacts with capital markets and secures funds for its projects. Through its *human resource* function, the business interacts with labor markets and obtains labor services of managers and employees. Through its *purchasing* function, the business interacts with suppliers and acquires capital equipment, facilities, manufactured parts and components, resources, energy, and land. Through its *engineering* and *research and development* functions, supplemented by training and license acquisitions, the business interacts with the market for intellectual property and creates its technology.

Third, the business must manage its use of inputs and *coordinate* its many functions. The company's *operations* function is responsible for productive activities that transform inputs into goods and services for customers. The company's *accounting* function is responsible for

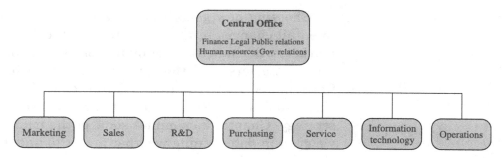

Figure 5.3: Company divided into functional units with central office services.

tracking internal allocation of funds as well as the company's payments and receipts. Finally, through its *public relations* and *legal* functions, the company interacts with the general public, government officials, regulatory agencies, and the judicial system.

Managers face the problem of coordinating the company's functional tasks and allocating them within the organization. The traditional solution has been to create separate divisions along functional lines, that is, to divide the organization by functions with a central office coordinating all of the functional divisions, as in Figure 5.3. A functional organization should generally be used by small businesses or by single-product firms. A functional organization can also be used by specialized business units within multiproduct firms.

The functional organization has the advantage that each area, such as sales or purchasing, operates as a department. The operations division includes all the personnel required to carry out manufacturing or other business processes. A functional division of labor within the company eases communication within the division. Members of the marketing division are likely to have similar professional training and perspectives. Also, specialization allows members of functional units to improve their skills by focusing on their functional task. Functional units can achieve some economies of scale and scope. Thus, centralizing the purchasing function may be less costly than dispersing purchasing personnel.

The departments of the modern business school continue to reflect functional divisions: accounting, finance, marketing, organizational behavior, human resources, and operations management. Organizational behavior applies psychology and sociology to the study of management and personnel problems. Some schools have a separate management strategy department. Economics often is associated with one of the functional departments (such as finance or operations management) to fit it in the basic framework.

The functional business organization tends to become unwieldy and inefficient for multi-product companies. Because functional units are focused on special tasks, the functional organization is not customer-oriented when the firm operates multiple businesses. The sales division is less able to focus on customer needs. Managers and employees in each of the functional areas have a tendency to view value creation entirely from the perspective of their own area. For example, engineers and scientists focus on products that satisfy technological criteria for performance but may not be tailored to market requirements. Henry Mintzberg calls these types of companies machine bureaucracies having "non-adaptive structures, ill-suited to changing their strategies". Mintzberg laments that this form "remains a dominant structural configuration — probably *the* dominant one in our specialized societies."[2]

In a multi-business firm, the functional organization is often highly centralized, with many levels in the hierarchy and a great deal of authority at the center of the organization.[3] The functional divisions generally cannot talk to each other directly but must work through the center of the organization. A communication between two functional departments generally must pass through the central office. A response requires a second trip up and back down the hierarchy, when sent through regular organizational channels. Thus, the functional organization often is not efficient at transmitting information.

For example, suppose that the operations division makes two products, say cars and trucks. The purchasing department must buy parts for the operations division, such as tires or spark plugs. If a manufacturing unit wishes to obtain more parts, it must send a message to the top of the hierarchy requesting the parts, and the center passes the message back down to the parts divisions. If cars and trucks use different parts, central purchasing adds an unnecessary layer of bureaucracy that complicates communication. If cars and trucks use similar parts, there are some benefits from central purchasing, such as a common information system and coordination of interaction with suppliers. One solution is to have central purchasing of some items that are used to manufacture both products and decentralized purchasing of parts that are used only in the manufacture of one of the products.

Like many companies, Xerox was a multi-business company that had a functional organization into the 1990s. Chairman and CEO Paul Allaire noted "in a functional organization, you need many staff functions to make things work. To ensure that manufacturing, for example, was hooked into sales, both had to be hooked into corporate." He continued:

"It breeds dependence and passivity. In a functional organization, there is a natural tendency for conflicts to get kicked upstairs. People get too accustomed to sitting on their hands and waiting for a decision to come down from above. Well, sometimes the decision does come down. But sometimes it doesn't. And even when it does, often it comes too late because market conditions have already changed or a more nimble competitor has gotten there first. Or maybe the decision is simply wrong — because the person making it is too far from the customer."[4]

The complexity of communication in a functional organization creates many problems. The marketing department cannot effectively convey customer information to the company's design and manufacturing arm, nor is it sensitive to the production constraints of the company. Marketing must do what it can with the company's product offerings, "selling what we have", rather than adapting products to customers. Similarly, finance must raise funds for all of the company's activities and may have a hard time distinguishing the company's many activities and their economic returns. None of the divisions are a self-contained business; they are all cost centers for the company.

In a functional organization, the costs of offering individual products are difficult to analyze since the functional areas are cost centers. For example, a single operations division mingles the cost of manufacturing the company's outputs. The company's marketing unit combines the cost of marketing the firm's products. Also, capital costs are not separated, so the value added of each activity cannot be determined. Accounting rules can allocate costs to products, but these rules are often arbitrary and do not reflect incremental resource costs of each product. As a result, costs associated with outputs under the accounting rules may not provide managers with sufficiently accurate information to make decisions about what products to produce.

Traditional management focused on production rather than markets. The functional organization was primarily designed in service of a large operations division, often further subdivided into several distinct units representing various stages in the production of parts and final product assembly. The basic organization chart not only indicates supervision and control, but also the nature of work carried out by the branches of the organization. Later work on classical management recognized the increasing complexity of the tasks faced by large corporations and placed greater emphasis on management rather than on production activities. Modern management has recognized the necessity of customer-focused organizations that could engage in strategic competition. This approach requires understanding how functional activities create value.

The Value Chain

The *value chain* is a way to think about how the firm's functional divisions work together to create value. Michael Porter observes that

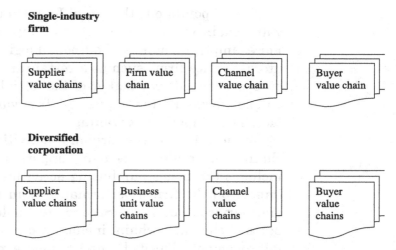

Figure 5.4: Michael Porter's value system for a single-industry firm and a diversified firm.

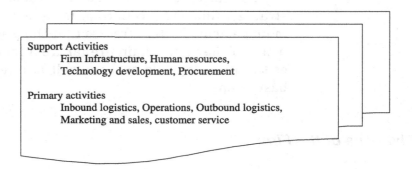

Figure 5.5: Michael Porter's generic value chain.

"every firm is a collection of activities that are performed to design, produce, market, deliver, and support its product" (see Figure 5.4). Porter defines the value chain of the firm as having five main components: inbound logistics, operations, outbound logistics, marketing and sales, and service (see Figure 5.5). First, inbound logistics includes material handling, inventory control, transportation, and contacts with suppliers. Second, operations includes activities associated with transforming inputs into the final product form, such as machining, packaging, assembly, equipment maintenance, testing, printing, and facility operations. Third, outbound logistics collects, stores, and distributes the product to customers, including warehousing, material handling, delivery, order processing, and scheduling. Fourth, marketing and sales activities are means for inducing and enabling buyers to purchase the firm's product, including advertising, promotion, sales force management, quoting prices, channel selection, and channel relations. Finally, service enhances or maintains product value, including installation, repair, training, parts supply, and product adjustment.[5]

Porter points out that in different industries, certain parts of the value chain will be more vital to competitive advantage than others. For example, inbound and outbound logistics are important for distributors. Service firms such as a restaurant or retailer will depend more on operations. For a maker of industrial or office equipment, service will be crucial. Porter emphasizes that value activities are the building blocks of competitive advantage.

In addition to the primary activities that comprise the value chain, Porter identifies four support activities. First, procurement includes the firm's purchasing activities, which often have a greater impact on the firm's performance than the direct costs of the goods and services obtained. Second, technology development affects all parts of the value chain, including product development and process improvement. Third, human resource management consists of the activities involved in recruiting, hiring, training, development, and compensation of personnel, often key to the firm's competitive advantage. Finally, firm infrastructure consists of general management, strategy, finance, accounting, legal, government affairs, and quality management. Porter stresses that the primary and secondary components of the value chain are far from independent. The effectiveness of the firm depends on how well it integrates and coordinates the basic steps.[6]

5.3 *Divisions of the Firm*

The organization can be divided into specialized units in two basic ways. Recall that the manager asks *what* the firm produces and *how* the firm produces it. Accordingly, the organization may be divided on the basis of what it produces or on the basis of how it produces. If the organization is divided on the basis of what it produces, its divisions correspond to its final goods and services or lines of business. If the organization is divided on the basis of how it produces, it is divided based on functional activities, with departments for marketing, sales, finance, operations, human resource management, purchasing, and so on. Companies often have hybrid organizational structures that are composed of both business and functional units.

In companies with business units, the manager also considers the allocation of functional activities between the company's central office and its divisions. Functional activities that are concentrated at the central office are said to be *centralized* while other functions such as sales and R&D that are located in the company's business units are said to be *decentralized*. Typical functions that are centralized at the corporate level include finance, accounting, public relations, and legal services.

Purpose of Organizational Structure

Managers design the company's organizational structure to enhance implementation of the company's strategies. The organization is divided into specialized units to provide focused tasks and to improve efficiency of operations. The manager's internal analysis evaluates the company's divisions and how the organization allocates activities between the central office and the divisions.

The design of an organization should not be haphazard or an accident of history, although every organization has some characteristics that are simply the result of inertia on historical events, and the process of evolutionary change. Rather than leaving things to chance, top managers must engage in conscious organizational design. Managers employ organizational design as a means of aiding strategy implementation. The internal analysis of company divisions and allocation of responsibilities prepare the way for organizational adaptation and evolution.

Changes in the company's goals are likely to be reflected in the company's organization. If the company is organized by lines of business, then deciding to enter new markets may require expanding the activities of some organizational units or creating new units. Similarly, exiting markets may require narrowing the focus of some organizational units or closing some units.

Deciding to offer new products and/or to shed some products often entails different types of business processes. Thus changes in the company's functional units will result, whether these units are company-wide or contained within lines of business. For example, a decision to stop manufacturing a product and to start offering customer services will require shedding operations and purchasing units and establishing or expanding service units. A decision to offer differentiated products rather than generic products may require the addition of a product design or R&D unit. Thus, the goals of the firm also affect the company's functional units.

Outsourcing decisions will likely affect the nature of the company's functional units. Switching to an outside supplier of parts will entail divesting or closing the company's parts manufacturing unit and changing the activities of related functions such as purchasing and human resource management.

Strategic shifts may be accompanied by changing the fundamental nature of the organization, favoring business units or functional units. The divisions of the organization and the allocation of tasks among those divisions will have important consequences for the allocation of authority and responsibility in the organization. The effectiveness and structure of the organization play important roles in determining what activities it can be expected to carry out in the future. Moreover, the flexibility of the organization affects the company's ability to change in response to new goals and strategies.

The manner in which companies divide their tasks has been evolving dramatically as new forms of organization are being devised. Organization design has proceeded through several substantial revolutions: Companies have moved from functional divisions to product-based divisions and then to business-based divisions. This significant shift in perspective is reshaping companies as never before.

The Product-Based Firm

One of the most important developments in the history of management has been the creation of the multi-divisional company, with divisions based on products. This form originated with DuPont and General Motors in the 1920s. Its key idea is that each of the product divisions generally has functional divisions. That is, in comparison with the functional organization, finance, personnel, purchasing, engineering, operations, and sales are split up and moved into the product divisions. Of course, all companies differ to the extent that functional departments are decentralized.

Alfred D. Chandler documents the historic developments at DuPont.[7] After its reorganization in 1921, DuPont was partitioned into five divisions: explosives, dyestuffs, pyralin, paints and chemicals, and fabrikoid and film. Some functions at DuPont remained centralized, including finance, advertising, corporate services, and some purchasing and engineering.[8] The clear advantage of this organizational form is that each of the manufacturing divisions tailors its personnel and much of its purchasing to reflect their changing individual needs. The increased autonomy of the divisions reduces the need for coordination with other divisions and allows the company to decrease the size of the central office.

Alfred P. Sloan, chairman of General Motors, presented a study of the company's organization in 1919 that was adopted as a plan of reorganization for the company the following year.[9] The plan, which was to become influential for the organization of large-scale companies, sought to bring some central coordination to the very decentralized General Motors company.[10] General Motors set up its well-known product divisions: Chevrolet, Pontiac, Buick, Oldsmobile, and Cadillac, and part makers such as Delco. The product divisions addressed market segmentation as well.

Sloan emphasizes the importance of determining the rate of return to individual divisions as a guide to the corporation's "strategic investment" decisions. He sets out general principles that recognize the trade-offs between central control of the company by the chief executive officer and independence of the company's divisions. General Motors adopted its bonus plan based on stock ownership as a means of creating incentives for executives to consider corporate profit rather than divisional profits.

The decentralizations General Motors and DuPont carried out in the 1920s are important to the history of management because they use organizational design changes to create incentives for management performance. Peter Drucker, perhaps the most prolific management author, observes that "Sloan's federal decentralization is still the best structure for the big multi-product company", but adds that newer multi-market businesses require different approaches.[11]

In the multi-divisional firm, some functions are carried out centrally (for example, finance and accounting, research and development, human resources, legal affairs) whereas others are carried out by all divisions (for example, marketing, manufacturing, and purchasing), as Figure 5.6 illustrates. Each division thus is a replica of the unified firm structure, with the exception of centrally administered tasks. The central functions sometimes are referred to as "cost-centers" and the product divisions as "profit-centers". This approach, while an improvement on the centralized organization, is fundamentally limited since the product groupings sometimes bear little resemblance to the firm's markets.

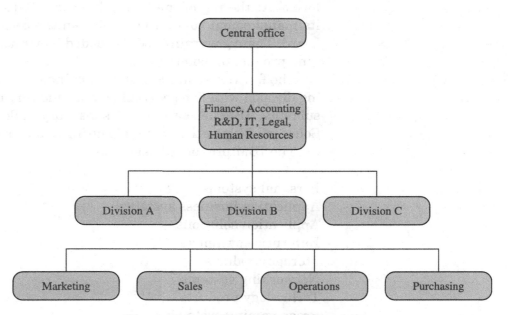

Figure 5.6:   The multi-divisional firm.
Some functional activities carried out by the central office and other functional activities carried out by the divisions.

Exhibit 5.1   IBM's Organization

Many companies have had difficulty shedding the functional or product-based forms. For example, IBM has long tried to escape but continued to have a hybrid organization. By the mid-1980s, IBM was still a highly centralized company, with four basic groups (communications, storage, products, and technology), a large U.S. marketing and services divisions, and two international divisions, one for the Americas and the Far East, and the other for Europe, the Middle East, and Africa.

In the late 1980s, John F. Akers, chairman of IBM, carved out product divisions (personal systems, application business systems, enterprise systems, technology products, programming systems, and communications systems). He retained a separate marketing arm with application solutions as a sub-unit that would provide business customers with integrated hardware and software systems to address their need.

At the beginning of the 1990s, IBM still found itself tied to the declining market for giant mainframe computers, which were fast becoming obsolete. Moreover, it has missed the market in laptops, desktop PCs, and workstations, as far smaller competitors earned greater profits.

In 1991, Akers proposed a reorganization of IBM that still had elements of both the functional and product-based organizational forms. At the top of the hierarchy were IBM's board of directors, its management committee, and then the corporate management board. These governing bodies headed two functional divisions and nine product divisions.

The functional divisions were the finance division and a marketing division with five geographic units and service and maintenance subunits. IBM also set up a subsidiary called the Employment Solutions Corporation to handle hiring and recruiting personnel.

The IBM product divisions were

Personal systems
Application business systems
Application solutions
Enterprise systems
Storage products
Networking systems
Technology products
Programming systems
Printer products.

(Continued)

Exhibit 5.1 *(Continued)*

The idea was to have the nine product divisions function with some autonomy. Akers suggested that the company would provide investors with financial performance data on each of its businesses.[12] He suggested that reorganization could turn "battleship IBM" into a fleet of nimble "destroyers".[13] Unfortunately, this laudable goal was undercut by the need to share marketing resources. The direct sales staff had traditionally focused on mainframes and minicomputers. Now they were expected to draw from the product divisions to provide customers with integrated solutions. Yet, IBM continued to have difficulty coordinating product design, production, and marketing.

After replacing John Akers as head of IBM two years later, Lou Gerstner decided to retain as a single unit the company's 40,000-member sales force. While retaining a geographic segmentation, the sales force would be divided further into product and industry specialists known within IBM as "fighter pilots".[14] Industry specialists will provide consulting services along with software and hardware from IBM and other companies, making them value-added resellers. Although the division of IBM's sales force by industry specialists represented an improvement over the previous organization, the unified sales force limited IBM's ability to focus on target markets.

The unified marketing force was a relic of the functional organization. The solution was not necessarily to carve up and distribute the sales force to the company's product divisions. Doing so would help toward the more decentralized multi-divisional form. However, IBM needed to create divisions that corresponded more closely to the customer groups it served and then to distribute its sales personnel into these new businesses. The company's organizational structure was a vertically integrated company that constrained the lines of business from competing with similar producers due to their reliance on the IBM marketers as their main sales channel. As IBM moved toward a consulting and service-oriented company, the industry expertise of the unified sales force offered some advantages.

The Business-Based Organization

A natural way to divide the organization is along the lines of its businesses. This type of division further decentralizes the company, shifting its functions, especially marketing, sales, and operations, to the level of the individual business. The company is decentralized much more than the multi-divisional product-based organization, allowing all of the functional areas in the individual businesses to be much closer to their customers. Marketing and sales are tailored to the needs of the business, rather than simply acting as promotion arms of the company's operations and engineering functions.

This type of organization enhances communication — since information from the field has less distance to travel, it can be acted upon within the business division. Directives and management objectives within the business can be communicated more effectively as well since the business managers can stay closer to the front line. Within the business, attention is focused on producing and marketing a set of products, so managers are not swept up in the full range of company activities.

By the early 1970s, businesses began to shift away from a focus on long-term planning based on simple growth projections. Strategic planning began as the result of intense competition and the increasing importance of marketing. This new concern was reflected in a reorganization at General Electric. GE classified parts of the organization into strategic business units (SBUs) based primarily on segmentation of the product market (see Exhibit 5.2).

Exhibit 5.2   General Electric and Strategic Business Units

At GE in the mid-1970s, the Corporate Executive Office included staff for strategic planning, finance, personnel, legal, administration, development, and technology. Research and Development, with 650 scientists and 1,200 support personnel, was attached to the central office.[15] The company was organized into nine "groups": Aerospace Business, Aircraft Engine Business, International and Canadian, Special Systems and Products, Components and Materials, Consumer Products, Industrial and Power Delivery, Major Appliance Business, and Power Generation Business.

Each of the groups was split into "divisions". For example, the Consumer Products Group was divided into seven operating companies: Home Entertainment Business, Housewares Business, Lamp Business, General Electric Broadcasting Company Inc., General Electric Cablevision Corp., General Electric Credit Corp., and Tomorrow Entertainment Inc. The Aerospace Business was split into Aerospace Programs Relations, Aircraft Equipment Products, Electronic Systems Products, Re-entry and Environmental Systems Products, and Space Products.[16]

(Continued)

> ### Exhibit 5.2 *(Continued)*
>
> General Electric in 1980 was composed of 350 businesses grouped into 43 strategic business units. William E. Rothschild, a manager of Corporate Strategy Development and Integration at GE, identified four main criteria for an organizational division to be an SBU:
>
> - It must serve an external market.
> - It should have a clear set of external competitors.
> - It should control its own destiny.
> - Its performance should be measured as a profit center.[17]
>
> The SBU must be defined on the basis of a set of customers rather than an internal supplier serving the company itself. The set of competitors that the SBU is trying to surpass in the market should be identified. The SBU should choose its own products, pricing, and suppliers, although it can draw on company resources such as a common sales force or R&D. Finally, the SBU must be judged as an autonomous business, in terms of profits, with proper accounting for the use of company resources.
>
> By the beginning of the 1990s, through acquisitions, divestitures, and reorganization, Jack Welch had slimmed down the GE structure to a dozen businesses, all of which reported directly to him. GE dropped air-conditioning, housewares, consumer electronics, semiconductors, and mining and acquired NBC, RCA, and Kidder Peabody. GE's business divisions included Aircraft Engines, Appliances, Capital Services, Lighting, Medical Systems, Broadcasting, Plastics, Power Systems, Industrial Systems, Transportation Services, and Global eXchange Services. These divisions were likely to change again as GE changed its focus to emphasize the provisions of services and to reduce the company's reliance on manufacturing.

The SBU organizational form represents an important development that firms can use to achieve better communication with the firm's distributors, retailers, and final customers. The SBU form is a significant improvement over the product-based company. It divides functional areas among the individual business units and gears them to serving the customers of that business unit (see Figure 5.7).

Partitioning the organization is a difficult but necessary task that reflects the manager's perception of the firm's customers. The divisions may be simply geographic to handle regional differences in customer groupings (see Figure 5.8). In some cases, the groupings may reflect

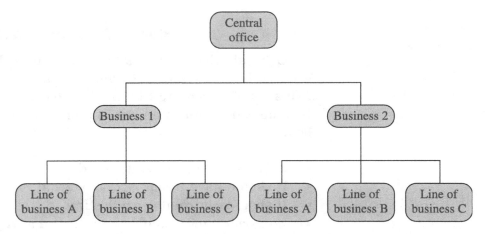

Figure 5.7:   Business units in a diversified firm.

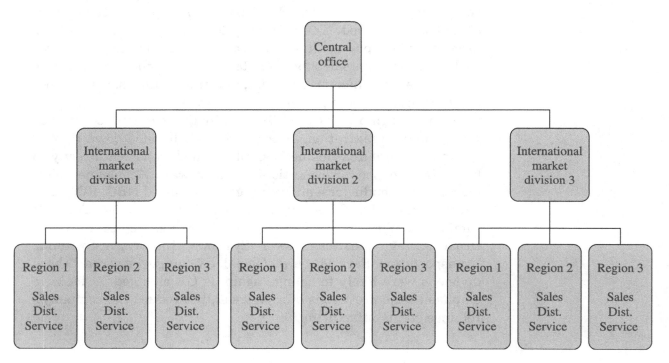

Figure 5.8:   Company with geographic divisions.

broad customer classes, such as industrial, commercial, and residential customers. Deciding how to partition the organization is a crucial component of innovative strategy. Dividing the organization in terms of market segments fosters improved communication and learning about the firm's product markets.

Within individual businesses, the functional divisions can remain. Thus, finance, personnel, purchasing, and R&D are in close contact with the relevant upstream markets while maintaining a focus on the requirements of their SBU's market. The SBU also fosters closer coordination between operations, purchasing, sales, marketing, and other functional areas.

The SBU form is important whether the firm sells directly to individual consumers or through wholesale or retail distributors. Wal-Mart divides its individual stores into 36 merchandise departments that are centrally managed for the entire chain and correspond to strategic business units. These merchandise units are able to communicate customer requirements to the suppliers of the merchandise in the department. Moreover, department managers effectively monitor sales and respond to customer needs.

The SBU form is also useful for firms with business customers. Long-term relationships with business customers often involve multiple contacts between the marketing, operations, engineering, and other departments of the supplier firm and those of the customer firm. Coordination of these interactions within the supplier firm is greatly facilitated when each SBU has its own functional departments.

5.4 *Performance*

A key step in any internal analysis is an evaluation of performance. Managers gather data on how the company has performed in the past and attempt to estimate future performance. Projected performance measures are useful for evaluating whether the organization is achieving the company's goals. Performance measures are needed to select business strategies and to adjust organizational structure. The manager evaluates the desirability of continuing the company's activities. In addition, new activities may call for redefining the company's performance measures.

Measuring Performance

Because the manager seeks to maximize the value of the firm, performance measures should provide information about how the company's goals and strategies contribute to value. In conducting the internal analysis, the manager should consider how the firm has performed relative to the goals that it has set in the past. This information is useful in determining whether the company's organizational abilities and choice of market opportunities are a good fit. The manager uses the performance information to estimate how the firm's abilities and choice of market opportunities will contribute to its value in the future.

Performance measures for strategic decision-making are necessarily forward-looking. Maximizing the value of the firm does not mean earning the highest possible profit in any given year, but rather earning the highest net present value of a stream of earnings. Looking at net present values of cash flows takes into account long-term strategies, not just short-term fixes. For purposes of strategic decision-making, the stream of earnings must be considered rather than just quarterly or annual returns. The appropriate rate of interest used to discount profits reflects the company's market cost of capital, which in turn depends on the company's degree of risk to investors. This approach takes into account the time-value of money.[18]

Companies should not rely solely on accounting measures of revenues, costs, depreciation, and cost of capital. Accounting and financial information such as measures of costs and revenues shown in the annual statement reflect past performance. This information is valuable to identify where the company stands. Past business performance provides information for the decision maker, but future performance is affected by changes in operating efficiency, customer demand, supplier costs, technological change, and competitor actions.

To choose the company's goals and strategic actions, it is necessary to develop measures that look ahead to the future effects of strategic decisions.[19] Managers should develop strategic accounting systems that give them information on their economic performance and allow comparisons with the performance of competitors. Managers must resort to formal and informal estimation procedures. Business plans consist of detailed projections of costs and revenues. Projections about rates of growth of company costs, revenues, and profits are ways of describing future performance estimates in comparison with already observed performance measures. Choosing actions to maximize the value of the firm requires estimates of the impact of the firm's activities on the present discounted value of future cash flows.

Revenues include all of earnings from sales and other sources of income. The company should take into account all of its costs, including all of its expenditures for production facilities and other capital equipment; labor costs; purchases of manufactured inputs, raw materials, and services; and technology expenditures, such as license fees and royalties.

The company's costs also include the *opportunity cost of capital*, whether investment is generated internally from cash flow, from sales of equity, or through borrowing. The opportunity cost of capital is the returns that would be obtained from investing in the best alternative opportunity. This is a market test that evaluates the best alternative opportunities for investing the money used by the company. The company measures its returns against market rates of return on capital. The company is creating value for its shareholders if and only if its revenues cover all costs, including the cost of capital.

Example 5.1   Calculating Value

A company is considering a two-year project that requires an investment of 600 in the first year. The project has operating costs of 200 in both the first and second year, and revenues of 1,300 in the second year. Suppose that the appropriate cost of capital is 10 percent. What is the value of the project? The cash flow in the first year is −800, that is, investment costs of 600 plus first-year operating costs of 200. The cash flow in the second year is 1,100, that is, earnings of 1,300 minus operating costs of 200. The present discounted value of cash flow in the second year is 1,100 divided by (1 + 0.10), which equals 1,000. The value of the project is 200, which is the sum of the present value of cash flows for the two years. The manager should approve the project because it creates value.

Concern over measuring value added is reflected in attempts by companies to estimate the returns to capital.[20] Companies such as Coca-Cola, AT&T, CSX, Quaker Oats, Eli Lilly, Briggs & Stratton, Georgia Pacific, and Tenneco have used these types of measures in evaluating their strategies. Surprisingly, some companies have not properly taken into account the costs of funds obtained from the company's shareholders.[21] *Fortune* magazine ranked top U.S. corporations in terms of estimated increments in market value, and Coca-Cola, General Electric, Wal-Mart Stores, Merck, and Microsoft topped the list.[22] These companies achieved additional profits by noticing the capital tied up in expensive productive facilities. John Snow, CEO of rail carrier CSX, took account of the capital invested in locomotives, containers, and railcars, and observed that "how we use capital determines market value."[23] Because of the risks of equity capital, the cost of equity is estimated to be about six percentage points higher than the long-term government bond rate, say 12 percent when the bond rate is at 6 percent.

Example 5.2   Opportunity Cost

An international company is considering whether or not to operate a production facility that it owns in another country. After one year's time, the production facility is expected to generate revenues of 1,000. After one year's time, the company will incur an operating cost of the production facility of 450. Suppose that the appropriate cost of capital is 10 percent. Another company has offered to purchase the production facility for 525. Should the

(Continued)

Example 5.2 *(Continued)*

company sell the production facility? At first glance, it might appear that the company should operate the facility, since it will earn 550 (1,000 minus 450), which is greater than the opportunity cost of holding on to the facility. However, discounting future returns, the company will only earn a present value of 500. Thus the company will obtain greater value from selling the facility. Put differently, after one year, the company will create value of 550. If the company sold the facility for 525 and the investor earned 10 percent, the investor then would have 577.50 (which is 525 times 1 + 0.10). Holding on to the facility the company owns is not free. The manager should take into account both the opportunity cost of the facility and the cost of capital.

Capital not only includes the funds tied up in equipment and real estate; companies also have contingent capital such as some forms of business insurance and the costs of hedging risk in commodity and international currency markets. Investment costs also include working capital, such as cash, inventories, and receivables. By reducing inventories through just-in-time techniques and improved market information, companies cut their costs of working capital. Product manufacturing can be closely tied to customer orders — suppliers can deliver straight to the production facility and plants can ship the products as soon as they are completed. In this way, manufacturers such as Ford Motor Company and assemblers such as Gateway and Dell Computer reduce working capital. American Standard set a goal of zero working capital to be achieved by adjusting production and inventories to demand fluctuations.[24]

Robert Goizueta, CEO of Coca-Cola, had an embroidered cushion in his office that read "The one with the biggest cash flow wins."[25] He boosted Coke's share price by a factor of almost 15 by shedding some less profitable businesses, reducing the number of plants for producing concentrate, and boosting productivity in the remaining plants. Generally, companies select capital expenditures based on the trade-off between the costs of obtaining investment capital and the returns to additional investment. Capital investment can be increased as long as the company's value is increased.

Strategy and Performance

At the corporate strategy level, top managers consider the contribution of the company's individual business units to the value of the company.

It is important for managers to try separating out the value contribution of individual businesses. This is often a difficult question because the corporation's business units share corporate resources so that their individual contributions are difficult to determine. It is sometimes useful to decentralize corporate resources to make individual businesses act more like profit centers, even if this means losing some of the advantages of centralized services.

Corporate managers should not assemble a portfolio of businesses to pool the risks of different industries. Investors are better able to decide what collection of companies should go in their portfolios. Top managers should consider divesting business units that do not contribute value to the corporation. The collection of businesses assembled by a company should be those that generate greater value when they work together than they could as separate companies. Some sharing of resources or management must occur that makes it worthwhile to combine the businesses. Diversified companies such as GE and ABB must demonstrate to shareholders that combining their diverse businesses adds value.

At the level of the individual business, managers should choose combinations of lines of business that maximize the value of the company. They should examine the set of goods and services provided to customers and evaluate the performance of the company's lines of business individually and in various combinations to determine whether they are contributing value to the business. Companies can divest themselves of lines of business, invest further in individual lines of business, or enter new lines of business.

The business manager also needs to determine what goods and services the company should provide in the future. Recall that the goals of the firm specify what markets the firm should continue to serve and what new markets the firm should enter. These choices require estimating future performance. The manager seeks to estimate what markets will generate the most profits for the company.

Companies evaluate what markets to enter and exit on the basis of their expected profitability. To properly evaluate the performance of the company's lines of business, it is necessary to give them profit-and-loss responsibility. The company should exit lines of business that are not expected to generate a positive net present value. The company should continue to operate existing lines of business or establish new ones that are expected to generate a positive net present value. Accordingly, expected performance of lines of business is crucial in determining the boundaries of the firm.

When lines of businesses share costs, it is necessary to make sure that each line of business makes a positive incremental contribution to the company's performance. Moreover, each combination of lines of business must make positive incremental contributions to the company. Without checking for incremental contributions, the company may show a profit, but unprofitable lines of business reduce overall performance. Proper evaluation of the performance of lines of business requires careful identification of costs and revenues that are attributed to them. At the same time, companies should not rely on arbitrary rule-of-thumb allocations of joint costs and overhead that cannot be properly attributed to individual lines of business.

Example 5.3   Identifying Incremental Returns

A company has two lines of business. The first line of business has revenues of 600 and attributable costs of 200. The second line of business has revenues of 400 and attributable costs of 500. The two lines of business have joint overhead costs of 75. Is the overall business profitable? Should the company continue both lines of business? In this example, the overall business is profitable since it earns 225. However, looking only at overall profitability would be misleading. The company should divest the second line of business because it does not contribute to the value of the firm. The first line of business operated alone would earn 325.

Example 5.4   Sharing Joint Costs

A company has two lines of business. The first line of business has revenues of 600 and attributable costs of 400. The second line of business has revenues of 800 and attributable costs of 750. The two lines of business have joint overhead costs of 200. The company uses an accounting rule of thumb to allocate joint overhead costs equally between the two lines of business; that is, each business is allocated 100 of the overhead. Should the company continue both lines of business? Using the accounting rule of thumb, the second line of business looks unprofitable; however, since the cost assignment is arbitrary and the two lines of business make positive incremental contributions, the company should continue both lines of business.

In evaluating the performance of the company for strategic purposes, managers should be careful to identify costs and revenues that attach to alternative decisions. This means that they should not consider costs that are irreversible, known as *sunk* costs. The costs of continuing an activity are the costs that are avoidable. Costs already incurred that cannot be recovered are not relevant to strategic performance measures even though they have affected performance to date. Irreversible investment must be considered before incurring the costs. This distinction causes a difference between entry and exit decisions.

Example 5.5 Sunk Costs

A company is contemplating providing a new product. The company expects revenues of 250 and operating costs of 200. The company would have to incur entry costs of 100, which are nonrecoverable. Should the company provide the product? The answer is clearly no, since the costs of entry exceed the net returns.

Consider a different situation. The company has already incurred entry costs of 100, which are nonrecoverable. After entering the market, the company discovers that revenues will be 250 and operating costs will be 200. Should the company provide the product? The answer is yes since sunk costs are written off and the net returns to providing the good are positive.

As managers receive additional performance information, they not only revise their choice of goals and strategies, but also consider changes in the company's organization. In evaluating the company's functional activities, the manager compares the expected performance of the company's functional units with market alternatives. The manager may choose to outsource parts production or some human resource management, depending on the contribution of these functions to the value of the firm. Also, the manager will consider the contribution of the company's resources and competencies in determining whether to emphasize existing abilities or to develop new ones.

The success of the company's strategy must be measured against that of competitors. Managers compare the firm's performance in terms of sales and profits with that of its competitors. This comparison provides a yardstick with which to evaluate performance of the company's investment projects and management decisions. Although all companies are different, many of the same forces affect all of the firms in an industry. They potentially have access to the same customers, they often compete for employees and suppliers in the same input markets, and they must solve similar technological and operational problems. The rewards go to the managers of the most successful firms. Competing to be the best motivates managers and employees to try harder, creating the discipline necessary to continually improve performance.[26]

The relative performance of companies in their industries often is reflected in the cost of capital. Investors wish to put their money in companies that consistently top their competitors. The competitive advantage of these companies reflects superior organization, better products, and managerial leadership. Investors cannot determine whether the company is maximizing its profits, but it is fairly easy to determine performance in comparison to competitors. Companies that consistently deliver additional value to shareholders will benefit from a lower cost of capital.

The company's performance measures are tied to its goals and strategies. The manager's review of the firm's past performance helps in the choice of new goals and the strategies to achieve them. The manager's estimates of future performance are also necessary to determine the company's goals and strategies. The performance measures for the business are closely tied to the evaluations of performance used to provide incentives for managers and employees. The next chapter considers the manager's internal review of the company's abilities and incentives.

5.5 *Overview*

The manager's internal analysis of the company's organization complements the external analysis of the company's actual and potential markets. The purpose of the internal analysis is twofold. First, the manager seeks to determine desirable goals and strategies given the existing features of the organization. In this way, the internal analysis is a critical input into the process of formulating goals and devising strategies to attain them. Second, the manager seeks to evaluate how the organization should be changed to accommodate the company's goals and implement strategy. Thus, the process of specifying goals and strategies and information obtained from the internal analysis are inputs into the process of organizational design.

The internal analysis begins with the company's organizational structure and performance. The company's organizational structure describes a number of important features. The boundaries of the firm reflect the range of the company's goods and services and its collection of functional activities. Specifying the company's divisions and the extent to which activities are decentralized to them determine the assignment of tasks within the company. The scale and consolidation of productive activities and the effectiveness of management affect cost efficiencies. The company's performance and estimated projections of future performance are crucial aspects of strategy. The manager evaluates how the firm's activities perform in comparison to market alternatives and uses prospective returns to evaluate its market entry goals and to compare alternative strategies.

Questions for Discussion

1. The four main aspects of internal analysis are organizational structure, performance, resources, and competencies. For each of these aspects, provide an example of specific information about the organization that a manager might need to know.
2. Select two companies within the same industry. Do the two companies make similar or different decisions about whether to perform a particular functional activity within their organization?
3. Select two companies within the same industry. Do the two companies provide similar or different types of customer service?
4. The five components of the value chain are inbound logistics, operations, outbound logistics, marketing and sales, and service. For a specific company, describe the activities that correspond to these components of the value chain.
5. What are some of the similarities and differences between product-based and business-based divisions of the organization?
6. For a specific company, discuss whether the company is organized along functional lines, product-based divisions, or some other organizational structure. What are some advantages and disadvantages of that company's organizational structure?
7. If a manager were to emphasize market share over profitability as a long-term measure of success, what effect would that performance criterion have on the activities of the organization?
8. Why should the manager evaluate the organization's performance in terms of the value of the firm?

Endnotes

1. Kenneth R. Andrews in his classic *The Concept of Corporate Strategy* (1971, Homewood, IL: Irwin) emphasized that strategy involves identifying the firm's strengths and weaknesses and the opportunities and threats presented by the business environment.
2. Henry Mintzberg, *The Structuring of Organizations* (Englewood Cliffs, N.J.: Prentice-Hall, 1979), pp. 346–347, emphasis in original. Mintzberg discerns five structural configurations: simple structure, machine bureaucracy, professional bureaucracy, divisionalized form, and adhocracy. For each type of organization, he assigns a different coordinating mechanism, a key part of the organization, and a type of decentralization. In the "simple structure," which occurs in some entrepreneurial organizations, organization is centralized with direct supervision by the "strategic apex." The professional bureaucracy is common to universities, hospitals, and public and not-for-profit agencies. The divisionalized form is the type Chandler described in his study of the DuPont company. Finally, Mintzberg characterizes the "adhocracy" as a decentralized

organizational structure influenced by an emphasis on expertise and advanced automated systems and designed to respond to a rapidly changing environment.

3. The functional organization has been called the unitary-form (U-form) firm by Oliver E. Williamson, *Markets and Hierarchies* (New York: Free Press, 1975).

4. Robert Howard, "The CEO as Organizational Architect: An Interview with Xerox's Paul Allaire," *Harvard Business Review*, September–October 1992, pp. 106–121.

5. Michael E. Porter, *Competitive Advantage* (New York: Free Press, 1985), p. 40.

6. Ibid., pp. 40–43.

7. Alfred D. Chandler, *Strategy and Structure* (Cambridge: M.I.T. Press, 1962). Chandler terms this type of company the multi-divisional, or M-form firm.

8. Ibid.

9. See also Peter F. Drucker's study of General Motors, in which he states that "decentralization . . . is not considered as confined to top management but a principle for the organization of all managerial relationships," *Concept of the Corporation* (New York: John Day, 1946), p. 49.

10. See Alfred P. Sloan, Jr., *My Years with General Motors* (New York: Doubleday, 1963).

11. Peter F. Drucker, *Management: Tasks, Responsibilities, Practices* (New York: Harper & Row, 1973), p. 520.

12. John Markoff, "The First Draft of IBM's Future," *New York Times*, December 6, 1991, p. C1.

13. John Markoff, "IBM Announces a Sweeping Shift in its Structure," *New York Times*, November 27, 1991, p. 1.

14. "At IBM, More of the Same-Only Better?," *Business Week*, July 26, 1993, p. 78–79.

15. Mintzberg, *The Structuring of Organizations*, p. 410.

16. Ibid., p. 411.

17. William E. Rothschild, "How to Ensure the Continuous Growth of Strategic Planning," *Journal of Business Strategy* 1 (Summer 1980), pp. 11–18.

18. The manager considers the present value of the company's stream of cash flows, discounted at the appropriate rate of interest. If the interest rate is r, and expected profit that will be earned in the year t is E_t, the present value of profit is

$$V = E_0 + \frac{E_1}{(1+r)} + \frac{E_2}{(1+r)^2} + \frac{E_3}{(1+r)^3} + \cdots$$

The profit stream can be divided into the present value of profit during the planning period, plus the residual value, which represents the net present value of profit in future years. Suppose that there is a five-year planning period, including the current year, and suppose that V^* represents the residual value. Then, the present value of profit is

$$V = E_0 + \frac{E_1}{(1+r)} + \frac{E_2}{(1+r)^2} + \frac{E_3}{(1+r)^3} + \frac{E_4}{(1+r)^4} + V^*.$$

19. The problems of using accounting methods as a guide to corporate strategy and its limitations for measuring economic profits are well known. See H. T. Johnson and R. S. Kaplan, *Relevance Lost: The Rise and Fall of Managerial Accounting* (Cambridge: Harvard Business School Press, 1986); Jeremy Edwards, John Kay, and Colin Mayer, *The Economic Analysis of Accounting Profitability* (Oxford: Clarendon Press, 1987); and Alfred Rappaport, *Creating Shareholder Value* (New York: Free Press, 1986).

20. For a discussion of how to apply the standard measures of economic profits to accounting data see John Kay, *Foundations of Corporate Success: How Business Strategies Add Value* (Oxford: Oxford University Press, 1993). The use of accounting data to estimate annual operating profit net of capital cost and prospective earnings explains the interest in a concept such as economic value added (EVA) and in its extension, market value added (MVA). The terms EVA and MVA are due to the financial consulting firm Stern Stewart. See G. Bennett Stewart, III, *The Quest for Value* (New York: HarperBusiness, 1991). EVA refers to after-tax net operating profit in a given year minus a company's cost of capital in that year. MVA refers to the company's capital from all sources of debt and equity as well as the capitalized value of R&D investment, minus the current value of the company's stock and debt. See also Anne B. Fisher, "Creating Stockholder Wealth," *Fortune*, December 11, 1995, pp. 105–116, and Shawn Tully, "The Real Key to Creating Wealth," *Fortune*, September 20, 1993, p. 38.

21. Fisher, "Creating Stockholder Wealth."

22. Ibid.

23. Tully, "The Real Key to Creating Wealth," p. 39.

24. Shawn Tully, "American Standard: Prophet of Zero Working Capital," *Fortune*, June 13, 1994, pp. 113–114.

25. Robert Goizueta summed up profit maximization with a basic analogy: "When I played golf regularly, my average score was 90, so every hole was par 5. I look at EVA like I look at breaking par. At Coca-Cola, we are way under par and adding a lot of value." Tully, "American Standard," p. 45.

26. The incentive effects of tournaments have been widely studied. See Edward Lazear and Sherwin Rosen, "Rank Order Tournaments as Optimum Labor Contracts," *Journal of Political Economy* 89 (1981), pp. 841–864, Oliver D. Hart, "The Market Mechanism as an Incentive Scheme," *Bell Journal of Economics* 14 (Autumn 1983), pp. 366–382, and Barry J. Nalebuff and Joseph E. Stiglitz, "Prizes and Incentives: Towards a General Theory of Compensation and Competition," *Bell Journal of Economics* 14 (Spring 1983), pp. 21–43.

CHAPTER 6

ORGANIZATIONAL ABILITIES AND INCENTIVES

Contents

The company implements strategy by acting through its employees. The manager depends on the efforts of the people in the organization to get things done. This means that the organization must be ready, willing, and able to execute the strategy. The manager's internal analysis evaluates the abilities of the organization and its willingness to carry out the strategy.

The *abilities* of the organization are its resources and competencies. The company's *resources* refer to tangible physical assets such as plant and equipment and more complex information resources such as brands, intellectual property, organizational capital, and reputation. The manager examines whether the company's goals are attainable given the company's existing assets and resources. The manager determines whether the company's goals are suitable for the company's resources. If the company's resources are highly valued in the marketplace, the company should attempt to realize the greatest value for its resources. This may require reorienting the company to make the best use of the resources or divesting resources that are not useful. If the company's resources are not appropriate to achieving its goals, the company may need to acquire additional resources.

The company's *competencies* refer to its technological expertise, process capabilities, functional skills, and other firm-specific abilities. These closely related terms pertain to activities, routines, or operations in which the firm excels in comparison with other companies. These activities distinguish the company from its competitors and guide the company in choosing what markets to serve. If the company's skills are exceptional and in high demand, the company chooses markets on the basis of where it will obtain the highest returns. If the company's skills are not exceptional or not in high demand, the company must attempt to develop new competencies and capabilities that are responsive to market opportunities and constraints.

The organization may have the ability to perform effectively, but it must also have the willingness to carry out the company's strategy. The company delegates authority to its managers and employees to carry out operations within the firm and to form business relationships with customers and suppliers. *Delegation of authority* refers to the allocation of responsibility and power within the organization both in terms of organizational units and personal relationships. By delegating authority to a manager or an employee, the company shares information and allows members of the organization to act for the company. The assignments may be specific operational tasks or generally defined management, sales, purchasing, and R&D activities. The manager examines the allocation of authority within the company to determine whether employees are empowered to carry out the strategic tasks.

Incentives refer to the motivation of managers and employees to perform their assignments. Incentives include not only monetary rewards for performance, such as salaries, bonuses, or stock options, but also employee morale, recognition, trust, training, and motivation. The manager's internal survey examines whether incentives are appropriate to the desired activities and whether employee interests are aligned with those of the company.

Chapter 6: Take-Away Points

The manager's internal analysis examines organizational abilities and incentives:

- The manager should examine the company's resources, both tangible and intangible assets, and evaluate the company's competencies, such as routines, skills, and processes.
- The manager determines whether the organization's abilities are matched to its goals and strategic activities.
- The manager considers the extent to which authority is delegated to employees and the incentives of employees to act in the interests of the company.
- The manager determines whether the organization provides the right incentives for employees and managers to carry out the company's strategic activities.

6.1 *Resources*

The internal analysis evaluates the company's resources. Broadly speaking, the resources of the firm are assets owned by the firm that are potentially useful in producing value. The great 19th century economist Alfred Marshall considered the success of companies to be related to their "internal economies", which are "those dependent on the resources of the individual houses of business engaged in it [the industry], on their organization and the efficiency of their management."[1] Edith T. Penrose was one of the first writers on management to emphasize the importance of company growth. She identified the resources of the firm and the competence of management as important factors in determining the company's growth potential.[2] Firm-specific resources are identified as sources of competitive advantage.[3]

Valuing Resources

The resources of the firm include its balance sheet assets and information assets (see Table 6.1). Balance sheet assets are listed in the annual report of publicly-traded companies. These include current assets: cash and cash equivalents, accounts receivable, inventories, prepaid expenses, and deferred income taxes. Balance sheet assets also include long-term assets: financial investments, property, plant and equipment, and capitalized costs. Companies also have balance sheet liabilities such as accounts payable, income taxes payable, payroll and benefit-related liabilities, retirement liabilities, and long-term debt.

Company resources further include information assets that are more or less intangible and often not part of the balance sheet. Companies have intellectual property such as brands, trademarks, copyrights, and patents, (see Table 6.2). They also have designs, blueprints, formulas, research results, and other technological information.

Table 6.1: Resources of the firm.
Some information assets may be explicitly included in the balance sheet.

Balance Sheet Assets	Information Assets
Current assets	Intellectual property
Cash and cash equivalents	Brands
Accounts receivable	Trademarks
Inventories	Copyrights
Prepaid expenses	Patents
Deferred income taxes	Designs
Long-term assets	Blueprints
Financial investments	Technology
Property	Business reputation
Plant and equipment	Market information
Capitalized costs	Organizational information
	Human capital

Table 6.2: Top 30 companies ranked by numbers of patents for 2006.

Rank	Company	Number of Patents
1.	International Business Machines Corp.	3,621
2.	Samsung Electronics Co. Ltd.	2,451
3.	Canon K K	2,366
4.	Matsushita Electric Industrial Co. Ltd.	2,229
5.	Hewlett-Packard Development Co. L.P.	2,099
6.	Intel Corp.	1,959
7.	Sony Corp.	1,771
8.	Hitachi Ltd.	1,732
9.	Toshiba Corp.	1,672
10.	Micron Technology Inc.	1,610
11.	Fujitsu Ltd.	1,487
12.	Microsoft Corp.	1,463
13.	Seiko Epson Corp.	1,200
14.	General Electric Co.	1,051
15.	Fuji Photo Film Co. Ltd.	906
16.	Koninklijke Philips Electronics N.V.	896
17.	Infineon Technologies AG	890
18.	Texas Instruments Inc.	880
19.	Siemens AG	854
20.	Honda Motor Co. Ltd.	778
21.	Sun Microsystems Inc.	776
22.	Denso Corp.	732
23.	NEC Corp.	728
24.	LG Electronics Inc.	694
25.	Ricoh Co. Ltd.	693
26.	Eastman Kodak Co.	688
27.	Sharp K K	665
28.	Broadcom Corp.	660
29.	Cisco Technology Inc.	649
30.	Robert Bosch GMBH	646

Source: "Patenting by Organizations 2006", U.S. Patent and Trademark Office (USPTO) Report, March 2007, http://www.uspto.gov/go/stats/topo_06.pdf.

The business reputation of companies among their customers, suppliers, partners, and investors is a resource as well. Companies also possess many types of market information about customers, suppliers, competitors, and partners.

What is the value of the company's resources? Companies provide estimates of the value of balance sheet items. These estimates are generally accurate for current assets such as cash, accounts receivable, or inventories. However, for assets such as plant and equipment, book value is generally calculated as purchase cost net of depreciation. Such values generally depart from market values since depreciation schedules are somewhat arbitrary accounting conventions. Companies estimate the useful economic lives of an asset and then depreciate a fixed amount per year. For some assets, companies use an accelerated depreciation schedule to attempt to account for technological obsolescence, which is by nature difficult to predict. For example, Cisco Systems uses straight-line depreciation for manufacturing facilities and laboratory equipment but accelerated depreciation for certain high-technology computer processing equipment. Moreover, as market conditions change, the value of assets can depart significantly from their purchase cost.

The best measure of the value of assets such as plant and equipment is their market value, that is, what the assets would be worth if they were to be sold. This generally differs from the value of the asset to the company, which is the incremental value the company produces due to ownership of the asset. The value of an asset to the company can exceed its market value when the company is employing the asset in a highly productive manner. Conversely, assets have a higher market value when they could be used more productively by others. Companies sometimes pay a premium above asset value when they acquire other companies. Such a premium is referred to as *goodwill*, which is treated as an asset after the acquisition.

The balance sheet asset referred to as *capitalized costs* denotes expenditures that are added up and treated as an investment in a tangible asset. For example, Lucent generally treats research and development costs as operating expenses. However, the company capitalizes labor expenses, related overheads and other costs that are incurred in developing software that can be sold, licensed, or leased. The cost of producing such an asset is unlikely to correspond to its market value. The revenues obtained from selling software might be greater or less than the costs incurred in producing it.

Many types of information assets may not appear on the company's balance sheet. Evaluating information assets is considerably more difficult than evaluating balance sheet assets. Sales of patents, brands, or technology give some indication of their market value. However, this need not indicate the value of these assets to the firm. By effectively managing the marketing and sales of products and services, companies can reap rewards that exceed their market value. For example, Microsoft obtained significant returns by retaining ownership of its disk operating system (DOS) software and licensing its use to IBM and other computer makers. Other types of information assets such as marketing knowledge about customers and procurement knowledge about suppliers is difficult to value. In addition, companies have all kinds of informal information about production processes and business practices. Companies have significant amounts of information about themselves, including information about the company's culture, organizational structure, and operations.

One of the most important types of internal knowledge is information about the characteristics of employees, known as *organizational*

Table 6.3: Assets and market value of the 10 largest U.S. companies in 2007.

Market value as of March, 28, 2008 measured by the number of common shares outstanding times the price per common share.

Company Name	Assets ($ millions)	Market Value ($ millions)
1. Wal-Mart Stores	163,514	208,730
2. Exxon Mobil	242,082	455,929
3. Chevron	148,786	175,479
4. General Motors	148,883	10,568
5. ConocoPhillips	177,757	120,284
6. General Electric	795,337	365,582
7. Ford Motor	279,264	12,337
8. Citigroup	2,187,631	109,255
9. Bank of America Corp.	1,715,746	169,115
10. AT&T	275,644	227,305

Source: *Fortune* 500's annual ranking of America's largest corporations (2008), http://money.cnn.com/magazines/fortune/fortune500/2008/full_list/.

capital. The company realizes the value of organizational capital when it assigns its employees to tasks within the organization that makes the best use of their combined skills. Employees' abilities, training, and expertise are the company's *human capital*. Organizational capital is used to effectively employ the human capital of the firm's employees.

One way of measuring the overall value of the assets of the firm is to look at the market value of the firm itself. The value of the firm reflects investor expectations about the future stream of cash flows, that is, the earnings obtained from use of the company's assets. The market value of firms differs substantially from balance sheet assets. Moreover, market value can be greater or less than balance sheet assets. Table 6.3 lists the balance sheet assets and market value of the 10 largest U.S. firms in terms of annual revenues. General Motors' balance sheet assets are over 14 times greater than its market value, suggesting that the assets are significantly overvalued. In contrast, Wal-Mart stores' market value is about 27 percent higher than the value of its balance sheet assets, suggesting a high valuation for the company's information assets. A more accurate picture of the company's equity is based on assets net of liabilities. However, the company's market value reflects information that investors have about the company's anticipated performance.

Resources as a Guide to Strategy

Having examined the firm's balance sheet assets and information assets, to what extent are existing resources a guide to strategy?

Clearly, managers must make the best use of the assets they have in hand. Making the best use of assets may require *employing* the assets to produce goods and services or manage transactions. Alternatively, the company may *divest* assets whose market value is greater than their value to the firm. The company may *write off* or dispose of assets that are not useful to the company and have little market value. The company can choose to *develop* or *acquire* new assets. As Michael Porter observes, "The presence of resources/activities within the firm that are rent-yielding is likely to reflect past managerial choices."[4]

Balance sheet assets such as property, plant, and equipment are routinely bought and sold. Intellectual property such as brands, patents, and trademarks are also readily bought and sold. Alternatively, companies create their own brands, trademarks, and patents. They consciously gather organizational information. Other less tangible information assets such as human capital are developed through hiring, training, and mergers with other companies.

Unique resources can be a source of competitive advantage for companies, particularly information resources such as patents or brands. The economic value of such an asset is the additional value of goods and services that results from the asset. Patented production methods can enhance the value of a product, and a consumer brand such as *Oreo Cookies* can enhance the revenues obtained from a product area. Information resources have value whether or not they are tradeable. For example, a production method may be firm-specific and hence not tradeable whereas a brand can be licensed or traded. The value of information assets is reduced or eliminated if its features can be copied by competitors, if substitute goods and services can be produced using comparable assets, or if they are surpassed by technological change.[5]

C. K. Prahalad and Gary Hamel see the company as a portfolio of competencies and core products rather than a portfolio of businesses that are related in terms of product markets. They define *core products* as "components or subassemblies that actually contribute to the value of the end products."[6] In their view such critical components should not be outsourced because they "contribute to the competitiveness of a wide range of end products. They are the physical embodiment of core competencies."[7] They suggest that the company should centrally produce or develop core products and the associated technological skills that underlie those products rather than decentralizing them to the company's business divisions. They further argue that top management should reallocate essential personnel and skills across the company's divisions, just as capital is allocated across divisions.

The value of assets depends on market alternatives, as emphasized in Chapter 2. The manager determines the *incremental* earnings that will be obtained by employing the resource; that is, the difference between earnings with and without the resource. If incremental earnings are negative, the asset should be divested or written off. If incremental earnings are positive, the manager compares those incremental earnings with the returns from divesting the asset. The manager should consider employing the asset only if using the asset yields greater incremental returns than divestiture.

The resources discovered by the internal analysis can only be evaluated in the context of alternative market strategies. The manager compares the use of its resources with alternatives that include developing and acquiring new resources. Michael Porter observes that in formulating strategy "stress on resources must complement, not substitute for, stress on market positions."[8]

6.2 *Competencies*

The manager's internal analysis examines the competencies of the organization. Competencies are particular types of information assets that have received a great deal of attention in management. Competencies refer to abilities of the firm in many different activities, including research, development, manufacturing, transportation, and transactions. Such abilities are organizational in nature; that is, they depend on many individuals and are therefore firm-specific and difficult to transfer.

In his masterful business history, *Scale and Scope: The Dynamics of Industrial Capitalism*, Alfred D. Chandler examines the growth of the modern industrial enterprise in the United States, Britain, and Germany. Chandler emphasizes that while economies of scale and scope in both production and distribution permitted the modern industrial enterprise to grow, achieving those economies depends on management of the organization.[9] Chandler notes that "such economies depend on knowledge, skill, experience, and teamwork — on the organized human capabilities essential to exploit the potential of technological processes."[10]

Chandler introduces the notion of *organizational capabilities*, which include both *strategic capabilities* and *functional capabilities*. Strategic capabilities are the managers' abilities to enter into growing markets and to exit from declining ones more quickly and effectively than competitors.[11] Functional capabilities include product innovation, process innovation, marketing, purchasing, and labor relations.[12]

Companies that rely heavily on strategic capabilities are likely to develop innovative entry strategies. Venture capital firms and leveraged-buyout companies specialize in investing in other companies. These companies must continually study markets, identify worthy prospects, and monitor their progress. For example, the venture capital firm Kleiner Perkins Caufield & Byers was one of the founding investors in Amazon.com. Among the many companies in which it has invested are America Online, @Home Network, Excite, Healtheon, Netscape, Ascend Communications, Cypress Semiconductors, LSI Logic, VLSI Technology, and Sun Microsystems.

Start-ups and entrepreneurial companies are also likely to exhibit strategic capabilities. For example, Amazon.com expanded from its initial position as an Internet-based retail bookseller, establishing other e-commerce stores offering recorded music, electronics and software, toys and videogames, hardware, and auctions. In addition, Amazon.com owns a share of online retailer Drugstore.com.

Companies with functional capabilities tend to be manufacturers or services providers. American Standard, a leading producer of plumbing products, air-conditioning and heating systems, auto components, and medical systems, identified functional process integration as its core competence. The company viewed itself as a "global pioneer in applying Demand Flow® Technology, a customer-responsive business system, to its business operations". The company applies its proprietary technology for "integrating and synchronizing work processes in a continuous flow". The process is intended to enhance product quality, speed up product design, and fulfill orders more rapidly. The company extended its process to administrative functions and restructuring its organization to manage its businesses according to processes rather than functions.[13]

Managers who can identify the firm's competencies can apply the information to organizational design and focus the firm's efforts on the appropriate products and functional activities. C.K Prahalad and Gary Hamel define *core competencies* as "the collective learning in the organization, especially how to coordinate diverse production skills and integrate multiple streams of technologies."[14] David Teece calls competencies "integrated clusters of firm-specific assets" and states that "the firm is a repository for knowledge ... embedded in business routines and processes."[15] George Stalk, Philip Evans, and Lawrence E. Shulman of the Boston Consulting Group define *capabilities* as "a set of business processes strategically understood."[16]

Advocates of organizing the firm around its competencies or capabilities point to successful firms that focus on specific activities.[17] Honda is successful due to its competency in manufacturing engines. Wal-Mart is said to owe its success to its capability of cross-docking at its distribution centers.[18] The problem with such examples is that the successful companies are identified after the fact. It is highly likely that such companies engage in valuable and efficient activities. Managers choosing what activities to focus on must identify what efforts will be successful in the future.

Identifying core competencies and capabilities is useful as a way of discarding the unnecessary tasks of the company, while keeping essential functions in-house and outsourcing other functions. However, managers still face the difficult task of identifying the company's core competencies or capabilities. Should competencies be the company's *previous* technological strengths or should they reflect *prospective* market opportunities? Simply identifying areas of technological expertise is not sufficient because the company's skills may not have high market value. Doing what you do best as a guide to choosing markets to serve is a prescription for failure if the market has little demand for those particular skills.

Computers and word-processing software have replaced the typewriter, which was originally patented in 1868. The typewriter-maker Smith Corona Corporation introduced its first portable word-processor, the PWP 9500NT, in the mid-1980s. This was not enough to retain customers switching to PCs. The company filed for bankruptcy in 1995 despite long experience selling typewriters and a decade of offering word-processors. The company had not recognized that it was providing services to customers seeking to create letters and other documents, so it hung on too long to its core competencies rooted in the old typewriter technology rather than adapting to new methods of creating printed matter. Had it adapted to new technology, the company could have used its experience, brand-name, and marketing knowledge to continue providing word-processing services.

Companies that have identified core competencies often divest lines of business involving products and services that lie outside the firm's area of expertise. They concentrate their efforts on businesses that are identified with their core competencies. This type of focus is valuable if the company has matched its abilities to its market opportunities. The company should not choose its lines of business solely on the basis of technology or organizational abilities. Companies should add lines of business that are expected to make positive contributions to value and to divest lines of business that are not expected to do so.

Companies that have identified core competencies are also urged to *outsource* functional activities that are outside their main area of expertise and conversely to retain those functional activities that are closely tied to areas of expertise. Core competencies are highly useful in leading managers to evaluate the firm's extent of vertical integration. Again, however, just doing what you do best is not sufficient as a guide to outsourcing. Outsourcing functional activities only makes sense if a suitable market alternative is available that provides greater value than internal sourcing. As with choosing lines of business, the decision as to whether or not to carry out functional activities must be subject to a *market test*. The manager judges what the firm does best and what is best left to others by comparing its performance with market alternatives.

Exhibit 6.1 Honda

Honda's considerable expertise in designing and manufacturing engines is well-known. Honda has applied this proficiency to a wide variety of products including motorcycles, cars, outboard motors, lawn mowers, and snow blowers. Honda's engine virtuosity has been often identified as an illustration of core competencies.[19] Difficulties Honda encountered and the measures the company took to overcome them provide a cautionary tale.

The history of Honda Motor Co. includes many highlights of engine design and manufacturing. The company was incorporated in 1948 after having introduced its first product, the A-type bicycle engine, the previous year. In 1949, the company produced its first motorcycle, the two-cycle, 98 cc Dream D-type. A couple of years later, Honda created the four-cycle, 146 cc Dream E-type motorcycle. American Honda Motor Co. was established in 1959 with a motorcycle store in the United States and European Honda GmbH (now Honda Deutschland GmbH) was set up in Hamburg, West Germany, in 1961. Honda's first sports car (S 500) and light truck (T360) appeared in 1963. Honda began marketing its portable generator in 1965. It introduced its CVCC low-emission automobile engine system and its Civic compact in 1972. In 1976, Honda presented the Accord CVCC (1600 cc), and production of the Civic series reached 1 million units. Lawn mower production at Honda Power Equipment Mfg. began in the United States in 1984, and engine production began there two years later. In 1988, Honda began making its low-emission, high-performance VTEC engine.[20]

In 1991, Honda's founder Soichiro Honda died and shortly thereafter, the company began to encounter financial difficulties. According to *The Economist*, "Rumors were circulating that the company — maker of the best-engineered Japanese cars — was about to be absorbed by its dreaded rival, Mitsubishi Motors. Bills were piling up from the financial follies of the 1980s, the company had no new models, and the economy was heading downhill fast."[21]

In 1990, Nobuhiko Kawamoto took over as president and chief executive officer. As *The Economist* observed, "Honda was in thrall to its engineers — a priesthood housed in separate premises, untroubled by the dismal businesses of marketing and accounting, with a say in everything from piston size to pricing. With the economy soaring in the late 1980s, it hardly seemed to matter that customers were not lining up to buy Honda's flights of engineering fancy."[22]

(Continued)

Exhibit 6.1 (*Continued*)

Recognizing Honda's problems, its new management took a different tack. Kawamoto had to reorient the company's organization toward its markets rather than its engineering core competencies. He put on hold the company's Formula One motor racing division as of 1992, despite its prestigious string of victories, to demonstrate the change in priorities. Mr. Kawamoto added merit-based pay and promotion in place of the company's seniority system. Furthermore, he raised the profile of market research within the company, resulting in new product design methods. Based on customer inputs, the company introduced a series of recreational vehicles such as the Odyssey van, the CR-V and the Step Wagon, which used an existing car platform and power train, rather than something designed from scratch.[23] The new models boosted Honda's domestic and international performance in the 1990s.

The Honda experience showed both the advantages and limits of a technological orientation rather than a customer focus. Beginning in 2000, the company planned a return to Formula One racing. The company's new CEO, Hiroyuki Yoshino, stated that "building on a solid tradition of technological innovation, Honda strives to provide its customers with ever more exciting and convenient new products, while constantly challenging existing parameters of quality, safety and environmental responsibility."[24]

A market test of functional activities compares the performance of intermediate processes within the firm to outside market alternatives. Companies should take on only those activities that the company can perform relatively better than market alternatives. Managers must evaluate how well the company performs each of these activities, taken singly and in various combinations.

Richard J. Schonberger suggests that vertically integrated companies link business functions to create a "chain of customers" within the organization, until the final customer is served. This does not necessarily mean that the company is divided into separate value chains or pipelines but rather that the company should align resources by the way goods and services flow.[25] At Harley Davidson, the motorcycle company, this meant reorganizing the engine and transmission plant where several hundred machines had been grouped by function. The machines were grouped into cells to produce shafts or gears, so that workers were able to see where the work went next and to complete the production of components.[26]

The idea that internal functions and processes serve internal customers is useful in understanding that outside suppliers could serve internal customers. The quality of service and costs of internal function and process should be compared to market alternatives. Also, managers should carefully evaluate the transaction costs and organizational costs of internal services and compare them with the cost of procuring services from other companies. The choice of internal processes depends on their performance relative to the best market alternative. Companies have learned that practically any internal process can be outsourced. Gone are the days when companies tried to go it alone by being fully vertically integrated and performing almost every step from product design to manufacturing to distribution. Any combination of product design, input manufacturing, or final assembly can be carried out by others. Many companies manage their supply chain by forming partnerships with other companies. Management of information technology and human resources can be outsourced to others. Information gathering and analysis can be outsourced to management consulting companies. Evaluating the performance of functional activities is an important component of the manager's internal analysis.

6.3 *Delegation of Authority*

For tasks involving any degree of decision-making, the manager must *delegate authority* to some degree to the employee. Managers' time is limited — they cannot do everything and be everywhere at once. Moreover, managers must rely on others with specialized skills and knowledge. Ultimately, top managers must seek the help of middle managers and employees to get the job done. Delegation requires granting the employees some degree of independence and the right to use their judgment. The manager must be willing to cede some control to the employee, making a trade-off between the advantages of getting more done and the costs of having less control. The internal analysis examines existing delegation relationships within the company.

Delegation and Performance

Ultimately, the effectiveness of the company's organization depends on how well the company's employees get things done. Formal structures matter, but how relationships work in practice matters more. Figure 6.1 emphasizes that implementation of the company's strategy involves delegation of authority to the company's employees. The manager is responsible for formulating the company's goals and strategy and communicating them clearly to the company's employees. The manager delegates authority to the company's employees who are then charged

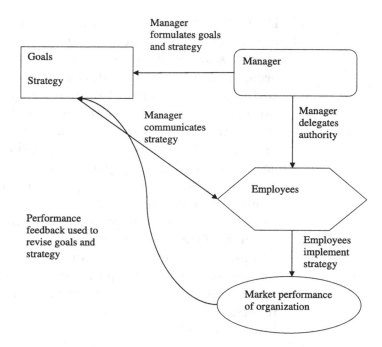

Figure 6.1: Implementing strategy involves delegation of authority to the company's employees.

with achieving the company's goals by implementing the strategy. As part of the internal analysis, the manager monitors the effectiveness of the organization in executing the strategy and uses the information to update the company's goals and strategies.

The manager's internal analysis reviews the company's vertical divisions. How many levels of management are there in the company as a whole and within its divisions? What is the span of control of its managers; that is, how many people report to each manager? What are the formal methods of communicating directives to subordinates and of receiving information from front line personnel? In addition to the company's formal lines of authority, the manager is concerned with the structure of informal relationships within the organization. What are the lines of communication in practice? How is power shared within the organization? What are the critical elements of corporate culture?

Delegation of authority occurs within the company's hierarchy. The relationship of delegation is sometimes referred to as the *principal-agent relationship* (see Figure 6.2). The principal delegates authority to the agent. The agent acts on behalf of the principal. The company may have middle-level supervisory relationships in which the principal monitors supervisors who in turn monitor agent performance. Such relationships exist throughout the company hierarchy. The internal analysis examines the general nature of the company's principal-agent

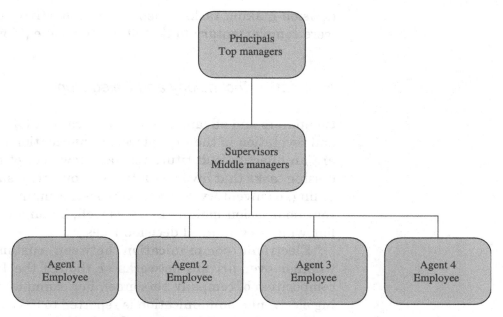

Figure 6.2: **Principal-supervisor-agent relationships in a company hierarchy.**

relationships. Who has the authority or will be given the authority to carry out the tasks at hand?

In the traditional centralized firm, divided on the basis of functional areas, it was possible to create a formal "command and control" structure, with managers clearly defining tasks for subordinates in the hierarchy. The assembly line concept was not confined to manufacturing, but extended across the organization. Managers divided many jobs into discrete components to take full advantage of specialization and division of labor. Employees were rewarded for well-defined performance measures based on completing specific tasks effectively or rapidly.

To some extent, even such tightly controlled activity is a form of delegation. Employees are given the responsibility for completing the task. They have some latitude in their choice of effort and level of dedication that they bring to the task. Companies can hire supervisors to monitor performance and ward off shirking, and they can offer rewards, such as promotion, bonuses, or piece rates, to stimulate employee effort. Even in such a constrained environment, employees may have better information than their supervisors about their own level of effort and interest, the types of skills that they bring to the task, and even about the quality of the output.

Such limited delegation is only useful for mechanical types of tasks. However, many of those tasks can be automated by mechanical equipment, a process that has continued unabated since the industrial revolution. Automation freed personnel for data processing and decision-making tasks, raising the productivity of labor services that were complementary to that of mechanical equipment.

Information Technology and Delegation

Computers and advances in communications have brought about a second revolution in the workplace, an information revolution. Computers are an effective substitute for vast new sets of mechanical data processing tasks that involve routine computation, such as record keeping, billing, or inventory control. Moreover, computers substitute for many routine decision-making tasks as well, by mimicking past decisions or following programmed decision trees.

Electronic communication between customers and suppliers, whether over private networks or though the Internet, changes the composition of company personnel. For example, one company employing electronic communication is reported to have reduced its data entry staff from 700 to 7.[27] Essentially, clerical personnel are removed from certain parts of the communication loop. Software that connects applications within the organization, such as order entry and processing, known as "bridging software," further reduces the costs of communication and data transfer across the organization. At least one-quarter of the cost of executing business transactions has been found to be data processing by personnel.[28] Electronic communication with suppliers and customers can cut the cost of sending a bill of lading from an estimated $13 to $1.50.[29] These costs are significant for an organization such as Sears Roebuck and Co. whose merchandise group deals with 5,000 primary suppliers and processes 21 million purchase orders annually.[30]

Many of the tasks now performed by computers were carried out not just by lower-level employees but by middle managers who kept track of production or marketing data, consolidated reports, and passed them on to upper management. Advances in computer applications have been one of the driving forces in company "downsizing". Companies terminated the positions of employees engaged in routine information processing, and those of their supervisors as well.

Any task requires employees to put in some effort. But the new tasks demand more than what has been called sweat equity. The enhanced productivity generated by computers and telecommunications technology increases the demand for higher skilled workers — not only workers with computer skills, but workers who have skills that computers have not replaced. The result has been a redefinition of employment as companies demand creativity, responsibility, and additional training from employees. As a consequence, delegation of authority in many organizations has increased as companies demand new types of performance from employees, and in return offer employees greater independence. Employees have become empowered by technical necessity.

Shoshana Zuboff coined the term *informate* to describe information-based technological change in the workplace: "Activities, events, and objects are translated into and made visible by information when a technology *informates* as well as *automates*."[31] According to Zuboff, although automation replaces human capacity,

> "when the technology also informates the processes to which it is applied, it increases the explicit information content of tasks and sets into motion a series of dynamics that will ultimately reconfigure the nature of work and the social relationships that organize productive activity."[32]

The smart machine fundamentally changes both manufacturing and office work by creating the need for smart workers:

> "The informated organization ... relies on the human capacities for teaching and learning, criticism and insight. It applies an approach to business improvement that rests upon the improvement and innovation made possible by the enhanced comprehensibility of core processes. It reflects a fertile interdependence between the human mind and some of its most sophisticated productions."[33]

Moreover, delegation of authority and increased information available to employees changes organizational relationships. Zuboff reminds us that

> "managing in an informated environment is a delicate human process. The ability to use information for real business benefit is as much a function of the quality of commitment and relationships as it is a function of the quality of intellectual skills."

Workplace information systems are not simply tools that increase productivity. Because computing power and data storage opportunities are so significant, the nature of the employment relationships is changed fundamentally.

The increasing knowledge shared with employees throughout the organization as the result of computers and the enhanced interaction using innovative communication technology has changed the configuration of the organization itself. Information technology represents the automation of routine tasks just as the industrial revolution brought mechanization to many production processes. One set of tasks that has become automated is the processing and exchange of information within the organization and between organizations, a set of processes formerly carried out both by clerical staff and management. Another important set of tasks undergoing automation has been transactions with customers and suppliers. Electronic commerce involves the substitution of computers and communications technologies for some types of labor services used to handle some purchasing and sales activities.

The signs that information technology has changed the workplace are both simple and substantial. In the business office, computers adorn every desk, with executives and staff alike engaged in word-processing, memo writing, data analysis, and sending memos by e-mail. The basic desk has long been replaced by the all-purpose cubicle, that confined world satirized mercilessly in the comic strip, Dilbert. The traditional distinction between executive and staff functions has begun to disappear, with executives handling some communications and monitoring real-time company information. According to *Fortune*, the number of secretaries has declined as companies have reduced the ranks of middle managers, with some of the remaining staff taking on selected administrative tasks formerly performed by executives.[34] An asset management firm near Philadelphia, SEI Investments, eliminated its secretaries and found that productivity increased.[35]

The physical separation of clerical and executive personnel has diminished apace, just as functional personnel are working side by side. Steelcase Inc., the office furniture manufacturer, commissioned a book titled *Office Access* which challenges conventional assumptions about office arrangement.[36] No longer do titles and organizational status determine location, size, or furnishings; rather, functional areas are physically proximate, and executives are at the center of the action.[37] Instead of allocating office space for productive isolation, the office is being reconfigured to encourage working together. Moreover, rather than viewing common areas as "wasted space", the guidebook urges office designers to promote employee interaction: "Good conversation breeds good ideas, and serendipitous conversation breeds the best ideas of all."[38]

At the same time as some companies redesigned their offices to encourage joint work, other companies explored ways to push employees out the door. As the work place evolves, many companies allow or encourage employees to interact with customers and coworkers online. Employees are using company data networks for messaging, meetings, scheduling, and virtual teams.

At AT&T's Sacramento, California, office, sales employees "were given laptop computers, cellular telephones and portable printers and told to create 'virtual offices' at home or at their customer's offices."[39] According to *The Wall Street Journal*, even the meager comforts of the cubicle are outdated: Companies like IBM, Ernst & Young, and Dun & Bradstreet "have begun taking away desks and offices to wring more value out of rented space and to compel employees to spend more time with customers."[40] Ernst & Young started a "hoteling" program in Chicago by taking away the desks of 500 accounting and management consultants below the senior management level. To work at the office for any given day, they must first call the company's concierge to reserve an office where they can park their files and receive calls or clients.

The *business document* continues its fundamental transformation, becoming a vehicle for collaborative work. The electronic document is a method of facilitating joint work and information exchange. A group of people within a business or negotiating a transaction between businesses can jointly write a report or contract even though they are at different locations or working on the document at different times. This ease of interaction can improve productivity and reduce costs of meetings and travel. It also potentially eliminates organizational boundaries that deter joint work.

Documents have generally referred to such simple forms as orders and invoices, and standards for preparing such forms have been the focus of the Data Interchange Standards Association and similar groups. Electronic documents no longer are simply digital versions of paper text. They are used as a means of storing and retrieving information from central databases. The new document not only includes these forms but now encompasses reports, business plans, and other longer and more complex communications. It also contains elements from multiple media, including voice, images, technical designs, and video in addition to data and text. This solves the difficult problems involved in the storage, retrieval, and display of different forms of information. The document can be replayed on a computer and transmitted across electronic networks. It is not necessary to obtain players and transmission devices for each type of information.

The electronic documents are virtual, in the sense that they are compilations with access to data from other databases; they evolve over time as they are updated by contributors; and they contain identification mechanisms that route them across networks. Documents are no longer simply stored in electronic files, but are themselves containers for sets of files. Industry groups continue to develop standards for document management that will enhance their functions as communication and information retrieval devices.[41]

The lower costs of information processing equipment have made it possible for all employees to have direct access to computers, without excessive mediation by the company's information technology (IT) managers. Unlike the traditional mainframe in the glass room, personal computers are a tool that is close at hand. The computer serves as a communication device for originating and terminating data transmission over both the company's networks and external networks. Communication over electronic networks has become a vital tool for coordination and collaboration within companies. The accessibility of computing and network technologies confers greater autonomy to employees, which has been reflected in changing responsibilities and organization design.

The development of corporate intranets and increased business use of the Internet have changed the role of the personal computer in organizations. Central IT management and standardization of hardware and software create economies of scale. Central coordination enhances communication within the company and between the company and its customers and suppliers. On the other hand, however, the system of PC clients and central servers recalls the more traditional system in which employees operated terminals connected to a central mainframe. By standardizing IT, companies may limit the delegation of authority to employees by restricting employees' abilities to manage their own computer applications. IT managers must be careful to maintain the advantages of decentralized computing while they pursue centralized corporate IT.

6.4 *Incentives*

To get the company's managers and employees to work in the company's interest usually requires *incentives*, such as salary, bonuses, promotions, and perks. The internal analysis examines the nature of the company's incentive programs. The manager answers some of the following questions: How is employee performance evaluated? How is employee performance rewarded? What specific types of incentives does the company use?

Then, the manager's internal analysis provides a comprehensive evaluation of whether performance measures, rewards, and incentive plans correspond to the company's goals and strategies. Do the rewards provide the right incentives to perform effectively? For example, auto dealers traditionally reward sales personnel through a combination of base salary and commissions to encourage aggressive bargaining over the sales price and to aggressively push customers to purchase more expensive cars. Some auto dealers such as Car Max, which offer cars at fixed no-haggle prices, pay sales personnel a base salary and a flat commission fee to reduce pushy sales tactics.

The company's relationship with its employees is founded on *trust*.[42] Delegating authority to the employee cannot be accomplished effectively without the company and its managers depending on the honesty and integrity of the employee. Although some monitoring may be needed, observing the employee's every move would simply be too costly. Fully controlling employee action not only would harm morale but it would also negate the very benefits from delegation of authority. Excessive control over employees stifles the creativity and market responsiveness that come with independence of action.

A relationship of trust carries certain duties. The employee is expected to act in the interest of the company and to report information thoroughly and accurately. Lack of trust between the company and its employees has high costs for the organization. Employees may be discouraged from acting in the interest of the company. Managers may be reluctant to delegate authority.

The manager's internal analysis considers the level of trust in company relationships. Is there cooperation between employees and between managers? Do managers share information and authority with employees? Are employees accustomed to handling authority and responding in a reliable manner? Does the company's culture support trust? What is the overall state of company morale? In addition, the internal analysis should attempt to evaluate what effect the company's proposed goals and strategies will have on trust in company relationships. Also, the internal analysis should consider whether organizational relationships exhibit sufficient trust for employees to be expected to carry out the company's strategies and achieve its goals.

For delegation of authority to produce better results than more bureaucratic command-and-control approaches requires the design of incentives for performance. Thus, the design of incentives and the monitoring of performance are integral aspects of decentralization. This does not mean that the company should determine those incentive schemes centrally and then wait for employees to respond. The need for decentralization applies to the process of incentive design *itself*. Managers selecting the rewards, measuring performance, and awarding bonuses and raises should be close to the task at hand. Such proximity means that rewards will be targeted to the specific needs of the operating division and that performance reviews will be adjusted for market conditions.

It is difficult, if not impossible, to get incentives just right. The factors that make incentive schemes necessary are precisely those that impede their effectiveness. Almost any work that employees perform can be viewed as a delegated activity. The company grants its employees authority to act on its behalf, whether in purchasing, production, or sales. Generally, the person performing the task has the best information about his own efforts and abilities, the resources needed to accomplish the task, and other external factors that affect the outcome of the work. Correspondingly, managers are less well informed, even though they may have better information about the interaction between the employees under their supervision or a broader view of the combination of tasks. The quality of the information that the manager has, both before the incentive scheme is designed, and after it has been put in place, is crucial to its success.

The incentive scheme, as a mechanical process of tying rewards to performance, cannot be expected to anticipate every market condition or technical problem that the employee may encounter. To build such a comprehensive incentive scheme would be both cumbersome and prohibitively costly. As a consequence, good incentive schemes involve both objectively defined rewards, such as sales commissions, and subjective performance reviews by immediate supervisors, who can adjust evaluations to reflect market conditions. For this reason, the managers designing the scheme must be located close to the employees performing the task.

Rewards should be based on *performance*, not measures of effort. Yet it can be difficult to monitor performance. A major electronics chain store rewarded its employees based on the number of units they sold. This created an incentive to push the least expensive items in the store, and total sales fell. To correct this, employees then were rewarded commissions based on the dollar value of their sales. The result was that customers would enter the store and not be able to find a single salesperson — they were all clustered around the big-screen TVs, the highest-priced items in the store. Customers seeking to purchase a phone or radio, were steered toward more expensive products. As a result sales of these items declined. Because employees were so focused on high-ticket items, customer service declined overall, leading to dissatisfaction and reduced store traffic. The chain store eventually abolished sales commissions altogether, paying employees fixed salaries, although with mixed results.

Even if some dimensions of effort can be measured in some way, it is usually not desirable to reward effort. Since effort is difficult to observe, the employee will tend to respond to the effort measures, boosting their achievement to match the desired effort profile rather than focusing on the end result, such as sales or profitability. This phenomenon is similar to "teaching to the test," where a school institutes a test for pupils as a means of gauging teacher performance in conveying subject matter. This gives the teacher an incentive to drill the anticipated material that will be covered in the test to increase student scores and thus improve the teacher's rating. The end result can be that the pupils know much less about the subject than they would have in the absence of the test because class instruction is no longer designed to provide in-depth coverage of important topics.

The phenomenon of "teaching to the test" is familiar to practitioners of total quality management (TQM). By specifying some short-list of procedures or measures of effort, employees, whether individually or in teams, are given an incentive to concentrate on meeting those requirements rather than making sure that the characteristics of the product satisfy customer requirements. Indeed, simply satisfying measures of product quality such as extremely low failure rates need not increase sales if customers are equally interested in aspects of the product that were not addressed, such as the price or product design.

The connection between effort and performance is not direct. The value of sales depends not only on the effort and skill of the employee but on the quality of the product, the activities of competitors, and the preference of consumers. Changes in market conditions, costs, or demand will have an impact on rewards to the employee. The result is that rewarding on the basis of performance entails risk for the personnel. The greater the dependence of compensation on performance, the more the employee is subject to risk. Since employees are averse to risk, it is necessary to compensate them through increases in base pay for the risks that they bear. Otherwise, the employee will seek other market opportunities.

The designer of incentives faces the following trade-off. Stimulating effort requires increased pay for performance. However, increasing the dependence of pay on performance increases employee risk, and thus the amount of compensation for risk that is required. The result is that the manager must determine some intermediate amount of pay for performance where the marginal benefits of shifting risk to the employee equal the marginal cost of paying compensation for risk.

It is expensive to make incentive contracts fully contingent because they are costly to write and negotiate. As time passes, changing market conditions and evolving company strategies require changes in desired performance. Rather than continually revising incentive contracts, companies rely on management assessment of performance and interpretation of rewards after performance takes place.

The increased delegation of responsibility is reflected in the changing nature of compensation. At Nucor Steel, a leading advocate of incentive pay, bonuses exceeded half of salary according to chairman Ken Iverson.[43] In Chicago, the building materials company USG Corp. used 60 different incentive plans, with every employee receiving some form of incentive compensation.[44] A survey by the National Association of Manufacturers found that more than two-thirds of the manufacturing companies surveyed offered incentive compensation, performance bonuses, and other non-wage compensation.[45]

Incentive pay shifts a share of business risks onto employees. Employees are rewarded based on various performance measures rather than on specific actions or effort monitored directly by management. If the performance measures are chosen carefully, employees will "work smarter" rather than simply harder, doing what is most effective to boost performance. Incentive pay is costly to employers, who must boost pay to compensate employees for the additional risks they face, and who must institute performance monitoring to implement the plans.

However, incentive pay becomes worthwhile when it boosts employee productivity sufficiently to cover these additional costs. Incentive pay is a necessary part of the delegation process. As employees are granted greater autonomy of action, they know more about their activities and market conditions than their supervisors. Performance-based compensation is an important means of aligning employee interests with those of the company. The manager's internal analysis should include a broad evaluation of the company's incentive programs.

6.5 *Overview*

The manager's internal analysis is directed at making the best match between organizational abilities and market opportunities. The manager knows that achieving the company's goals depends on whether the employees can effectively implement the company's strategy. Accordingly, the manager's internal analysis examines the company's organizational structure and its performance, as considered in the preceding chapter, to determine whether the company is taking on the functional tasks for which it is best suited and to see how well the company is performing.

In addition, as the present chapter shows, the manager's internal analysis examines the organization's abilities and incentives. The manager determines whether the company has unique resources and competencies that would be suited to serving particular markets. The values of the company's balance sheet and information assets and the quality of its competencies depend on market opportunities. The manager attempts to realize the greatest market return from these resources and competencies. If the company's resources and competencies are not well suited to the company's strategic plans, the manager must alter the company's goals and strategies accordingly or the organization must acquire new resources and develop new competencies.

Even if the members of the organization have the right abilities for the company's plan of action, those employees must be given the authority to take action. Accordingly, the manager's internal analysis must survey the delegation of authority within the firm to make sure that employees have the necessary instructions, information, and resources to carry out their assigned tasks.

Moreover, employees must be motivated to act in the interest of the company. The manager's internal analysis considers the various types of incentives that employees face within the organization. Are the employees rewarded for individual initiative and for cooperative behavior? Do employees have incentives to increase sales, increase profits, improve product quality, or cut costs? Does the organization monitor the right types of performance measures in determining employee rewards? The manager seeks to determine whether employees will try to achieve the company's goals or whether their efforts will be directed at targets that are not consistent with those of the company or even at the employee's personal goals. As the company's strategies change, managers must tailor incentives to match new strategic directions.

Questions for Discussion

1. Give examples of assets for which the market value is generally close to the book value, and others in which the two may differ significantly.
2. Explain how the value of an asset to the firm might differ from its market value and from its book value.
3. Explain the difference between strategic capabilities and functional capabilities.
4. How have innovations in information technology and telecommunications changed the competencies that companies seek to acquire? Consider how these technologies might impact the competencies that companies seek in various industries (banking, consulting, retail).
5. Why is delegation of authority essential for the manager? What are the benefits and costs of delegating authority?
6. Have advances in computer technology and available software over the last two decades increased or decreased delegation of authority to employees?
7. You are the manager of a division of a major insurance company. A newly developed information technology system and related software allow the division to significantly lower its costs. The system will require laying off 100 employees. However, you need the help of all the employees in the division to install the new information technology system and adapt the division's processes. What incentives are needed to induce employees to undertake these tasks?
8. Consider incentives for sales personnel. How might sales commissions be structured to encourage employees to sell more units? How might sales commissions be structured to encourage employees to generate more revenue? Can these two types of incentives be inconsistent?
9. Consider a company that sells capital equipment to business and also provides a limited amount of related services to its customers. The management of the company decides to change its strategy by selling less equipment while providing more related services such as maintenance and production planning. How might the company change the incentives for its sales personnel?
10. A company has a policy of being a cost leader, producing a high volume of output that it sells at a relatively low price. The management of the company decides to change to a product differentiation strategy that will involve selling a high-quality product at a relatively high price. How might the company change its incentives?

Endnotes

1. Marshall, A., *Principles of Economics*, 8th ed., (London: Macmillan, 1890).

2. Edith T. Penrose, *The Theory of the Growth of the Firm* (Oxford: Basil Blackwell, 1959).

3. See Berger Wernerfelt, "A Resource-Based View of the Firm," *Strategic Management Journal* 5 (1984), pp. 171–180; Jay B. Barney, "Firm Resources and Sustained Competitive Advantage," *Journal of Management* 17 (1991), pp. 99–120; Katherine R. Conner, "A Historical Comparison of Resource-Based Theory and Five Schools of Thought Within Industrial Organization Economics: Do We Have a New Theory of the Firm?" *Journal of Management* 17 (1991), pp. 121–154; and Raphel Amit and Paul J. H. Schoemaker, "Strategic Assets and Organizational Rent," *Strategic Management Journal* 14 (1993), pp. 33–46.

4. Michael E. Porter, "Towards a Dynamic Theory of Strategy," *Strategic Management Journal* 12 (1991), pp. 95–118, at p. 108. The idea that a company's resources are largely the outcome of management choices has been expressed by many management scholars. See particularly Chester Barnard, *The Functions of the Executive* (Cambridge, MA: Harvard University Press, 1938); H. I. Ansoff, *Corporate Strategy* (New York: McGraw Hill, 1965); K. R. Andrews, *The Concept of Corporate Strategy* (Homewood, IL: Irwin, 1971); and R. E. Miles and C. C. Snow, *Organizational Strategy, Structure and Process* (New York: McGraw Hill, 1978).

5. For discussions of market imperfections and resource scarcity, see Dennis Yao, "Beyond the Reach of the Invisible Hand: Impediments to Economic Activity, Market Failures and Profitability," *Strategic Management Journal* 9 (1988), pp. 59–70; and Margaret A. Peteraf, "The Cornerstones of Competitive Advantage: A Resource Based View," *Strategic Management Journal* 14 (1993), pp. 179–191.

6. C.K Prahalad and Gary Hamel, "The Core Competence of the Corporation," *Harvard Business Review*, May-June 1990, pp. 79–91, p. 85.

7. Ibid., pp. 82–83.

8. Porter, "A Dynamic Theory of Strategy."

9. Alfred D. Chandler, *Scale and Scope: The Dynamics of Industrial Capitalism* (Cambridge, MA: Belknap Press of Harvard University Press, 1990), p. 8.

10. Ibid., p. 24.

11. Ibid., p. 24. David Teece defines *dynamic capabilities* as "the ability to sense and then to seize new opportunities, and to reconfigure and protect knowledge assets, competencies, and complementary assets and technologies to achieve sustainable competitive advantage"; David J. Teece, "Capturing Value from Knowledge Assets: The New Economy, Markets for Know-How, and Intangible Assets," *California Management Review* 40 (Spring 1998), pp. 55–78, at p. 72.

12. Chandler (1990) also includes organizational capabilities, the product-specific and process-specific "facilities for production and distribution acquired to exploit fully the economies of scale and scope." Organizational capabilities are exhibited by top managers "in their responsibilities for

coordination, strategic planning, and resource allocation," by middle managers in all the functional activities (production, distribution, purchasing, research, finance, and general management), and by lower management and the work force. Ibid., p. 24.

13. www.americanstandard.com/advmfg.html.
14. Prahalad and Hamel, "The Core Competence of the Corporation," pp. 79–91.
15. Teece, "Capturing Value from Knowledge Assets."
16. George Stalk, Philip Evans, and Lawrence E. Shulman, "Competing on Capabilities: The New Rules of Corporate Strategy," *Harvard Business Review*, March-April 1992, pp. 57–69.
17. See for example Gary Hamel and C.K. Prahalad, *Competing for the Future* (Cambridge, MA: HBS Press, 1994).
18. Stalk, Evans, and Shulman ("Competing on Capabilities") attribute Wal-Mart's success to its operations expertise as a means of giving attention to customer needs. "This strategic vision reached its fullest expression in a largely invisible logistics technique known as 'cross-docking.' In this system, goods are continuously delivered to Wal-Mart's warehouses, where they are selected, repacked, and then dispatched to stores, often without ever sitting in inventory." The benefits of such a system are evident: The costs of holding inventory are reduced and the composition of inventory can be better tailored to demand patterns, thus reducing the risk of stockouts or overstocks. The costs of such a system are the need to coordinate individual stores, distribution centers, and suppliers. Stalk, Evans, and Shulman emphasize the role of Wal-Mart's private satellite communication system that sends point-of-sale data to its thousands of vendors, 19 advanced distribution centers, and point-of-sale data gathering. They contrast Wal-Mart's fleet of 2,000 company-owned trucks with K-Mart's outsourcing of transportation as a key to the different performance of the two companies.

While operational investments are an important part of Wal-Mart's growth and performance, they do not tell the whole story, nor do they provide a complete description of Wal-Mart's organization. Rather, it is Wal-Mart's 36 product areas that allow the company to tailor its product offerings to the individual submarkets in which it competes. While cross-docking and other distribution technologies are the means to an end, Wal-Mart excels at interpreting demand patterns for its suppliers, choosing what customer markets to enter or exit, choosing product categories, and finding suppliers. Transportation is not necessarily intrinsic to these economic activities and appears far from essential.
19. See Prahalad and Hamel, "The Core Competence of the Corporation"; and Stalk, Evans, and Shulman, "Competing on Capabilities."
20. The information in this paragraph is drawn from the chronology given at www.honda.co.jp/english/history/1946-1959.html.
21. "The Trouble with Excellence," *The Economist*, July 4, 1998, p. 68.
22. Ibid.

23. Ibid.

24. www.honda.co.jp/english/company/index.html.

25. Richard J. Schonberger, *Building a Chain of Customers* (New York: Free Press, 1990), p. 34.

26. Ibid., p. 45.

27. Doug van Kirk, "EDI Is Coming to a PC Near You," *InfoWorld* 15, no. 33 (August 16, 1993), p. 15.

28. This estimate by the Electronic Data Interchange Association is in Don Tapscott and Art Caston, *Paradigm Shift: The New Promise of Information Technology* (New York: McGraw-Hill, 1993), p. 97.

29. Ibid.

30. Ibid.

31. Shoshana Zuboff, *In the Age of the Smart Machine: The Future of Work and Power* (New York: Basic Books, 1988), p. 10.

32. Ibid., pp. 10–11.

33. Ibid., p. 414.

34. Alan Farnham, "Where Have All the Secretaries Gone?" *Fortune*, May 12, 1997, pp. 152–154.

35. Ibid.

36. See Barbara Presley Noble, "At Work: It's 1993. Is Your Office Outmoded?" *New York Times*, January 24, 1993, p. 25. The article reports on the book *Office Access* published by the electronic publishing firm The *Understanding* Business (TUB), San Francisco.

37. Ibid.

38. Quoted in Noble, ibid.

39. Mitchell Pacelle, "Vanishing Offices," *The Wall Street Journal*, June 4, 1993, p. 1.

40. Ibid.

41. See Andy Reinhardt, "Managing the New Document," *Byte*, August 1994, pp. 91–104.

42. In legal terms, the relationship between company and employee, that is, between principal and agent, is a *fiduciary* relationship.

43. Pat Widder, "Shifting Pay Patterns Help Keep Lid on Inflation," *Chicago Tribune*, April 19, 1997, Business Section, p. 1.

44. Ibid.

45. National Association of Manufacturers, Board of Directors Survey, Washington, D.C., April 1997.

PART IV

COMPETITIVE ADVANTAGE

CHAPTER 7

COMPETITIVE ADVANTAGE AND VALUE CREATION

Contents

This chapter examines how companies create value in the market and how they capture it to increase the value of the firm. After completing the chapter, you will have an understanding of the connection between value creation and competitive advantage. Also, you will know the basic components of value and how value created is shared with customers and suppliers.

The *value created* by the firm equals the benefits the firm's customers receive minus the costs the firm's suppliers incur and minus the costs of using the firm's own assets. To increase value created, the company increases benefits to its customers, lowers costs of its suppliers, uses its resources more effectively, or combines suppliers and customers in new or more efficient ways.

The firm's ability to create and capture value depends on the strength of competition and the characteristics of the firm. In markets where customer demand outruns industry capacity, many firms can add value. In markets where industry capacity outruns customer demand, a firm must have a competitive advantage to survive.

The firm must share the value that it creates with its customers and suppliers. The share of the value that the firm is able to capture is the *value of the firm*. Value-driven strategy involves three basic rules. To attract customers away from competitors, the company must provide sufficient customer value as compared to rival firms. To attract key suppliers away from competitors, the company must offer sufficient supplier value. To attract investment capital in competition with other market investment opportunities, the company must increase the value of the firm for its investors. Understanding these three important rules provides managers with a consistent framework for designing and applying strategy.

To obtain a competitive advantage, the company must create greater total value than its competitors and capture the incremental value that it brings to the market. The *competitive advantage* of a firm equals the difference between the overall value created by the industry when the firm is in the market and the overall value that would be created by the industry when the firm is not in the market. Thus, competitive advantage is the extra value created by the firm.[1]

Chapter 7: Take-Away Points

The manager chooses strategies to carry out the company's goals and to gain competitive advantage:

- To attract customers away from competitors, the company must provide sufficient customer value.
- To attract key suppliers away from competitors, the company must offer sufficient supplier value.
- To attract investment capital in competition with other market investment opportunities, the company must increase the value of the firm for its investors.
- Total value created is the sum of customer value, supplier value, and the value of the firm.
- To obtain a competitive advantage, the company must create greater total value than its competitors.
- Competitive advantage often requires product innovation, process innovation, or transaction innovation.

7.1 *Creating Value*

Managers must pay close attention to value creation because it is the source of the company's potential profits. The company creates value by coordinating its purchases and sales transactions. The company generates value by providing products to customers, which it produces both by purchasing inputs from suppliers and supplying some of its own. The value the company creates is equal to the difference between the benefits the company's customers receive and the cost to the company's input suppliers, including the cost of the company's self-supplied inputs.

All value creation begins with the company's final customer. The customer receives some *benefits* from consuming a product provided by a company. The dollar measure of those benefits is the customer's *willingness to pay*, which is defined as the *maximum amount that the customer would pay for that product*. Accordingly, the customer's benefit is also referred to as the customer's willingness to pay. For example, if a customer is willing to pay at most $200 for a particular product, then that is the customer's benefit from consuming that product. The value created by the firm is necessarily limited by its customers' willingness to pay.

There is no free lunch. Providing a product that benefits customers necessarily requires costly inputs. The firm obtains various inputs from suppliers. The firm also supplies some of its own inputs, including information assets such as business methods, inventions, and market knowledge. For most productive inputs provided by the firm itself, the most accurate measure of cost is the *market value* of that input, which is simply the current market price of the input. For those inputs provided by the firm for which there is no readily available market price, it is necessary to estimate the market value.

The best estimate of the market value of an input is based on the opportunity cost of the input. Recall that opportunity costs are what the inputs would earn in the best opportunity forgone; that is, the return from the best alternative employment of that input. For example, if a company owns a plot of land that it could sell to another company, that is the opportunity cost of using the land. The cost of the entrepreneur's time and effort in starting a firm is what the entrepreneur could have earned in his or her best alternative occupation.

The costs incurred by the firm's suppliers are the purchase costs of all inputs including labor, natural resources, manufactured parts and components, technology, and capital equipment. Supplier costs further include the costs of all services obtained by the firm, including the costs associated with completing transactions, such as legal, accounting, marketing, and sales costs. The costs of the supplier also include the cost of capital whether that capital is obtained through debt or sale of equity.

Therefore, the *value created* by the firm equals the benefits obtained by the firm's customers minus the total costs of inputs provided by the firm and its suppliers. The principles of value creation can be illustrated with a basic example. A single customer representing a specific market segment is willing to pay a maximum of $200. Therefore, the most value that the company could create is $200. In serving the customer, the company employs some of its own assets that are valued at $80. The company also purchases inputs from a supplier which cost $50. The value created by the company's buy-and-sell transaction is the customer's net benefit net of the cost of using the firm's assets and the supplier's costs: $200 − $80 − $50 = $70.

When Michael Dell started his company, he observed that IBM sold a personal computer for $3,000 even though it contained $600 worth of parts.[2] He knew that the difference was too high, which suggested that

IBM's other costs were too high and that IBM's markups were too high due to lack of competition. Michael Dell realized that the costs of assembly and delivery could be lowered considerably, thus increasing the value created. Later, by operating online, Dell Computer dramatically lowered its transaction costs still further. By lowering markups, Dell Computer was able to share the increased value with its consumers and suppliers. For a $2,000 computer, Dell Computer's cost of goods sold was approximately $1,600 and its overheads (selling, general, and administrative costs) was about $220, so that the markup was about $180.[3]

Example 7.1 Value Created and Captured by a Computer Company

A computer company serves a customer who is willing to pay $2,500 for a computer. The computer company charges its customer a price of $2,200 per computer. The computer company purchases parts and services from suppliers (the box, monitor, speakers, hard-disk drive, memory, software, and assembly services). The cost to the computer company's suppliers of providing all these parts equals $1,200. The computer company pays its suppliers $1,750. The computer company also has overhead costs (selling, general and administrative) of $200. How much value is created by the computer company? How much of that value is captured by the computer company? The total value created by the computer company is $2,500 − $1,200 − $200 = $1,100. The total value captured by the computer company is the company's revenue minus its overhead costs and the payment to its suppliers, $2,200 − $1,750 − $200 = $250.

Value creation provides an important linkage between the steps of the strategy process. The manager's external analysis yields information about the company's customers and their willingness to pay for the company's products. The external analysis also gives the manager information about the company's suppliers and their costs. This information enters directly into the manager's consideration of what value the company creates.

The manager's external analysis also provides information about the company's competitors: their costs, their prices, and the products that they provide. This information is very useful in determining whether the company's transactions with its customers and suppliers create value in competition with other firms in the industry. Do the company's customers derive greater or lesser benefits from purchasing the products of competitors? Do the company's suppliers incur greater or lesser costs in serving competitors? These considerations will be important to the manager in evaluating what value the company adds to the market.

The manager's internal analysis is useful in determining what assets the firm has to offer the market place. He or she uses the combination of internal and external analysis to determine the benefits the company's assets add in serving customers and the opportunity costs of using those assets. The internal analysis yields information about what types of products the firm can provide to its customers. In addition, the manager determines what activities should be performed within the organization and what types of inputs will be procured from suppliers. Together, this information helps the manager determine the potential value that the company's products will create.

The concept of value extending from suppliers through the firm and on to its customers is related to but distinct from Michael Porter's concept of the value chain. The value chain refers to the firm's internal processes. As Porter observes, "Every firm is a collection of activities that are performed to design, produce, market, deliver and support its product." Porter emphasizes that each firm's value chain is embedded in a value system of activities composed of supplier value chains upstream and channel and buyer value chains downstream.[4] These external value chains complete the picture by including buyer benefits and supplier costs.

As part of the manager's external analysis, it is useful to understand the manner in which customers derive benefits from their products. This will help managers tailor their product accordingly. Although it is difficult to measure precisely what an individual customer is willing to pay for a good or service, some inferences are possible. Customers reveal something about their willingness to pay by their purchasing decisions. If customers pay $150 for a product, their willingness to pay is *at least* that amount, but it might be $175 or it might be $300.

Statistical techniques for estimating total market demand also provide information about the total willingness to pay of customers in the market. When market prices fluctuate and total customer purchases change, companies get some indication of price sensitivity and can estimate how much customers are willing to pay. Brokerage fees fell substantially after deregulation. Customers were willing to pay hundreds of dollars per trade before deregulation of brokerage fees in 1975, so it can be inferred that those customers viewed a trade as providing a benefit of at least that amount (at least when they made that trade). After deregulation, brokerage fees fell below that level but earlier rates provide some guide to customer benefits per trade. With the advent of Internet securities trading, many customers were willing to pay about $30 per trade. As competition intensified, Internet brokers began to charge $5 per trade or less. Those customers who traded online when fees were over $30 had benefits of at least that amount per trade. Customers attracted to online trading by the lower prices were likely to have benefits less than $30 and greater than $5.

More complicated inferences can be made by comparing bundles of products. Some customers trade with full-service brokerages at up to $150 per trade rather than with discount brokerages at $50 per trade. Those customers must perceive that they obtain benefits of at least $100 from the services, over and above trade execution. In the same way, customers who trade with a discount broker at $50, rather than going online at $5, obtain benefits from personal interaction at least equal to $45.

Also as part of the manager's external analysis, it is useful to understand the costs of the company's suppliers. This understanding will help managers to determine the types of products they should obtain from suppliers and the types of activities that the company will perform itself. Managers are able to obtain information about the costs of their suppliers, especially if suppliers are willing to share cost information. Industry cost estimates may be available if the suppliers employ standard production techniques. In addition, market prices for the products suppliers use allow inferences about supplier costs. Managers can combine data on prices and standard industry markups to make informed estimates of supplier costs. Customer benefits and the costs of the firm and its suppliers are the building blocks of value.

7.2 *Growth and Value Creation*

The company's ability to create value depends critically on industry conditions. The potential for value creation depends on how the growth of market demand compares with the growth of industry capacity. There are two main scenarios: If market demand outruns industry capacity, practically all companies can operate profitably and add value to the market. If, on the other hand, industry capacity outruns market demand, some companies may not be profitable, and companies must have a competitive advantage to survive. These two scenarios are illustrated in Figure 7.1.

Industry capacity refers to the total capacity of the companies operating in the industry. Market demand varies, of course, depending on price. To compare market demand with industry capacity, it is necessary to determine the price at which market demand equals industry capacity. That price is a critical determinant of the state of the industry. If the price at which market demand equals industry capacity is greater than the unit cost of the highest-cost company in the industry, then market demand is said to outrun industry capacity. If the price at which market demand equals industry capacity is less than the unit

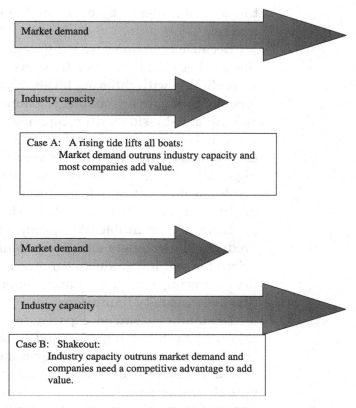

Figure 7.1: Growth and value creation: the two main scenarios.

cost of the highest-cost company in the industry, then industry capacity is said to outrun market demand.

Industry Demand Outruns Capacity

As President John F. Kennedy remarked: "A rising tide lifts all boats." A period of rapidly rising demand offers significant opportunities for value creation. When total market demand outruns industry capacity, it means that individual firms have encountered short-run capacity constraints and cannot serve the entire market. In consequence, industry capacity is scarce relative to market demand, so many companies in the industry can be profitable. This has the important strategic implication that individual firms can create value even without a competitive advantage.

Management strategy must adjust to take into account the relationship between market demand and industry capacity. With demand outrunning capacity, competition exerts less downward pressure on prices. The lowest-cost firms cannot use prices to expand market share beyond the limits of their productive capacity. Because capacity is scarce, even the highest-cost firms are able to operate. This means that the capacity offered by even the highest-cost firms adds value to the market.

Because companies with different cost levels can operate profitably, the market offers room for firms with a cost disadvantage. Firms can be profitable with differing levels of productive efficiency and different types of productive technology. Companies with obsolete plants might operate side-by-side with companies with modern facilities. The market can be served by many different types of firms, including larger established companies with scale economies and new entrants with the high costs that start-ups often encounter.

When demand outruns capacity, firms with products of differing quality can serve the market, so that even firms with lower-quality products are profitable. Also, multiple firms with similar or generic products can compete effectively. Moreover, firms that offer different prices for similar products can serve the market simultaneously. The firms offering price discounts cannot capture the market as long as they face capacity limitations.

Demand can outrun capacity for many reasons. The industry can experience relative high demand growth due to economy-wide growth. Increased consumer income, higher employment, and even tax cuts can stimulate the overall economy and lead to growth of demand in individual industries. Demand growth that individual industries experience may be simply the result of business cycle effects on the overall economy as felt in specific industries. Industry demand growth can occur due to shifts in population; for example, a housing demand boom occurred in the southwestern United States as more people moved there. Demographic changes such as the baby boom after World War II spurred economic growth that stimulated specific industries, such as education and housing. An aging population can stimulate demand for health care or retirement resorts.

Many types of factors stimulate demand growth in specific industries. A simple shift in customer tastes may favor a specific type of product. For example, changing lifestyles of working families led to increased demand for restaurant meals and prepared foods. The success of complementary products creates demand shifts. For instance, the increased use of personal computers stimulated demand for software applications such as spreadsheets and word-processing programs. The growth of the automobile industry spawned a wide range of other industries, including petroleum refining, automobile repair and service stations, motels, and companies producing parts such as tires and batteries.

Reducing barriers to international trade can induce a demand boom as companies take advantage of import and export opportunities. Companies discover new markets for their products abroad. Reductions in trade barriers often stimulate demand for imports of foreign music, movies, food, clothing, and electronics. Changes in government regulation also lead to demand booms, as happened when deregulation of telecommunications led to an explosion in the demand for phones, fax machines and other communication devices.

The introduction of new technologies has often led to demand booms as customers discover new products. Consider the opportunities created by the introduction of new modes of transportation from the railroad to the automobile to the jet airplane. New methods of communication including the telegraph, the telephone, the Internet, and the wireless phone have led to significant demand booms as customers adopt the new product. New technologies such as advances in information and communication can stimulate demand by changing the organizational structure of companies and the organization of markets. With less vertical integration and increased outsourcing, companies have increased demand for information processing and communications equipment.

Technology effects are often closely associated with new product introductions and help to explain the growth phase at the start of a product life cycle. Typically, the product life cycle involves four stages: introduction, growth, maturity and decline. Sales and profit rise during the introduction and growth stages and fall during the latter part of the maturity stage and the decline stage. The product life cycle has limited predictive powers because the turning point is difficult to observe and to predict. However, the product cycle does reflect an underlying pattern of firms' demand changes and capacity investments.[5]

Firms still have incentives to be efficient during a demand boom. Companies still earn greater profits by lowering their costs. Moreover, companies can stimulate sales through greater product quality. However, because competitive pressures are substantially reduced or eliminated during booms, it may be useful for companies to delay investments that have little short-run benefits, focusing their efforts on taking advantage of current sales growth. Managers may use the boom to prepare for future investment to reduce operating costs or enhance product features.

Scarcity of supply relative to demand can take many forms. Companies that supply natural resources, such as metals or energy resources, experience booms during periods of economic growth. Real estate demand booms result in sales or rental of all available space, rising prices, and new construction. Manufacturing industry booms lead to shortages of production capacity for manufactured inputs, components and final goods. The manufacture of computer notebooks has been hindered at various times by shortages in liquid crystal display (LCD) screens. Companies may benefit from industry-level scarcity of distribution capacity or shortages of sales personnel. Finally, there may be limits on all kinds of competencies, from technological knowledge to management skills.

Demand growth attracts inputs to the growing industry as demand for final products leads companies to expand production. Thus, demand booms attract funds from venture capitalists, banks and securities investors. Entrepreneurs are drawn to the opportunities a booming industry provides. Booms lead to higher wages and increased hiring, which in turn attracts managers and employees to the industry. Suppliers of parts, components, and services are attracted by the opportunity to serve companies in a booming market.

Why does demand outrun capacity? Part of the problem for companies responding to a demand boom results from lags in establishing capacity. The demand boom may come as a surprise so it takes the industry time to respond. When energy prices rise, it takes time to increase exploration and drilling for new petroleum and natural gas resources. When the demand for computers increases, it can take several years to build a new plant for manufacturing semiconductors. When the need for particular skills such as surgeons or computer programmers grows, long training times lead to delays in obtaining qualified personnel.

Because capacity is costly to establish, companies cannot adjust it to match demand fluctuations perfectly, so the market naturally experiences periods of capacity shortfall. Moreover, established companies experience *adjustment costs*, that is, increased costs associated with installing new capacity and integrating that capacity with existing operations. Competitive advantage is not essential to profitability when there are delays in new entry or expansion of firm capacity.

In some cases, the situation in which demand exceeds capacity can be sustained for a long time. Many industries have undergone long periods in which the number of firms grew steadily, including, for example, computers, lasers, automobile tires, gas turbines, and heat pumps.[6] However, the shortage of capacity creates incentives for established companies to expand their capacity and for companies to enter the market with new capacity, and competition intensifies.

Example 7.2   The Piano Boom

Consider a market for pianos in which there are three customers, each of whom can purchase at most one piano and each of whom has a willingness to pay of $200. There are two manufacturing firms, each of whom has the capacity to produce at most one piano. One firm has a cost of $50 and the other has a cost of $75. Do both firms supply pianos? How much value is created? Because there is excess demand relative to capacity, there is a boom in pianos. The consumers bid against each other for the scarce pianos so that the

(Continued)

> ### Example 7.2 (*Continued*)
>
> price of pianos is equal to $200. Both firms supply pianos and capture all of the value in the market so that they earn profits of $150 and $125 respectively. The total value created is equal to $150 + $125 = $275. Notice that even though one firm has a cost advantage, both of the firms in the industry are able to operate profitably due to limited capacity and relatively high demand.

> ### Example 7.3   The Piano Boom with Diverse Customers
>
> Consider a market for pianos in which there are three customers, each of whom can purchase at most one piano. The customers are willing to pay $200, $175, and $140 respectively. As in the previous example, there are two manufacturing firms, each of whom has the capacity to produce at most one piano. One firm has a cost of $50 and the other has a cost of $75. Since the industry has a total capacity of two, the customer with the lowest willingness to pay does not obtain a piano. Consumers bid against each other for the scarce pianos so that the price of pianos is just enough to exclude the third consumer, that is, just above the willingness to pay of the excluded consumer of $140. Both firms supply pianos and earn profits of $140 − $50 = $90 and $140 − $75 = $65 respectively. The total value created is equal to $200 + $175 − $50 − $75 = $250.

Industry Capacity Outruns Demand

When industry capacity outruns market demand, the strategic situation changes considerably. The industry must shed its excess capacity because it is usually very costly for companies to carry excess capacity. As a result, individual companies need to reduce their productive capacity or exit the market altogether. Some companies may merge as a means of lowering costs as well as retiring duplicative parts of their capacity. The main strategic implication is that companies need a competitive advantage to be profitable and survive. Therefore, it is often the case that when industry capacity outruns market demand, companies must have a competitive advantage to create value in the market.

Industry capacity can outrun market demand for many reasons. Economy-wide forces discussed previously, such as changes in income, employment, or other business cycle effects, can slow the growth of demand or produce a decline in demand for a given industry.

During the Great Depression, half of the plants and more than half of the firms in the U.S. auto industry closed between 1929 and 1933. Companies with larger plants and larger organizations benefited from mass production economies and were most likely to survive the shake-out.[7] Excessive global capacity in the automobile industry at the start of the 21st century led to mergers and consolidation such as Daimler Benz's acquisition of Chrysler and Mitsubishi.

Even if industry demand continues to grow, an unexpected slow-down in demand combined with continued growth in industry capacity can cause companies' building capacity to overshoot. Industry-specific effects, such as changes in customer tastes, also can reduce industry demand below industry capacity. Industries that produce outmoded products experience declining demand. The industry in a specific country can experience declining demand as it competes with lower-cost or higher-quality imports.

Even with substantial demand growth, capacity can outrun demand as a result of discovering economies of scale and scope. If technological change allows individual companies to expand their capacity while realizing economies of scale, then a smaller number of companies can meet market demand. As companies expand their potential scale, competition will drive companies out of the market. Companies with a smaller, inefficient scale will be driven out by more efficient companies with greater scale economies. Companies that realize scale economies will compete on cost efficiencies.

In some industries, economies of scale are so significant that there is only room in the market for a small number of companies. For example, in large jets, the global market apparently can only support two companies, Boeing and Airbus. Their scale economies result from the complexity of designing, manufacturing and assembling large aircraft. In computer operating systems, the high fixed costs of writing software and the minimal costs of producing an additional copy of the software give Microsoft the scale economies to easily supply software to run 90 percent of personal computers. In retailing, large companies such as Wal-Mart and Target take advantage of scale economies in distribution networks, but the retail sector also includes many small specialized stores that offer convenience and service.

In some markets, including computer operating systems, the benefits customers derive from standardization create winner-take-all industries. However, in other markets, the benefits of variety create opportunities for many companies. Although there are benefits from standard formats for computer games, such as those offered by Nintendo and Sony, the need for many diverse games creates opportunities for a host of game designers.

When an industry is newly established, capacity tends to outrun demand. Joseph A. Schumpeter observed "the appearance of entrepreneurs in clusters"; that is, innovative start-up companies tend to enter the market at around the same time.[8] He attributes this clustering in part to the elimination of obstacles to entry by pioneers who show the way. Pioneers may inspire entrepreneurs to enter into a specific industry. Sometimes pioneers in one industry can encourage entrepreneurs to enter into many other industries if they demonstrate how to overcome general types of obstacles.

Entrepreneurs often enter an industry simultaneously in search of the potential rewards of establishing a successful business. It has long been debated whether entrepreneurs are rational profit maximizers taking a calculated risk or irrationally exuberant gamblers who overestimate the chances of success or who simply enjoy creating a new business. The end result is the same; when excess capacity enters a market, some businesses will succeed and others will fail and exit the market. The growth of market demand and the entry decisions of firms determine whether the rewards of the winners outweigh the costs to the losers. Bringing innovation to market requires risk taking.

Capacity is likely to outrun demand in new industries due to technological uncertainty. If production technologies are not well understood, each entrant will try out a different production process. More capacity enters the market than required to satisfy demand because each entrant believes that its variant of the production technology has a chance of being the best. As production begins, companies discover the cost and efficiencies of their technology, and competition weeds out the inefficient processes.

Similarly, when new types of products are introduced, each entrant will test a different type of product. Excess capacity enters the market because each entrant believes that their product has a chance of proving to be superior to those of its competitors. As customers try the products, companies discover how customers view various product features, and competition drives out inferior products.

When capacity outruns demand, a shakeout results. A *shakeout* refers to a substantial reduction in the market share of some firms in an industry or the exit of some of the firms in an industry as a result of competition, leading to an overall reduction in the number of firms. The pattern of growth, rapid decline and leveling off has been observed in many new industries.[9] Shakeouts have characterized the early history of a wide range of industries. Hundreds of car companies entered the market at the beginning of the 20th century, but a combination of cost differences and product differentiation weeded out most of them.

The link between technological change and market structure is illustrated by the U.S. tire industry. In the tire industry, innovations in both production methods and product features led to a shakeout. The tire industry experienced steady growth in the number of firms for its first 25 years, reaching 274 firms, and then went through a significant shakeout with the number of firms declining by over 80 percent over a period of 14 years.[10]

Industries need not experience such life cycles. Steven Klepper suggests that shakeouts need not occur in industries with certain critical features. First, shakeouts need not occur where there is a separation between those firms that develop and sell process technologies, and those firms that apply the technologies to the design and manufacture of final products. An example of this situation is the chemical industry, in which independent companies that were process specialists entered the market by the end of World War II.[11] Second, shakeouts are less likely in those industries where there is a separation between firms that develop product innovations and license them, and firms that produce the final products. Finally, shakeouts are less likely in industries where final demand is heterogeneous and fragmented so that smaller companies serve market niches.[12] In the turbo-prop engine industry from 1948 to 1997, the market structure remained relatively stable with generalist companies coexisting with specialist companies. In the turbo-prop engine industry market, demand segmentation combined with a lack of significant increasing returns to scale in R&D and manufacturing helped to prevent the shakeout.[13]

The effects of industry excess capacity can be mitigated by transaction costs. In markets that are subject to imperfect information about prices and products, customers must search among companies to find the best deal. Because search is costly, companies have localized market power that allows them to remain profitable. Customers may pay a higher price or settle for a lower-quality product because of the costs of continuing to search. Even if a company does not provide the best deal in the market, it creates value by offering customers product availability or immediacy, so that its customers can avoid the costs of additional search. Firms with high costs or less attractive products thus provide value to their customers by reducing their search costs. Accordingly, imperfect information allows value creation even without competitive advantage.

The advantages conferred by imperfect information need not be sustainable. Companies can find ways to improve customer information. Through marketing and sales techniques, companies can communicate with customers about better deals or reduce consumer search costs. As customers are better able to find the companies offering the best deal, companies again need to have a competitive advantage to remain profitable.

As total industry capacity outruns demand, the least efficient firms or the firms with the least attractive products are generally forced to exit the market. However, there is still room for a diverse set of firms to operate. The capacity limits of firms operating in the market continue to allow less efficient firms or firms with less attractive products to operate profitably. This explains in part the great diversity of firms within any given industry. Limits on individual firm capacity imply that supplying products to the market creates value.

When industry capacity outruns customer demand, a shakeout of industry capacity follows. Some firms will exit the industry by closing capacity. Other firms will merge to consolidate their capacity and close down inefficient capacity. The second-largest disk-drive manufacturer Maxtor bought Quantum, and Hitachi bought IBM's disk-drive manufacturing business, for example. Companies need to have a competitive advantage to continue operating in the industry and survive the shakeout. Those companies with a competitive advantage are sometimes able to grow their capacity at the same time that the industry is shedding its less efficient capacity. Moreover, having a competitive advantage enables the company to pursue its strategy through the ups and downs of business cycles.

Example 7.4   Dot-Com Shakeout

Suppose that there are four firms with competing websites offering to provide a particular product. Suppose that each of the firms has a capacity of 100 units, so that total industry capacity equals 400 units. The four firms have unit costs of $20, $30, $40, and $50 respectively. At a price of $32, market demand equals industry capacity of 400 units. At a price of $45, market demand equals 300 units. What happens next?

Clearly a shakeout must occur, since the industry cannot support the highest-cost firm with cost of $50. Industry capacity has outrun market demand and the highest-cost firm must exit the market. After a shakeout, three firms will be operating in the industry and total industry capacity will be 300 units. Now, market demand has outrun capacity and all three firms in the industry can operate profitably, each selling 100 units at a price of $45.

Next consider what would happen if each of the firms were able to expand their capacity to 200 units without any change to its unit costs. Then, all but the two lowest-cost firms would exit the industry. Each would sell 200 units at the price of $32.

7.3 *Competitive Advantage and Value Creation*

In markets where capacity exceeds demand, value creation generally requires competitive advantage. A company with a competitive advantage consistently outperforms competitors, that is, it earns greater economic profits. To achieve competitive advantage, companies seek the best match between organizational abilities and market opportunities. Few, if any, competitive advantages can be sustained indefinitely, so the company must continually seek opportunities to create the most value.

Companies tend to differ in terms of production methods, product features, brand names, locations, and many other aspects. The critical differences that determine success or failure are the sources of competitive advantage. The company's earnings are limited by its competitive advantage. It can obtain no more than the additional value it creates over and above that of its competitors. The earnings realized by the firm depend on the share of additional value that the firm is able to capture. Therefore, competitive strategy requires both value creation relative to competitors and capturing a portion of that value through relationships with suppliers and customers.

Managers must monitor the overall state of the industry, comparing industry capacity to market demand. They must also closely examine their company in comparison with competitors to distinguish those features that are critical to success. *To outbid competitors for customers, suppliers, and shareholders, the firm must create total value at least as great as its competitors.*

The notion of creating value provides insight into the sources of competitive advantage. Value creation has three aspects: the benefits received by customers, the costs incurred by the company and its suppliers, and the particular combination of customers and suppliers. Since the total value created by the firm also equals customer willingness to pay minus the costs of using the firm's assets and the costs incurred by suppliers, achieving a competitive advantage means that the firm must either increase customer benefits, lower supplier costs, or discover innovative transactions.

Accordingly, there are three sources of competitive advantage: (1) *Cost efficiencies* that make more efficient use of the firm's assets and supplier inputs or that lower supplier cost; (2) *Product differentiation* to raise customer benefits; and (3) *Transaction innovations* that lower the costs of transactions or that create new combinations of customers and suppliers. The three types of competitive advantage are

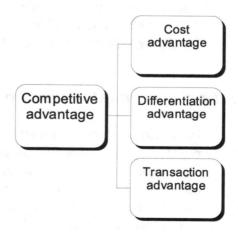

Figure 7.2:   The three sources of competitive advantage.

called cost advantage, differentiation advantage, and transaction advantage as seen in Figure 7.2. Alternative strategies for creating value are associated with each of these alternatives.

Competitive advantage equals the difference between the value created by the company and the potential value created by its competitors. When market demand outruns industry capacity, competitive advantage increases the value added by the company and also increases its potential profits. When industry capacity outruns market demand, competitive advantage also ensures that the firm will survive.

Example 7.5   Competitive Advantage

Consider a market with two suppliers, one customer, and two firms each offering a pair of jeans. Each supplier provides the fabric to make one pair of jeans and has a cost of $10. Each firm has costs of $20 to make the fabric into jeans, including the opportunity costs of its designers and tailors. The customer will only buy one pair of jeans. The customer has a willingness to pay of $120 for the first firm's product and a willingness to pay of $170 for the second firm's product. The customer will transact with the second firm and the second firm will transact with one of the suppliers. The total pie is $170 − $10 − $20 = $140. Without the second firm, the players would have a pie of $120 − $10 − $20 = $90. So, the second firm's competitive advantage is $140 − $90 = $50, which is just the additional amount the customer is willing to pay for that firm's product.

Given this information, it is possible to calculate the value captured by the second firm. Since there are two identical suppliers, they will bid against each other until they receive a breakeven payment of $10 and supplier value equals zero. The second firm can charge the customer no more than $80. Otherwise, the first firm would be able to attract the customer at a price of $30, enough to cover its payment to the supplier of $10 and its additional cost of $20, and still offer a customer value of $90. So, the value captured by the second firm is $80 − $10 − $20 = $50, which exactly equals its competitive advantage.[14]

7.4 *The Components of Value*

There are three components of value: customer value, supplier value, and the value of the firm. Customer value is the customer's benefit net of the expenditure to purchase the product. Supplier value is the supplier's revenue net of cost. The value of the firm is the share of value created that is captured by the firm (see Figure 7.3).

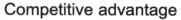

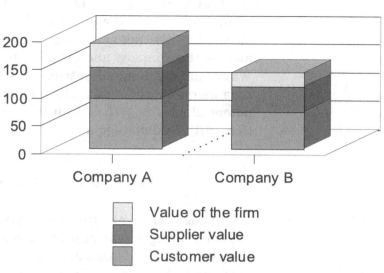

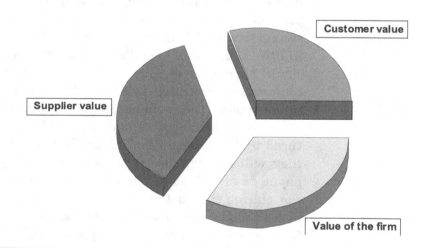

	Company A	Company B
Value of the firm	40	25
Supplier value	55	45
Customer value	90	65
Total	185	135

Competitive advantage of company A: 185 − 135 = 50

Figure 7.3: Total value created by the firm is the sum of customer value, supplier value and the value of the firm.

Example 7.6   A Guide to Value Creation

Customer willingness to pay for the firm's product or service:	$200
Asking price of the firm's product or service:	$110

Customer value: $200 − $110 = $90

Supplier costs:	$15
Bid price offered to the firm's supplier:	$70

Supplier value: $70 − $15 = $55

Cost of using the firm's assets	$10

Value of the firm: $110 − $70 − $10 = $30
Asking price − bid price
− cost of the firm's assets:

Value created by the firm:

Customer value + supplier value
+ value of the firm: $90 + $55 + $30 = $175

Customer willingness to pay
− supplier costs − cost of the
firm's assets: $200 − $15 − $10 = $175

These two methods of calculating value creation are equivalent.

Exhibit 7.1   Global Manufacturing: "My First Tea Party Barbie"[15]

Mattel captures value of approximately $1 per doll:

Retail price	$10.00
Shipping, ground transportation, marketing, wholesale margin, retail margin	$ 7.00
Overhead and management (Hong Kong)	$ 1.00
Materials (Taiwan, Japan, U.S., Saudi Arabia, China)	$ 0.65
Labor (Asia)	$ 0.35

Mattel earnings per doll: $ 1.00

Key locations for production of Barbie Doll:

El Segundo: Mattel Inc.
U.S.: Cardboard packaging, paint pigments, molds.

(*Continued*)

> ### Exhibit 7.1 (Continued)
> China: Factory space, labor, electricity.
> Saudi Arabia: Petroleum.
> Hong Kong: Management, shipping.
> Taiwan: Refines petroleum into ethylene for plastic pellets for Barbie's body.
> Japan: Nylon hair.

Customer Value

IBM states that its goal is to "strive to lead in the creation, development and manufacture of the industry's most advanced information technologies, including computer systems, software, networking systems, storage devices, and microelectronics. We translate these advanced technologies into value for our customers through our professional solutions and services businesses worldwide." How do companies create value for their customers?

The customer derives *net benefits* known as customer value or consumers' surplus from purchasing the good. *Customer value* is equal to the customer benefits minus the price the customer pays for the good. Both price and benefits are elements of customer value.

For example, a customer buys a toaster and receives benefits equal to $120, so the customer has a willingness to pay that amount for the toaster. If the customer pays only $90 for the toaster, then the customer receives a net benefit. The *customer value* is $30, which is willingness to pay of $120 minus the purchase price of $90. Another way to look at this is to observe that the customer would have been willing to pay $30 more than the purchase price, so the company's price and product delivers a customer value equal to that amount.

To attract customers, *the firm must create customer value that is at least as great as that offered by competitors*. A competitor's product or service may provide the customer with a different combination of price and product features. To bid away the customer from competitors, the firm must offer a better deal.

There are exceptions to the rule. Market frictions may permit firms to deliver less value than competitors and still attract customers. If competitors face capacity constraints and entry barriers, the firm will be able to attract customers without necessarily outperforming competitors. If customers are not well informed about alternatives or face costs of switching to another firm's products, a firm may continue serving customers while delivering lower value. However, such advantages are temporary and likely to be quickly eroded by competition.

To raise customer value relative to its competitors, the firm has three options that correspond to Michael Porter's well-known generic strategies: product differentiation, cost leadership, and focus.[16] Consider each of these strategies in turn.

The *product differentiation* strategy involves raising customer willingness to pay by enhancing the quality and other features of the company's product relative to competitors. Some customers are willing to pay relatively more for the firm's product than they would for those of competitors. The firm creates greater customer value than its competitors by offering a good that generates additional benefits that are sufficient to outweigh price increases.

There are many product differentiation strategies. The benefits consumers obtain from using a firm's product or service can be increased by offering enhanced features such as quality or durability. Customer value can be enhanced by convenience in purchasing and by complementary services such as warranties or maintenance. Customer value can be increased by subjective features such as brand image, advertising, and aesthetic design.

Example 7.7   You Get What You Pay For

Suppose that there are two firms and one consumer. The consumer has a willingness to pay of $120 for the first firm's product. The first firm charges a price of $90. The consumer would obtain a surplus of $120 − $90 = $30 from buying the first firm's product. A second firm employs a product differentiation strategy, offering the customer a product that the consumer values at $150. Whether or not the second firm incurs some additional costs of producing a higher-value product, the second firm charges the customer a price of $110. Suppose that this price is sufficient to cover the second firm's costs of producing a higher-value product so that the second firm would be profitable. Which good would the customer choose? The consumer would obtain a surplus of $150 − $110 = $40 from buying the second firm's product. Accordingly, the consumer buys the second firm's product because the customer value created is greater. The second firm captures the customer by the higher-value product differentiation strategy.

Cost leadership means offering a lower price to customers relative to that offered by competitors for a similar product or service. Companies pursue low costs not just to operate efficiently, but to become effective price leaders, undercutting competitors' prices. If the firm's costs are lower than its competitors, then the firm can offer lower prices, higher quality, or both. When products are generic and rivals do not face capacity constraints, a successful low-cost strategy requires that the company have costs below all other firms in its industry.

Cost reductions are achieved through innovative best-practice organizational processes, with careful monitoring of purchasing expenditures, application of computer and communications technology in a cost-effective way, trimming of overhead costs, and efficient operations. Costs are lowered by investment in cost-reducing capital equipment when the reduction in operating costs outweighs the capital equipment costs. Cost economies can be realized from large-scale operations. Alternatively, cost reductions can sometimes be accomplished through outsourcing manufacturing and services when outside providers offer lower-cost alternatives. By lowering cost for the same quality level, the low-cost firm is able to undercut the price of competing firms. Cost advantage is considered in Chapter 9.

Another strategy is to lower costs by offering lower quality. Consumer value is increased if the price reductions outweigh quality reductions. By providing a basic product with cost savings, firms can offer consumers a better deal. An example of a basic service that some consumers perceive to be a better deal is the full range of generic in-house products offered by supermarkets. Generic pharmaceuticals are also better deals for some consumers who perceive the lower cost of the generic as outweighing the possible difference in characteristics with branded pharmaceuticals.

Example 7.8   A Better Deal

Suppose that there are two firms and one consumer. The consumer has a willingness to pay of $120 for the first firm's product. The first firm charges a price of $90. The consumer would obtain a surplus of $120 − $90 = $30 from buying the first firm's product. A second firm employs a cost leadership strategy, offering the customer a product that the consumer values at $100. The second firm charges a price of $55. Suppose that this price is sufficient to cover the second firm's costs of producing the lower-value product so that the second firm would be profitable. The consumer would obtain a surplus of $100 − $55 = $45 from buying the second firm's product. Accordingly, the consumer buys the second firm's product since the surplus is greater than if the consumer bought the first firm's product. The second firm captures the customer by the cost leadership strategy.

Michael Porter defines *focus* as achieving low cost or product differentiation for a particular buyer group, segment of the product line, or geographic market rather than for the industry as a whole.

Southwest Airlines focused on a geographic region and short point-to-point flights as a means of reducing costs. Even though it offered no-frills service and was based in secondary airports, Southwest Airlines enhanced quality relative to the limited set of competing alternatives by offering direct flights rather than flights requiring changing planes at large hub airports. The airline also offered better on-time performance and friendly service.

Example 7.9   A Tale of Two Cities

Suppose that there are two airlines providing service between two cities and one business passenger. The first airline is a major carrier and the second airline is a regional carrier. The major airline provides service between the two cities that requires the passenger to change planes whereas the regional airline provides direct service. Because of the business passenger's opportunity cost of time, the passenger values the major airline's service at $500 and the regional airline's service at $600. The major airline charges an airfare of $450 and the regional airline charges an airfare of $300. Since the regional airline offers better service at a lower price, the business passenger chooses the regional airline. To compete, the major airline must offer better service or a substantially lower airfare.

Example 7.9 illustrates a focused strategy of a regional carrier that both improves quality and lowers the price for a targeted segment. The problem of the regional carrier is extending the strategy that is effective for a market segment to multiple city pairs. The problem of the major airline is how to fend off many regional carriers that target specific city pairs. The major airline established its hub-and-spoke system as the least-cost way to serve a complex route structure. In the example, the major airline is vulnerable to competition in specific market segments. The major airline's predicament in the example illustrates the problem of being stuck in the middle.[17] To remedy this type of problem, a firm must undertake strategies designed to improve its product quality, to reduce its cost dramatically, or to focus on specific market segments.

United Airlines was stuck in the middle in competing with such regional carriers as Southwest. United established its Shuttle by United in 1994 as an "airline within an airline". The shuttle would operate low-cost flights on the West Coast in the United States operating as a short-haul carrier with flights of under 750 miles serving over 200 destinations. By the year 2000, United's shuttle service would be one of the largest regional carriers if it were split off from the parent company. The shuttle service attempted to avoid the problem of being stuck in the middle by mimicking the low-cost, direct-route focus of the regional carriers while retaining some of the benefits of a major airline. Among those benefits, the shuttle service originated and terminated travel to the major airline's routes and participated in its strategic alliances. Moreover, the shuttle benefited from the advertising, brand name, reservation systems, and frequent flyer programs of the major carrier.[18] Although United Airlines created the shuttle service to provide both a lower-priced and higher-quality service to its regional markets, it could not sustain the separate company. United ended the shuttle brand and folded the flights back into the parent company's offerings.

Supplier Value

Companies also create value for their suppliers. Suppliers provide essential parts, components, resources, technology, and services, and companies' success depends, to a great extent, on the quality of those goods. A company can reduce its procurement costs through long-term relationships with its suppliers. If suppliers have lower costs of serving a particular company because that company is easy to work with, the company can share in these cost savings. Companies have an interest in the profitability of their suppliers to assure continued reliability of service.

The company's suppliers include parts and components manufacturers; manufacturers of capital equipment; wholesale product sellers; service providers (accounting, legal, consulting, information technology); and technology licensing and R&D. The firm's employees and managers supply labor services to the firm.

Supplier value equals the difference between the payment the firm makes to the supplier and the supplier's cost. Supplier value is also referred to as *producers' surplus* or *operating profit*. Firms bid against each other for the products and services of their suppliers.

The problem of finding suppliers is different from that of attracting customers. Suppliers can serve multiple customers. Suppliers are also competing against each other to sell their products and services. Established suppliers can expand their capacity and new suppliers can enter the market if a need for additional products and services develops. For example, computer makers benefit from advertising that their computers feature Intel microprocessors. Intel can expand production capacity to meet the demand from computer makers. Competing suppliers such as AMD can supply substitute processors.

To attract key suppliers, *the firm must create supplier value that is at least as great as competing alternatives*. Generally, companies must offer value to their suppliers that is at least as great as the market alternative. However, when key suppliers have scarce capacity and offer unique products and services, companies compete more intensively to outbid each other for supplier products and services. To bid away the supplier from competing buyers, the firm must offer a better deal. To achieve greater supplier value, the firm must offer the supplier a higher price in comparison with competing buyers or it must purchase products and services that entail lower supplier costs.

Companies often compete for the best suppliers, particularly if suppliers differ in terms of their abilities, product quality, reliability, or access to resources. Companies bid for scarce supplies in markets for specialized parts and specialty chemicals and plastics. Even when supplier products and services are generic, such as industrial materials or standardized parts, companies must offer the going rate. Competition for supplies is evident in commodity futures exchanges, such as the Chicago Board of Trade, that are double auction markets in which both buyers and sellers bid competitively.

The fact that companies bid competitively for products and services became more evident with the advent of business-to-business electronic commerce. Transactions between businesses in the United States exceed $14 trillion per year with a small but quickly growing fraction of transactions taking place on the Internet.[19] On some of the new Internet exchanges, businesses bid competitively to purchase and to sell such diverse items as steel, auto parts and office supplies. Supplier margins depend in part on the extent of competition between buyers and the extent of competition between sellers.

There are exceptions to the rule, however. In some markets, suppliers with market power may price-discriminate across their customers by charging premiums to some companies and offering discounts to others. Conversely, customers with market power may extract concessions from their suppliers. Also, when market frictions are present, companies may end up paying different prices for productive inputs simply due to imperfect information and the costs of shopping for inputs.

One strategy for attracting suppliers is *cooperation*. The company gets better service and price by lowering the cost to the input supplier. There are several ways to lower supplier cost. First, the company can request a product or service from the supplier that is less costly to produce than products or services purchased by competitors. By designing parts and components that are easier to make, a manufacturer can obtain price concessions from its parts supplier. Also, the company can be easy to work with by limiting the number and frequency of design changes and service demands. Second, the company can share information with suppliers about its demand through electronic data interchange, thus allowing suppliers to hold lower inventories of finished goods and tailor their own parts orders and manufacturing to meet demand fluctuations more closely. Third, the company can automate its billing and invoicing with suppliers to lower their transaction costs. For example, Cisco lowered its procurement costs by linking to its many suppliers over the Internet, despite coordination problems, whereas Lucent was slower in moving its supplier contacts online.[20] Fourth, the company can share its technological knowledge, management, and technical personnel, and other resources with suppliers. Finally, buying in bulk can also lower supplier costs by reducing the number of orders and allowing the supplier to achieve economies of scale.

By lowering the costs to suppliers of meeting its needs, the company creates greater value even if it pays its suppliers less than competitors.

Example 7.10   Being a Good Partner

Suppose that there are two firms and one supplier. The supplier's cost of providing an input to the first firm is $50. The first firm offers the supplier a bid price of $110. The supplier obtains profit of $110 − $50 = $60 from supplying an input to the first firm. A second firm employs an input differentiation strategy by asking the supplier to manufacture an input that costs the supplier only $30 to produce. Whether or not the second firm incurs some additional costs in helping the supplier to lower its costs, the second firm offers the supplier a bid price of $100. If the supplier has a capacity constraint that only permits it to serve one firm, which firm would the supplier serve? The supplier would earn a profit of $100 − $30 = $70 from serving the second firm. Accordingly, the supplier would serve the second firm and earn a higher profit. The second firm captures the supplier by an input differentiation strategy. Notice that the strategy is profitable if the supplier's cost savings are less than the second firm's internal costs of being a good partner.

Another strategy for attracting a key supplier is *value leadership*. The company must use its critical inputs to create a product of higher value. Since the company values certain inputs highly, it is willing to pay more to attract those key suppliers. If suppliers have similar costs of serving a company and its competitors, they will be attracted by a higher bid price. By paying more to its suppliers, the company can also ask suppliers to provide higher-quality products and services. By offering above-market payments to suppliers, companies have higher expenditures than competitors but in some cases obtain greater benefits in terms of supplier loyalty and incentives to perform reliably.

The value-leadership strategy on the supplier side corresponds to the cost-leadership strategy on the customer side. The company should follow this strategy only if the benefits of offering customers a higher value product outweigh the costs. In such cases, it may be worthwhile pursuing long-term contracts or acquiring the suppliers outright.

Companies bid for suppliers only if the suppliers' inputs are perceived to confer a competitive advantage. To be able to outbid competitors for supplier products, a company must value the supplier's products more than competitors do. This requires the company buying the inputs to deliver greater value to its own customers or to have lower internal costs than competitors. Thus, paying more to suppliers must be accompanied by a product differentiation or price leadership strategy on the final output side. A firm that provides products that its customers value highly or that operates efficiently can then attract good suppliers by sharing with them some of the additional value it creates.

Example 7.11   Paying the Piper

Suppose that there are two firms and one supplier. The supplier has a cost of $50 to supply an input to the first firm's product. The first firm offers the supplier a price of $80. The supplier earns profit of $80 − $50 = $30 from serving the first firm. A second firm employs a value leadership strategy, offering the supplier a bid price of $100 and asking the supplier to provide an input that costs the supplier $60 to produce. Suppose that the additional value of the input to the second firm justifies the additional cost it pays. If the supplier only has the capacity to serve one firm, which firm would the supplier choose to serve? The supplier would earn a profit of $100 − $60 = $40 from serving the second firm. Accordingly, the supplier serves the second firm since it obtains a higher profit by doing so. The second firm attracts the supplier by the value leadership strategy.

Jordan D. Lewis suggests that General Motors' aggressive campaign to reduce payments to suppliers in 1992 discouraged many of the company's best suppliers and may have raised costs.[21] To retain important suppliers, companies must provide them with value at least as great as alternatives, either by lowering their costs or by paying enough to retain them.

Value of the Firm

The *value of the firm* is total value created net of customer value and supplier value. The value of the firm is the present value of customer revenues minus the present value of payments to suppliers and the costs of using the firm's assets. The firm attempts to *capture* a greater share of value by raising prices to customers or by lowering payments to suppliers or by using its assets more efficiently. Capturing a larger piece of the value pie depends on the strength of competition for customers and suppliers and the company's abilities.

The firm also tries to *create* greater value which increases the size of the pie. Creating greater value is accomplished in three ways: (1) operating more efficiently by providing customer benefits at a lower cost or by lowering supplier costs, (2) providing greater benefits to customers by improving products and services, and (3) developing innovative transactions that offer new value to the market. The three strategies for creating greater value correspond to the three types of competitive advantage: cost advantage, differentiation advantage, and transaction advantage.

The *value of the firm* equals the net present value of expected cash flows. It determines the value of the firm's debt and the value of owner's equity. Investors have many alternative opportunities for capital investment, so the company's owners and lenders must be compensated for the cost of capital. To attract investment capital, *the firm must provide sufficient value to investors so that they earn a rate of return at least as great as comparable investment alternatives.* Thus, to attract investment capital, managers try to maximize the value of the firm. Maximizing the firm's present value of expected cash flows assures that the firm is maximizing the owner's equity because the shares represent ownership of the firm's future stream of expected cash flows net of debt costs.

For publicly-traded companies, *shareholder value* is the market value of owner's equity. Since shareholders claim the residual earnings of the company, shareholder value is the net present value of expected cash flows minus debt costs. When managers make investment decisions, there is always an alternative of returning funds to the company's shareholders. Thus, the cost of capital represents the opportunity costs of capital to the firm's shareholders. Companies deliver value to shareholders in two ways, through dividend payments and through increases in the company's stock price. Increases in the company's stock price reflect investors' perceptions about the company's future prospects. Managers that fail to deliver shareholder value will find that the stock market will penalize the company's stock price. The lower the company's stock price, the greater the cost to the company of raising capital. Managers' performance is also monitored by the directors serving on the corporate board, who represent shareholder interests.

Because it is difficult to judge whether or not a firm is earning the highest possible profit, investors often evaluate the performance of the company in comparison to its competitors. In addition, investors examine the rate of return on invested capital in comparison with all other investment opportunities, adjusting for risk. Managers strive to achieve the greatest attainable shareholder value given market opportunities and the abilities of the firm.

The principles of management strategy are necessarily consistent with the principles of finance. The value of the firm reflects the market's evaluation of the long-term effectiveness of management strategy and the degree to which the company's competitive advantages can be sustained. Effective management strategy means only choosing projects with a positive net present value and, when choosing between projects, choosing those that have the greatest net present value. Managers should expect that capital markets will favor only those investment projects or mergers that add value. They should not expect investors to be misled for long by cosmetic actions that do not create value, such as shifting expenditures over time or across divisions of the company.

Mergers and acquisitions should only be pursued if combining the companies increases their total value. There are no financial benefits from combining two companies simply for size or for portfolio diversification. Investors could add the two companies to their portfolios at a lower cost than merging the two organizations. The combined companies must achieve greater net present value of expected cash flows by operating together than they could achieve separately. This requires sharing of assets or combining activities in a way that increases revenues or lowers costs.

Some management strategy recommendations place considerable emphasis on the firm's current earnings, the earnings per share of stock, the ratio of the company's stock price to earnings per share (P/E ratio), or the growth rate of earnings. Such accounting measures do not provide an accurate picture of the economic value created by the company.[22] Current earnings or earnings per share are only short-term indicators. Earnings may be useful in a very limited way as an indicator of past performance and as a possible predictor of future performance. Earnings growth, however, need not imply an increase in the value of the firm since it is also necessary to take into account the cost of investments and rate of return to the company's investments. Managers that invest in projects below market rates of return may increase earnings but they are destroying shareholder value. Shareholders would prefer that management not make such investments since shareholders could earn greater returns investing the money elsewhere and earning market rates. Moreover, earnings do not take into account the company's dividend policies and the riskiness of investments. Investments that are expected to yield below-market rates of return adjusted for risk will lower the value of the firm and result in a lower market price for the company's stock.[23]

Companies often use accounting measures such as return on investment (ROI) for strategic analysis, where ROI is measured as the ratio of net income to book value of assets. Such measures fail to capture fully the company's economic value. Net income is a short-term measure and strategies often play out over a longer period of time. Also, the ratio is highly sensitive to the company's depreciation and investment capitalization policies, which affect the book value of assets. The company's book value of assets is likely to depart from the market value of those assets. Thus, ROI does not provide a good indicator of the effects of strategy on the long-run value of the firm.

Managers should apply net present value (NPV) analysis to generate a more accurate long-term picture of how strategic decisions affect the company's value. Such analysis includes the opportunity cost of capital in strategic decisions, as discussed previously in Chapter 5. The manager should focus on the present value of expected cash flows in examining strategies. Using NPV analysis means that the manager considers the trade-off between current and future cash flows in comparison with the firm's cost of capital. In addition, the choice of the discount rate in NPV analysis is based on the risk involved, with a greater discount rate applied to riskier projects. The manager should incorporate market uncertainty into the NPV analysis in making estimates of revenues and costs.

Example 7.12   NPV Analysis Under Uncertainty

The manager has an investment project that costs $160 for the initial investment. The project only pays off in its second year. Due to market uncertainty the project can have two possible payoffs, either $330 or $110. The company's appropriate discount rate is 10 percent so that the present value of the project payoff in the favorable case is $330/(1 + .10) = 300$, and the present value of the project payoff in the less favorable case is $110/(1 + .10) = 100$. If the two outcomes are equally likely, that is, each occurs with likelihood of 1/2, the project's NPV is the expected present value of payoffs net of the initial investment cost: $-160 + 1/2 \times 300 + 1/2 \times 100 = 40$. Since the project has a positive NPV, the manager should go ahead with the project, at least if there is not the possibility of delaying the project.

Because managers face uncertainty, they must be able to respond flexibly to changes in customer preferences, supplier technologies, and competitor actions. Strategic flexibility is essential because it provides options to decision makers. Traditional analysis of net present value of lines of business should include attention to the value of such options.

Option value should be taken into account to avoid the pitfalls of traditional NPV analysis.[24]

Option analysis of the information presented in the preceding example suggests a different conclusion when the manager is able to delay the project and gather more information. The value of the project with the option to continue is greater than the value of going ahead right away because the manager waits to observe market conditions before investing and thus avoids the cost of investment in the less favorable state. This benefit outweighs the forgone benefits due to delay. This does not mean that waiting is always best; it means that, under some conditions involving significant uncertainty, a wait-and-see approach may be better than jumping the gun. What is more important is that the manager would be willing to pay for the flexibility to decide later.

Example 7.13   Wait and See

Consider the choice of the manager in Example 7.12 and suppose that the manager has the option of waiting. Delaying the project would allow resolution of the uncertainty. The cost of waiting is further delay of the benefits, but this cost is partly offset by delaying the investment cost as well. Is delay worthwhile in this case? By waiting until the uncertainty is resolved, the manager can decide whether or not the project is worthwhile. The manager would undertake the project in the favorable case since it has positive net present value equal to $300 - 160 = 140$. The manager would not undertake the project in the unfavorable case since it has negative net present value equal to $100 - 160 = -60$. Since the favorable and less favorable states are equally likely, the manager expects to carry out the project with a likelihood of 1/2, so the expected value of waiting to evaluate the project is calculated as 1/2 of its value in the favorable case, $1/2 \times 140 = 70$. Discounting the expected value of waiting gives the net present value of the project when the manager can learn about the state of the world and has the option to change: $70/(1 + .10) = 64$, rounded to the nearest dollar amount. This is greater than the NPV of the project without the option of waiting, so in this case a delay is worthwhile.

The value of an option to delay is significant when additional information can be gathered. The value of the option in the preceding example is the difference between the two project values, $64 - $40 = $24. Put differently, the manager would be willing to wait and invest later, even if the initial cost of investment was higher in the case of the delayed project. It turns out that the investment could be substantially higher and still allow the manager to prefer waiting. If the investment cost after delay were as high as $212, that is, $52 higher than the $160 without delay, the manager would still prefer to wait.[25]

The option value of waiting has some organizational implications. Suppose that the organization is not good at monitoring market conditions and makes decisions very slowly. The result of these bureaucratic inefficiencies is that the company will proceed along the lines of the traditional NPV analysis. To get the project up and running, the company will need to start its decision process early, and it will necessarily rely on projections of future market conditions. Since the company is not skilled at observing market change, it will base its decisions on these projections rather than gathering information just in time. In contrast, a company that monitors market conditions and reacts flexibly would choose to wait and act after market uncertainty is resolved.

Monitoring market conditions and organizational flexibility has its own costs, however. The higher returns realized by waiting in some cases imply that the manager would be willing to pay more for organizational systems that would allow monitoring of market conditions and improve decision-making. In the example, the manager would be willing to pay $24 for such organizational enhancements. The company might be willing to restructure company decision-making procedures or hire additional managers to carry out decision-making roles. Other costs could include investments in information technology and communications systems.

Alternatively, recall that the manager would be willing to incur greater investment costs for the purpose of waiting to observe market conditions. In the example, this additional investment equaled $52. Such additional costs might take the form of flexible manufacturing equipment, just-in-time inventory systems, and automated delivery systems that allow faster and more accurate responses to market conditions. Achieving flexibility requires investment in the necessary production facilities and information technology.

7.5 *Overview*

The firm's total *value created* is equal to the sum of customer value, supplier value, and the value of the firm. Thus to compete effectively, the firm must deliver greater customer value, supplier value and value of the firm than its competitors. To achieve this means that the firm must create greater total value, which is the difference between customer willingness to pay and the cost of inputs obtained from suppliers or supplied by the firm itself. Managers must consider these three components of value jointly, not separately. The difference between the value created by the company and its competitors is the firm's competitive advantage.

There are three sources of competitive advantage: Differentiation advantage allows the firm to offer customers products and services with greater benefits. Cost advantage allows the firm to lower the costs of using its own assets and to reduce its payments to suppliers by reducing the costs of delivering a particular benefit to customers or by reducing supplier costs. Transaction advantage allows the firm to discover new combinations of customers and suppliers that create greater value. Chapters 9–12 explore these aspects of competitive advantage.

Questions for Discussion

1. Consider the strategy of a movie rental chain.

   (i)   What might be customers' willingness to pay for a movie rental? How would you go about estimating willingness to pay? What is the cost per movie of the movie studios who are the movie rental chain's suppliers? How would you go about estimating their cost? What are the costs per movie of the movie rental chain? What is the value created per movie by the movie rental chain?

   (ii) Consider again the movie rental chain in (i). What are the retail and wholesale prices per movie? What factors affect those prices? What determines the consumers' surplus, producers' surplus, and value of the firm?

   (iii) Given the example of the movie rental chain in (i) and (ii), how might the movie rental chain increase the value it creates? How might the movie rental chain increase the value that it captures?

   (iv) The movie rental chain must decide between selling and renting digital video discs (DVDs). What factors should it take into consideration? Would the cost of capital affect this decision? How might competition affect this decision?

2. A company contemplating international expansion must choose between two target countries. The first target country offers revenues net of operating costs equal to 210 each year over two years. The second target country offers revenues net of operating costs equal to 300 the first year and 105 the second year. The firm's cost capital is 5 percent. Compare the net present values obtained from serving the two countries. Which target country should the manager choose? What does this example illustrate about the trade-off between current and future revenue?

3. The world steel industry faces substantial excess capacity. The world price of steel is about $300 per ton. A major steel producer has operating costs that are about $450 per ton; that is, about

$150 above the world price of steel. What should the steel producer do? Some of the options are as follows: continue producing steel in the hopes that the world price will rise, shut down its facilities and exit the industry, invest funds to modernize its facilities so as to lower operating costs, buy steel at the world price for sale to its customers, relocate the company's facilities in another country. What are some of the considerations that enter into the strategic decisions of the managers of the steel producer?

4. An electronics manufacturer that makes computer components has costs that are below those of some of its competitors but above those of other competitors. Due to a boom in demand, the electronics manufacturer is able to operate profitably at full capacity. Should the electronics manufacturer invest in facilities to lower its operating costs? What factors should the company's managers consider in making their decision?

5. Ten companies in a segment of the pharmaceutical industry are contemplating competing research programs to develop a drug to treat a particular medical condition. Some of the research programs will be successful in developing a treatment and obtaining approval by government regulators. The resulting drugs are likely to differ in terms of their effectiveness and side effects. Only one or two of the drugs are likely to be successful in the medical marketplace. How would an individual pharmaceutical company make a decision about whether or not to enter into a research program? How should managers think about the uncertainties of the projects? Would there be situations in which managers might stop their research projects before they are completed?

6. A manager finds that he can boost the company's current earnings by counting some orders as revenues as soon as the order is received rather than when the revenues are received. Does this approach increase the value of the firm? Should this approach increase the value of the company's stock?

7. A company producing a wide range of manufactured goods diversifies by acquiring another company producing appliances. The two companies will be operated independently and will not share resources, production facilities, marketing, or purchasing. The managers of the company proposing the acquisition argue that since the two companies will form a larger company they will be able to raise capital at a lower cost. Will the merger increase the value of the two companies? Will this type of merger increase the stock price of the two companies in comparison with the initial stock prices? Should the acquiring company pay a premium above the market price of shares for the company it seeks to acquire?

Endnotes

1. See Adam M. Brandenburger and Harborne W. Stuart, Jr., "Value-based Business Strategy," *Journal of Economics & Management Strategy* 5 (Spring 1996), pp. 5–24. The notion that value is the additional contribution of a player to a coalition in a game is familiar to students of cooperative game theory. See, for example, Richard Aumann, "Game Theory," in J. Eatwell, M. Milgate and P. Neuman, eds. *The New Palgrave: A Dictionary of Economics* (London: Macmillan, 1987), pp. 460–482. See also the discussion in E. Davis and J. Kay, "Assessing Corporate Performance," *Business Strategy Review* 1 (1990), pp. 1–16.

2. Michael Dell, "Netspeed: The Supercharged Effect of the Internet," address to The Executives' Club of Chicago, October 23, 1998, http://www.executivesclub.org/static/News98-99/Michael%20Dell.htm

3. For the year 2000, it was estimated that Dell Computer had cost of goods sold equal to 78.8 percent of operating revenue, and selling, general, and administrative cost of goods sold of 10.9 percent. See Lori Calabro and Gunn Partners, "Cost Management Survey: Bend and Stretch, Why the Best Companies Remain Focused on Cost Cutting — Whatever the Business Cycle," *CFO Magazine*, February 26, 2002, cfo.com

4. Michael E. Porter, *Competitive Advantage: Creating and Sustaining Superior Performance* (New York: Free Press, 1985).

5. See Gary Lilien, Philip Kotler and K. Sridhar Moorthy, *Marketing Models* (Englewood Cliffs, NJ: Prentice Hall, 1992), pp. 512–517.

6. Steven Klepper and Elizabeth Graddy, "The Evolution of New Industries and the Determinants of Market Structure," *Rand Journal of Economics* 21 (Spring 1990), pp. 27–44.

7. Timothy Bresnahan and Daniel F. G. Raff, "Intra-Industry Heterogeneity and the Great Depression: The American Motor Vehicles Industry, 1929–1935," *Journal of Economic History* 51 (June 1991), pp. 317–331.

8. Joseph A. Schumpeter, *The Theory of Economic Development* (New Brunswick, NJ: Transaction Publishers, 1997), p. 229.

9. Klepper and Graddy, "The Evolution of New Industries."

10. Steven Klepper and Kenneth L. Simons, "The Making of an Oligopoly: Firm Survival and Technological Change in the Evolution of the U.S. Tire Industry," *Journal of Political Economy* 108 (2000), pp. 728–760.

11. Ashish Arora, "Appropriating Rents from Innovation: Patents, Licensing and Market Structure in the Chemical Industry," *Research Policy* 26 (1997), pp. 391–403.

12. Steven Klepper, "Industry Life Cycles," *Industrial and Corporate Change* 6 (1997), pp. 145–181.

13. Andrea Bonaccorsi and Paola Giuri, "When Shakeout Doesn't Occur: The Evolution of the Turboprop Engine Industry," *Research Policy* 29 (2000), pp. 847–870.

14. This example is derived from Brandenburger and Stuart, who apply game theory to analyze the "added value" of any market participant. In particular,

a competitive firm's added value in the market is the value created by all the players in a market minus the value created by all the other players except the firm itself. If the firm in question has a competitive advantage over its rivals and if the rivals are displaced from the market by competing with the firm, then the firm's added value will equal its competitive advantage as I define the term. See Adam M. Brandenburger and Harborne W. Stuart, Jr., "Value-based Business Strategy," *Journal of Economics & Management Strategy* 5 (Spring 1996).

15. Sources of data: U.S. Commerce Dept., Chinese Ministry of Foreign Trade and Economic Cooperation, Mattel Inc., Hong Kong Toy Council. This example is presented in Rone Tempest, "Barbie and the World Economy," *Los Angeles Times*, Part A, p. 1, September 22, 1996, Home Edition. See also Robert C. Feenstra, "Integration of Trade and Disintegration of Production in the Global Economy," *Journal of Economic Perspectives* 12 (Fall 1998), pp. 31–50.

16. Michael E. Porter, *Competitive Strategy: Techniques for Analyzing Industries and Competitors* (New York: Free Press, 1980).

17. Ibid., pp. 41–44.

18. Information about Shuttle by United obtained from http://www.ual.com/airline/default.asp?section=shuttle.asp&SubCategory=Our_Services&destination_URL=/airline/Our_Services/shuttle.asp.

19. Transactions between businesses in 1999 in the U.S. were reported to equal approximately $14 trillion, of which approximately $90 billion were conducted over the Internet. See Louis Uchitelle, "It's Just the Beginning," *New York Times*, Special Section on E-Commerce, June 7, 2000, p. E1.

20. Steve Rosenbush, "Rosenbush Covers Telecommunications from New York, Commentary: How Lucent Lost Its Luster," *Business Week*, August 7, 2000.

21. Jordan D. Lewis, *The Connected Corporation: How Leading Companies Win Through Customer-Supplier Alliances* (New York: Free Press, 1995).

22. G. Bennett Stewart, III, *The Quest for Value* (New York: Harper Business, 1991).

23. Alfred Rappaport, *Creating Shareholder Value* (New York: Free Press, 1986).

24. The numerical example is due to Avinash K. Dixit and Robert S. Pindyck, *Investment Under Uncertainty* (Princeton: Princeton University Press, 1994).

25. The NPV of the project without delay is $40. The net present value of the project with delay and investment level I is NPV* = 1/2 × (300 − I)/(1 + .10). The highest investment level such that NPV* falls to $40 is I = $212. This can be easily verified by plugging $212 into the NPV* equation.

CHAPTER 8

TRANSACTION COSTS AND THE FIRM'S VERTICAL STRUCTURE

Contents

Executing the company's strategy requires completing a number of functional tasks: raising finance capital, hiring and managing personnel, obtaining technology, procuring inputs, designing products, managing operations, marketing, and selling products. The manager determines which of those tasks the company will perform within the organization and which of those tasks the company will rely on others to perform. The choice of what functional activities are carried out within the organization establishes the company's *vertical structure*.[1]

The manager's choice between in-house production and outsourcing is sometimes referred to as the *make-or-buy* choice. The company's degree of *vertical integration* represents the types of functional tasks the company performs in order to produce its final output. The extent of the company's vertical integration is determined by a number of fundamental management decisions. Should the company extend into wholesaling or retailing? Should the company be a parts supplier or a manufacturer? Should the company be involved in raw materials sourcing? Should the company deal with basic research and development? The degree of vertical integration depends on the ease of coordination with suppliers and distributors, the need to make capital commitments, the relative costs of market transactions and vertical consolidation, and knowledge required for production and distribution tasks. Companies have various alternatives to vertical integration including spot transactions, contractual relationships, alliances and joint ventures.

Managers further specify the company's vertical structure by establishing vertical levels of authority and assigning functional tasks within the organization. Managers determine whether functional tasks will be centralized or decentralized. Functional tasks are said to be *centralized* when they are performed at the corporate level. Functional tasks are said to be *decentralized* when they are performed within each of the divisions rather than at the central office. The vertical structure of the organization also depends fundamentally on the extent of delegation of authority to managers and employees. By delegating greater authority to employees, and to teams where appropriate, companies reduce transaction costs.

Chapter 8: Take-Away Points

The manager evaluates market transaction costs and organizational costs in establishing the vertical structure of the firm:

- The manager should consider both the direct costs and the transaction costs of market transactions, including sales, purchasing, and finance.
- The manager should choose between whether to carry out an activity or outsource, taking into account the transaction costs of using the market versus the governance cost of performing the task within the firm.
- The manager should consider the transaction costs associated with alternative modes of market entry, comparing spot transactions, contracts, licenses, alliances and joint ventures, mergers and acquisitions, and organizational growth.
- Managers should decentralize the organization by assigning functional tasks to market-based divisions, including marketing and sales, R&D, human resources, finance, purchasing, and operations.
- The manager's make-or-buy decision often depends on expertise of the company relative to potential suppliers and subcontractors.
- Managers should delegate authority to increase organizational flexibility and responsiveness to market conditions.
- Managers follow a series of basic steps in delegating authority: Share goals and strategies, communicate the required tasks, grant independence, inform, monitor progress, reward performance.

8.1 *Transaction Costs and the Make-or-Buy Choice*

Transaction costs are a critical element in the manager's design of the firm's vertical organizational structure. Although transaction costs are sometimes difficult to identify, managers ignore such costs at their peril. Generally, transaction costs are the *indirect* costs of the company's market transactions, including management and employee time, legal services, back-office operations, and information systems. Transaction costs affect management choices regarding modes of market entry and the firm's asset ownership.

Transaction Costs and Organizational Costs

In designing the organization, the manager must be concerned with total cost. For example, the company might obtain a component at a purchase price of $100, but it might spend an additional $20 in transaction costs to research the market and negotiate with the providers of that component. The manager should purchase the component if the organizational cost of producing the component in-house exceeds the total purchase cost of $120.

Consideration of transaction costs has a fundamental impact on strategic decisions. Dealing with multiple suppliers may lower costs through competition, but it may raise total costs because of the need to manage multiple relationships. Dealing with suppliers on a spot basis rather than through contracts may confer flexibility and lower short-term purchase costs, but it may entail higher transaction costs due to the need for repeated negotiation. A domestic supplier may charge more than an overseas supplier, but the total costs of dealing with the domestic supplier may be lower because of the costs associated with arranging the overseas purchase, including the costs of communicating with the supplier, scheduling transportation, obtaining import documentation, and converting currencies.

Transaction costs include many types of *management and employee labor costs* associated with obtaining financing, acquiring technology, hiring personnel, and finding and negotiating with suppliers of inputs and services. In addition, firms encounter transaction costs in dealing with customers, including many types of *marketing and sales costs* and all manner of *back-office process costs* associated with sales and purchasing, such as billing and invoicing. Transaction costs often represent a major share of the company's *communications and information technology expenditures*. Transaction costs also entail costs associated with *negotiating, writing, and monitoring contracts* with suppliers and customers and the costs of *accounting* needed to gather and provide information in the company's dealings with government, capital markets, and business partners.

Advances in telecommunications technology have drastically cut the costs of communication and information exchange between trading partners, which in turn has helped lower the costs of coordination both between companies and within organizations. The increasing reliance on outsourcing suggests that technology and market changes have lowered the transaction costs of market contracting relative to the costs of managing production within large vertically integrated organizations.

Companies lower transaction costs by exchanging data on demand and inventories with their suppliers, allowing for production and deliveries to respond rapidly to demand shifts. Companies can have just-in-time deliveries rather than holding large inventories. Companies can link electronically with their suppliers, drastically reducing the costs of ordering, billing and monitoring deliveries. By connecting internal computer systems to external buying and selling decisions, companies can update their inventory records and production decisions, achieving coordination in business-to-business transactions that was previously limited to production within the organization.[2]

Nobel prize winner Ronald Coase was the first economist to identify the importance of transaction costs in determining the structure of the firm. He recognized that companies must choose between making an input and buying it from others. Coase observed that companies choose between *making and buying* an input by comparing the costs of using the market with the costs of making the input:

> A firm will tend to expand until the costs of organizing an extra transaction within the firm become equal to the costs of carrying out the same transaction by means of an exchange on the open market or the costs of organizing in another firm.[3]

Companies choose to carry out productive activities that are less costly to handle within an organization as compared to market exchange. In competitive industries, the vertical stages of production are divided among firms in an efficient manner, that is, "in such a way that the cost of organizing an extra transaction in each firm is the same."[4]

Thus, the choice between whether to carry out an activity or outsource it depends in large part on a comparison of the transaction costs of the two activities. Whether an activity is performed in-house or contracted out, organization costs are associated with each activity. Managers must coordinate and supervise internal activities just as they must devote effort to coordinating and monitoring external suppliers and distributors. Manufacturing parts and supplies in-house brings advantages of greater coordination but comes at the cost of a large organization with a potentially unwieldy bureaucracy. Outsourcing services and procurement of parts and supplies reduces the size of the organization. If a company encounters problems of coordination with a supplier, it may need to carry inventories or it may experience production interruptions.

Modes of Entry

Transaction costs affect the firm's decision on how to enter new markets. The *mode of entry* refers to the extent to which a new entrant vertically integrates into production and distribution. At one end of the spectrum, an entrant can focus on carrying out spot transactions. At the other end, the entrant can be fully vertically integrated into production and distribution. In between, the entrant can pursue a variety of contractual alternatives. Licensing brands and technology provides entrants with a way of avoiding practically all integration into production and distribution. The range of alternatives is given in Table 8.1.

The manager begins by identifying the target market to be served. Then, the manager decides how the company's organization should be structured to serve its target markets. The question of modes of entry has particular importance in international business, where the target market generally refers to a particular country or geographic region.[5] The company can choose many combinations of entry modes for sourcing and serving, as represented in Table 8.1. The company's sourcing modes can be similar or different from its serving modes.

The company coordinates its sourcing and serving activities to create the greatest value added. The manager's decisions about vertical integration affect how the company sources from suppliers and how the company serves its customers. The company's vertical structure and its sourcing and serving strategies are closely connected (see Figure 8.1).

Companies that simply broker transactions are perhaps the least vertically integrated. Thus, auction houses such as Sotheby's and Christie's or Internet auctioneer eBay facilitate transactions without being a counterparty. Some wholesale and retail merchants engage in both spot purchases and spot sales. Companies often enter into longer-term commitments to suppliers, customers, or both.

Alliances and joint ventures provide additional ways of reducing vertical integration. For example, a joint venture between a manufacturer and a distributor avoids the need for either party to fully vertically integrate, while enabling both parties to coordinate their activities through contracts and informal agreements. By entering into

Table 8.1: Modes of entry.

Options	Sourcing	Serving
Spot transactions	Buy from others	Sell to others
Licenses and franchises	Purchase technology and brands from producers	Sell technology and brands to distributors
Contracts	Purchase contracts	Distribution contracts
Alliances and joint-ventures	Form alliance with suppliers	Form alliance with distributors
Growth	Establish manufacturing unit	Establish distribution unit
Mergers and acquisitions	Merge with supplier	Merge with distributor

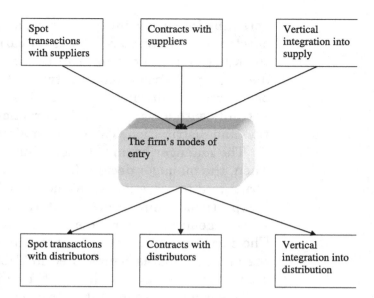

Figure 8.1: **The manager determines the company's modes of entry by deciding how best to source from suppliers and how best to serve customers.**

a joint venture, partner companies share the total costs of investment. In deciding whether to enter into an alliance or a joint venture, companies take into account the transaction costs of forming and maintaining these relationships.

Supply Chain

Managers in every type of company continually face make-or-buy decisions. Should the company manufacture a component itself or purchase it from another company? Should the company perform particular services or rely on external suppliers? The decision may be as simple as retaining a hospitality company to run the employee cafeteria. Alternatively, the make-or-buy decision may be critical, such as choosing a partner to manage all of the company's manufacturing and parts procurement.

Practically any company activity can be outsourced. The company's internal processes can be provided by service companies, including temporary staff services, human resource management, employee training, accounting, legal, information technology, purchasing, and facilities maintenance. Companies employ consulting services provided by such companies as McKinsey and Accenture as a means of outsourcing some types of management decision-making. Service providers offer both expertise and cost savings by achieving economies of scale from managing operations for multiple customers. Managers evaluate transactions with external service providers in comparison to administration of services within the company. Managers that choose to retain external service providers calculate that the total costs of contracting for services is less than the total costs of providing those services internally.

As the costs of communication and information processing change, companies reevaluate their make-or-buy decisions. Changing the vertical structure of a company can involve complex restructuring efforts. For example, General Motors Corporation spun off its Delphi Automotive Systems subsidiary, creating the world's largest maker of automobile parts with over 200,000 employees. The divestiture was part of a long-standing attempt to reverse vertical integration strategies dating back to before the Great Depression. Delphi division supplies hundreds of auto parts including seats, spark plugs, and dashboards.[6] The spin-off proposal came on the heels of a United Auto Workers strike at a Delphi parts plant and a non-Delphi parts plant that idled much of the company. Similarly, Ford spun off its Visteon auto parts manufacturing unit with over 82,000 employees. Thus, GM and Ford determined that their auto parts units and automobile manufacturing units would be better off dealing with each other through market transactions rather than through internal transactions.

Among the many advantages of vertically separating parts manufacturing and final product manufacturing is that parts makers can sell to many manufacturers, thus increasing economies of scale and scope and yielding greater returns to R&D. Also, the manufacturer is no longer tied to an internal supplier and can source from many parts makers, thereby gaining access to the best suppliers in terms of productivity, quality, location, and convenience. Competition allows parts makers and manufacturers to make the best match in the marketplace. Moreover, parts makers and manufacturers can seek new trading partners as technology and market conditions change. The significant reductions in vertical integration observed in the automobile industry typify a general restructuring of manufacturing for many other types of products.

In addition to relying on independent parts suppliers, companies can outsource their procurement and manufacturing functions. Outsourcing the management and operation of the production sequence itself, known as *supply chain management*, is gaining increased popularity. The advantage of supply chain management is that the company can choose to carry out in-house only those activities that contribute the greatest value added relative to market alternatives.

Solectron Corporation is the largest provider of electronic manufacturing services to original equipment manufacturers (OEMs) in the consumer and business electronics markets. Solectron's international manufacturing network coordinates production for companies such as Hewlett Packard, IBM, Ericsson, and Cisco. Solectron's integrated supply chain management includes not only production but also product design, product introduction management, material procurement, distribution, and warranty and repair services.[7]

The company that manages supply chains for others creates a virtual company linked by communications and information technology systems. Many of the advantages of supply chain management result from decreased product cycle times, which allow companies to adjust production levels more closely to observed demand, thereby reducing inventories. The reduced cycle time provides the company with flexibility and rapid response. Moreover, the company that manages the supply chain offers flexibility by assembling customized teams of suppliers for each major downstream customer.

Intermediation and supply chain creation constitute a major change from the traditional value chain view. According to that view, the company controls part of the value chain by carrying out production within the organization and owning the critical technologies and capital equipment. With intermediation and supply chain creation, on the other hand, owing productive facilities is no longer essential. The supply chain manager focuses on creating innovative transactions rather than acquiring production processes and equipment. Similarly, the companies that are customers of supply chain managers can reduce their own level of vertical integration, allowing them to focus on areas where they add the most value.

The vertical span of the organization does not mean that the company provides each stage in the production sequence within that span. Instead, the company may select crucial segments and outsource those in between on the vertical chain. In other words, the company can leap over critical steps in the production sequence through outsourcing.

Exhibit 8.1   Li & Fung

To illustrate how supply chain management works, consider Li & Fung, Hong Kong's largest export trading company, which works with 7,500 suppliers in over 26 countries. The company creates customized supply chains for each of its customers, which generally are American and European retail companies.

According to Victor Fung, grandson of the founder and company leader along with his brother William, the company's novel concept is "breaking up the value chain" through dispersed manufacturing. Li & Fung provides the upstream functions of design, engineering, and production planning. Then, it organizes and outsources the middle stages of production, including raw material and component sourcing and managing production by contractors

(Continued)

Exhibit 8.1 *(Continued)*

in factories around the world. Then, Li & Fung provides the downstream functions of quality control, testing, and logistics, as well as the marketing and sales interaction with its customers, the retail companies.[8]

Li & Fung takes advantage of low-cost labor by assembling a vast workforce in many countries, numbering perhaps a million people. The company's competitive advantage stems from its efficiency in coordinating the large number of suppliers along its supply chain. The company's suppliers might produce only one component of the finished product, for example the zipper or the shell of a parka. Li & Fung determines the best sources for the individual components and coordinates design, final assembly, and quality control.

Li & Fung structures its organization to maintain personal relationships in its dealings with its many suppliers, delegating considerable autonomy to the company's buyers. Victor Fung colorfully illustrates the company's intermediation role:

> "I have a picture in my mind of the ideal trader for today's world. The trader is wearing a pith helmet and a safari jacket. But in one hand is a machete and in the other a very high-tech personal computer and communication device. From one side, you're getting reports from suppliers in newly emerging countries, where the quality of the information may be poor. From the other side, you might have point-of-sale information from the United States that allows you to replenish automatically."[9]

Li & Fung orchestrates its vast set of suppliers to create the best mix of activities for each item it produces. Only 2,500 of Li & Fung's 7,500 suppliers are active at any one time,[10] allowing the company's suppliers to pursue other manufacturing projects and reducing excessive reliance on Li & Fung orders.

Other companies in Hong Kong and throughout Asia employ the supply chain structure of companies like Li & Fung. Coordinating virtual supply chains through networks of alliances provides productive efficiencies and fast production cycles. By relying on its network of outside suppliers, Li & Fung achieves flexibility that might not be possible with vertical integration into manufacturing.

Distribution

A company that manufactures a product must choose how much to be involved in distribution. The company may rely entirely on independent wholesalers and retail distributors. Alternatively, the company may choose to perform some wholesale services and deal directly with retailers. Finally, the company may vertically integrate into distribution by performing both wholesale and retail distribution activities.

The manager compares the total cost of distribution activities with the total cost of relying on external distributors. The cost estimates are adjusted for the effectiveness of alternative modes of distribution. The manager compares the organizational costs of establishing and operating wholesale and retail distribution units with the direct costs of hiring outside distributors plus the transaction costs involved in dealing with them. These transaction costs include the costs of searching for outside distributors, negotiating and contracting with the distributors, and monitoring their performance.

Many grocery manufacturers control their distribution systems by contracting with brokers that act as their agents.[11] Companies such as Castle & Cook Inc., CPC International, Kellogg's Mrs. Smiths products, Sara Lee Corp., H. J. Heinz (StarKist, Ore Ida), and Hormel & Co. sell through brokers and ship directly to retail warehouses.[12] For these companies, the transaction costs of dealing with independent brokers offers advantages over the transaction costs of using internal distribution personnel to deal directly with retailers.

In contrast, other companies have discovered that being closer to the retail level creates greater returns because the company has greater access and interaction with customers. Many of the major grocery manufacturers have traditionally chosen captive distribution systems.[13] Kellogg, Quaker Oats, Campbell Soup, and many others operate a direct sales force and ship directly to retail warehouses. Campbell sells and distributes *Pepperidge Farms* bakery products through direct store delivery on company-owned and independent routes.[14] Procter & Gamble operates a combined sales force for all seven of its divisions and ships directly to retail warehouses.[15]

Many distributors are vertically integrated companies. Retail chains, such as Wal-Mart and Target, engage in extensive wholesale operations to serve their retail outlets. These companies obtain cost efficiencies from establishing and operating central warehouses, inventory management, and transportation systems and coordinating these operations with a network of retail stores.

Media conglomerates are examples of vertically integrated companies owning both production and distribution outlets. Media companies such as AOL-Time Warner, Disney, Viacom, Bertelsmann, and News Corporation seek cost efficiencies from vertical integration of content production and distribution. The formation of such media companies suggests that the organizational costs of allocating content within a company may be less than market transactions costs for certain types of content. For example, the German media conglomerate Bertelsmann produces content through its ownership of book publishers (including Random House, Bantam Doubleday Dell, Ballantine Books, C. Bertelsmann Verlag, Plaza & Janés, and Editorial Sudamericana) and distributes the content in part through book clubs with 29 million members and online booksellers BarnesandNoble.com and bol.com.[16] In addition, Bertelsmann operates Arvato, a printing and publishing services company. Similarly, Bertelsmann's RTL Group is a media production company that creates thousands of hours of programming in 35 countries each year (including such shows as Baywatch), an independent wholesale distributor of distribution rights to media content, and a retail distributor of content through its 24 television stations with 100 million viewers and its extensive online presence.[17]

Media managers debate whether companies obtain greater market power by owning content, such as films and television programs, or by owning distribution systems, such as cable television, broadcasting, and satellite systems, but the issue may be transaction costs. Michael D. Eisner, chairman of the Walt Disney company, is said to feel pressure to own additional distribution systems, even though the company already owns broadcaster ABC and various cable television channels such as Lifetime, Disney channel, and ESPN. In the United States, customers spend about $2 billion per year renting Disney home videos of which Disney receives only about one-third.[18] This gives some indication of the transaction costs in media markets. The question is whether vertical integration into distribution would increase efficiency of transactions or entail additional costs from managing a vertically integrated company.

Another aspect of vertical integration of production and distribution is that companies gain market knowledge by wholesaling or retailing their own products. A wholesale or retail strategy requires developing organizational units for marketing, distribution, and sales. Carrying out the wholesale function entails employing a sales force for promoting the products to retailers or contracting for the services of professional sales representatives. The company must create a wholesale distribution network with warehouses or direct distribution systems and computerized tracking methods for products, or rely on independent wholesalers. Companies that seek to obtain market knowledge through vertical integration determine that the transaction costs of obtaining and applying such knowledge through market relationships is greater than internal learning.

Licensing

Companies can minimize their extent of vertical integration by *licensing*. Companies license patents, technology, trademarks, brands, business methods, product designs, software, transaction methods, and business franchises. They can choose to license their technology to manufacturers who are then responsible not only for production but also for sales. By licensing technology, companies avoid investments in production or distribution. Licensing technology is becoming increasingly common in international business, allowing companies to sell their technology abroad without the need to develop manufacturing facilities or distribution arrangements in other countries.

Licensing potentially confers a number of competitive advantages. The company participates in domestic and international markets for intellectual property. Increased sales from licensing technology generate incremental returns to the company's innovation efforts. By investing more in R&D, rather than on manufacturing and distribution activities, the company benefits from scale and scope economies in R&D and gains expertise in creating intellectual property. Conversely, the company lowers its capital commitment to physical plant or marketing thereby lowering its capital costs. Technology can be adapted to changing market conditions more rapidly through licensing than through vertical integration since clients using the technology often are more familiar with specific applications. Companies licensing technology can also provide value-added services. IBM, as the leading world supplier of technology for application-specific integrated circuits, emphasizes speed to market for customers. According to IBM, licensees can obtain "the same system-on-a-chip (SOC) expertise, extensive core library, and methodologies that we use in our own products."

Licensing arrangements allow manufacturers and distributors to partner with developers lacking the production and marketing capabilities. Pharmaceutical companies outsource R&D by contracting with biotechnology companies. For example, one of the best-selling diagnostic tests in the United States was a blood test for hepatitis-C marketed to hospitals and blood banks by Johnson & Johnson and Abbott Laboratories. Johnson & Johnson received a royalty from Abbott and split its profits and the royalty with Chiron Corp., the developer of the test.[19]

Franchising is an important type of licensing that creates a process for rapid growth and lowers the necessary capital outlay by relying on the investment of franchisees. The franchise company benefits from the incentive effects of franchise ownership because small business owners are able to intensively monitor performance. Leading franchisers include McDonald's, Burger King, Yogen Früz Worldwide, Subway, Baskin Robbins, GNC Franchising, KFC, and Dairy Queen.

Companies are increasingly turning to international licensing of technology and brands, which offers the advantage of avoiding product import and export transaction costs, currency risks, transportation costs and tariffs, and other costs of trade regulations. Companies do not have to incur the costs of establishing manufacturing or distribution facilities abroad, where they have less knowledge of the business environment. Licensing can reduce the risks of contracting for manufacturing and distribution with unfamiliar international partners. Moreover, companies avoid investment abroad where trade regulations imposed by foreign governments can reduce value.

Companies engaged in international licensing face various challenges. They face the problem of protecting brand equity overseas where there are additional costs of monitoring the performance of license holders. Licensing also entails transaction costs, including the costs of negotiation and enforcement of contractual agreements and currency risk. Companies must also deal with different levels of legal protections for intellectual property such as patents and trademarks across countries. Licensing technology abroad can increase the risks of imitation and copying. Companies engaged in international franchising incur additional costs of finding franchisees abroad and monitoring their performance.

Companies that provide services also transfer knowledge and training without vertical integration. Consulting companies in engineering, management, electronic commerce, and information technology offer various types of expertise without making investments in the industries that they advise. The shift of companies such as General Electric and IBM into services represents such transfers of business and technical knowledge.

Exhibit 8.2 Distribution Joint Venture

Hutchison Whampoa based in Hong Kong, Anda based in Shekou, China, and Tibbett & Britten Group from the United Kingdom formed a joint venture in the People's Republic of China, named Hutchison Tibbett & Britten Anda. The joint venture was the largest retail logistics contractor in China that provides both warehousing and distribution services.

The partners to the joint venture offered complementary skills and resources. Hutchison Whampoa provided knowledge of China and Chinese market connections, as well as access to its Yantian Port facilities and Shanghai container terminals. Anda provided a trucking network and access to its Shekou container terminal. Tibbett & Britten Group in turn provided knowledge of contract logistics services and supply chain management. The joint venture shipped products sourced from Chinese manufacturers to retail sector customers in Europe and North America. Hutchison Tibbett & Britten Anda also provided access to the Guanlan Retail Consolidation Centre which, according to the company, is the largest warehouse and container handling facility in South China.

(Continued)

> **Exhibit 8.2** *(Continued)*
>
> The partners already operated a joint venture that runs distribution for Hutchison Whampoa's supermarket group Park N' Shop. In addition to the main partners, the joint venture was a vehicle including secondary joint venture partners. Air Tiger Express added air freight services and Orient Express Container Co. Ltd supplied ocean freight services.[20]

8.2 *The Firm's Vertical Structure*

The advantages and disadvantages of vertical integration stem from the costs of market transactions in comparison with the costs of expanding the organization. This section examines how transaction costs affect the manager's vertical integration decisions.

Coordination

Companies face the difficult task of coordinating many disparate activities. Purchasing, production, and distribution must match demand to avoid the costs of holding unsold inventories or missed sales from insufficient inventories. Parts and components must meet specifications for products to be manufactured. Product designs must match customer needs to compete effectively.

These types of coordination may be costly to achieve through arm's-length transactions with suppliers due to transaction costs. If the information systems of a company and those of its potential suppliers cannot be linked electronically, coordination advantages may accrue from in-house production. Improvements in information technology, particularly in software that allows for companies to coordinate their production requirements, potentially reduces the advantages of vertical integration.

If a firm operates in a market with high variability of customer demand, it may be easier for managers to adjust internal production levels by work orders rather than for purchasing managers to adjust orders from outside suppliers through contractual arrangements. Long-term relationships with suppliers mitigates the transaction costs of changing orders. Suppliers who are accustomed to order variability may offer flexibility and lower costs. A company that relies on internal production capacity to deal with variable demand will inevitably carry costly excess capacity. An outside supplier can avoid some of the costs of excess capacity by pooling orders from many sources. Thus, the decision to use the company's own capacity to deal with highly variable demand depends on the costs of holding excess capacity as compared with the costs of adjusting outside orders. CEMEX handles this problem by manufacturing about half of its cement requirements and relying on external suppliers for the other half.

Some managers favor vertical integration for strategic reasons. Some suppliers are concerned that they may not sell all of their output unless they own the distribution channel while some distributors are concerned that they may not have adequate product availability if they do not own manufacturing facilities. Such concerns may not justify the costs of operating a vertically integrated company. If supplies are generic, owning supply capacity offers distributors little advantage. If product supplies are differentiated, distributing the company's own products will not yield a competitive advantage unless those products yield greater benefits for consumers than those of competitors. The competitive advantage comes from the superior products, not from the vertical integration of production and distribution. Accordingly, distributors also may not gain an advantage from owning supply capacity if products are differentiated. Conversely, suppliers gain little advantage in owning distribution. If specific distribution channels have a competitive advantage over others, that advantage is not likely to be enhanced by vertical integration into supply.

The major oil companies such as BP, Exxon-Mobil, and Royal Dutch Shell have traditionally been fully vertically integrated from the oil field to the final customer. They engage in exploration, production, transportation, refining, chemicals manufacturing, wholesaling and retail distribution of refined petroleum products. Each of these activities and other functional services within the major oil companies could operate as a large-scale standalone business. The major oil companies operate internationally, exploring for crude oil and natural gas in fields around the world and serving final customers in many countries. The vertically integrated structure of these companies suggests that transaction costs of international coordination have been greater than the costs of operating vertically integrated companies. However, as the costs of international communication fall and as trade agreements lower tariff and non-tariff barriers to trade, the major oil companies will contemplate restructuring their organizations.

In some markets, regulation created conditions for vertical integration. Electric utilities such as Pacific Gas and Electric or Commonwealth Edison have traditionally spanned their entire market range: generating power, transmitting power as wholesalers, and distributing power to residential, commercial, and industrial customers. Their vertically integrated structure was in large part due to incentives from regulators who monitored costs and service quality. As electric power markets have opened to competition, forces have driven electric utilities toward vertical divestiture, and competitors offered unbundled power or acted as marketers and brokers without owning any facilities at all. Other companies have specialized in building or operating generation facilities to supply power to competitive wholesale markets. Large vertically integrated energy conglomerates discovered and spun off a variety of new businesses that provide back-office services, billing, marketing, or sales functions as standalone businesses. Competition has spurred companies to recombine and restructure their extensive vertically integrated organizations to accomplish strategies targeted on market segments.[21]

Companies require *flexibility*, that is, a rapid response to changes in production technology, supplier performance, or customer preferences. In some cases, vertical integration into manufacturing or distribution may improve flexibility if companies can achieve internal coordination at lower costs than coordination with their suppliers. However, outsourcing increases flexibility in a number of ways. By relying on outside suppliers for parts, components, and resources, managers can concentrate on customer service. Specialized suppliers can shorten product cycles, allowing the company to adapt to changes in customer preferences. For example, supply chain manager Li & Fung offers faster product cycles to major retail chain stores, allowing the stores to respond quickly to fashion trends and to adjust inventories more accurately.

Moreover, by dealing with outside suppliers, the company has the option to switch to new suppliers who offer more up-to-date equipment, parts, or product designs. In the computer industry, companies such as IBM were vertically integrated, producing their own software, processors, memory, and other computer components. As the industry developed, companies such as Dell relied on outside suppliers to obtain the best software and computer components.

Often, a higher degree of vertical integration creates a company bureaucracy that slows down the company's response to market conditions and hinders flexibility. Managers are busy supervising a larger vertically integrated company and their concern with activities within the organization may take time away from external analysis.

Many CEOs at the largest public companies are struggling to keep pace with organizations that have rapidly expanded through mergers and acquisitions. Referring to the large organizations operated by new behemoths such as DaimlerChrysler and Citigroup, *The Wall Street Journal* asks "Can anyone run these monsters?" Extensive overseas operations, armies of employees, and a deluge of market data put strains on management decision-making at companies such as General Motors and Procter & Gamble. A. G. Lafley, CEO of Procter & Gamble with well over 100,000 employees, acting on concerns about the speed of decision-making, has attempted to shed vertical layers in the company and urged managers to focus on the company's core businesses.[22]

Capital Commitment

Many vertically integrated companies undertake extensive capital investment to establish manufacturing facilities for parts and components, assembly plants, and distribution facilities. For example, vertically integrated petroleum companies invest in equipment for oil exploration and production, tankers and port facilities, refining plants, and retail outlets. Vertical integration does not in itself increase capital investment in an industry but concentrates that investment within large firms that must raise substantial amounts of capital and allocate that capital within the organization. What are the advantages and disadvantages of combining investment in multiple stages of production within a single firm?

If a company can allocate capital within its organization at a lower cost than capital markets, then there are returns to vertical integration. For such returns to materialize, it is necessary both that managers of the company are better informed than investors about the performance of its business units and that managers allocate capital more effectively than capital markets would. This may be the case for start-up companies if managers are effective venture capitalists. Problems arise, however, if internal capital allocation covers up the performance of the company's activities, allowing the good performance of some units to cover up the poor performance of others. This suggests that internal allocation of capital is not a sufficient reason for vertical integration. Accordingly, if a company's business units do not belong together for strategic reasons, internal capital allocation cannot justify vertical integration.

Large conglomerates in the form of industrial groups are common in many countries: the large industrial conglomerates in Japan and Korea, the grupos of Latin America, the hongs of Hong Kong, and the business houses of India and Malaysia. Tarun Khanna and Krishna Palepu suggest that focused strategies may not fit in emerging markets if the large, diversified industrial groups fill institutional gaps that result from incomplete financial markets, imperfect information in product markets, and inadequate legal systems. If financial markets do not function adequately, industrial groups raise capital from earnings and from domestic and international investors and allocate the funds through internal capital markets.[23] However, as financial institutions strengthen in emerging economies, the value added by the formation of such industrial groups is likely to diminish substantially. The economic problems encountered in many developing countries can be traced to inefficient capital allocation within the industrial conglomerates.

Transaction-specific investment provides another potential justification for vertical integration. Some economists and management authors favor vertical integration because of the so-called *hold-up* problem.[24] The problem, also referred to as contract risk, is generally described as follows. If a supplier invests in special-purpose equipment designed to produce a particular part for a customer, known as *transaction-specific investment*, there is a risk that after the investment is made, the customer will renegotiate its contracts and lower the payments to the supplier or equivalently increase performance demands. In this case, the customer is said to behave in an *opportunistic* manner, taking advantage of the supplier's investment commitment. Anticipating this situation, the supplier will avoid making such transaction-specific investment and the result will be that outsourcing partnerships will be inefficient in comparison with in-house production. One potential solution to the hold-up problem is the vertical integration of the supplier and the customer. This allows the vertically integrated company to fully appropriate the benefits of the transaction-specific investment.

However, vertical integration is not necessarily the only solution or even the preferred solution to the hold-up problem. First and foremost, contract law supports the formation and enforcement of efficient contracts. A simple contract clause that requires the customer to pay for any investment that is specific to serving that customer is a very direct way to solve the problem of tailored investment. In addition, the supplier and the customer may write a legally enforceable contract that rules out such renegotiation that would prevent either one of the parties from obtaining the expected benefits from the transaction. Even in the absence of contracts, long-term relationships between buyers and sellers build trust and create incentives for the parties to avoid opportunistic renegotiation. Competitive market forces also reduce the likelihood of hold-ups. Informal networks of buyers and suppliers deter opportunistic behavior because companies need to maintain a trustworthy reputation to continue transacting within the network. Generally, buyers and sellers are restrained from engaging in contractual hold-up by the forces of competition.

Managers contemplating full vertical integration rather than market contracts must choose between alternative methods of expanding the company. The company can vertically integrate by growing the organization internally, acquiring the necessary technology, personnel, and facilities. Alternatively, the company can expand through mergers and acquisitions (M&A). The decision on how to expand the organization is a different type of make-or-buy decision. Managers must compare not only the direct costs of growth with those of M&A, but also the substantial transaction costs of the alternatives.

Knowledge and Market Relationships

Independent suppliers and distributors, by serving multiple manufacturers, can increase output levels and offer greater variety of goods and services than in-house suppliers and distributors. The additional volume or variety of output allows them to achieve greater economies of scale and scope than could specialized internal company units. Accordingly, vertical integration into manufacturing and distributing means giving up some of the cost economies of scale and scope. These additional costs must be traded off against potential cost economies that might arise from vertical integration.

The make-or-buy decision often depends on the company's expertise relative to potential suppliers and subcontractors. If a company has greater technical knowledge than potential suppliers, it may wish to make the parts itself. If suppliers have the necessary experience and R&D, it may be necessary to buy the parts or services.

The company must also decide on where to focus its efforts and attention. Even if the company is more skilled at certain manufacturing tasks than potential suppliers, it may still choose to outsource those tasks if it has a comparative advantage in focusing on other activities. In its make-or-buy decision, the company chooses to concentrate on those activities that yield the greatest relative value added based on the company's skills, while relying on suppliers for other activities. In such a situation, the company may enhance the performance of suppliers by sharing knowledge and production techniques through long-term partnerships and joint ventures.

Managers face a trade-off in obtaining technological knowledge. By relying on external suppliers of designs, software, equipment, parts, or components, companies rely on their suppliers' technological knowledge. An advantage of using external suppliers is that the company can turn to those suppliers that have the most highly developed expertise and the most advanced R&D capabilities. The corresponding disadvantage is that the company does not develop its own expertise and R&D. Managers need to evaluate the transaction costs of dealing with technologically skilled external suppliers. They should compare these costs with the transaction costs of acquiring knowledge through product licensing, hiring trained personnel, establishing of company research facilities, and acquiring companies with technological knowledge.

The role of knowledge in the vertical integration decision is illustrated by Intel's development of the Itanium microprocessor, which has a new architecture able to handle 64-bit processing. Intel, in partnership with Hewlett-Packard (HP), designed the chip for high-end servers used in corporate networks and the Internet. The project began with HP's design of an advanced processor architecture called WideWord. HP already built processors for its own servers and workstations, but HP was reluctant to incur the full costs of designing a new chip, which would run between $1 billion to $2 billion. Accordingly, HP decided to limit its vertical integration into chip design by becoming a partner with Intel and relying on Intel's design capabilities. Further,

HP chose not to vertically integrate into production of the chip since Intel had substantial expertise in chip fabrication and the costs of building a manufacturing facility could exceed $1 billion.[25] IBM, which also builds its own chips, including the Power PC processors for servers and workstations, chose not to vertically integrate into either the design or manufacturing of the Itanium chip. Instead, IBM invested in the Itanium project and opted to purchase the chip from Intel for use in IBM products.[26]

Market knowledge is an important consideration in the decision to enter into distribution. If a company has greater familiarity with customers than outside distributors, it may chose to distribute its own products. When the company decides that its efforts are best employed in distribution, it may choose to exit other activities. A company such as Nike concentrates on design and marketing while outsourcing most of its manufacturing. Conversely, if external distributors have greater knowledge of customers and distribution methods, it may be necessary to contract with those distributors. However, even if the company has greater market knowledge than external distributors, it may still choose to contract with others for distribution if the company can add greater value by focusing its efforts on other activities such as R&D or manufacturing.

One of the advantages of vertically integrating into distribution is a closer relationship with final customers. By relying on other companies for distribution, a company obtains information about customer preferences only indirectly. Some manufacturers survey their final customers to better understand how customers use their products. Companies that rely on outside distributors attempt to overcome this information problem by distributing some portion of their output directly to obtain a window on the market. After an early period of separating their Internet sales strategies from catalog and store sales, many companies moved to an integrated strategy. For example, Staples folded its Internet unit Staples.com back into the company after discovering that sales per customer were much higher for shoppers that visited the company's three channels (in-store, catalog, and online) as compared with customers that visited only the store or used both the catalog and the store.

8.3 *Decentralization of Functions*

Decentralization entails moving responsibility for functional areas out of the corporate central office and into the company's market divisions. By decentralizing, the company reduces the need for hierarchical layers, with heads of market divisions reporting directly to the corporate office. Managers assign functional tasks to the divisions, including marketing and sales, R&D, human resources, finance, purchasing, and operations. Functional tasks tend to be decentralized when the company seeks to make divisions bear the responsibility for their own costs and revenues. Some of the tasks that tend to be performed centrally by the corporate office include finance, public relations, and legal and government relations.

As the basis for horizontal divisions of companies has evolved, so has the structure of the corporate hierarchy. Companies have cut levels from their organizations and drastically pared the ranks of middle management. As they downsize, companies are also decentralizing by shifting tasks from the central office to the division level. Managers increase their effectiveness by delegating authority and sharing information appropriately.

The benefits of decentralization are many. The costly overheads of the central office bureaucracy are reduced. Managers of the firm's divisions are able to respond quickly to market conditions, satisfying customer needs or anticipating competitor strategies. Decentralization avoids the delays associated with sending information and proposals up the corporate hierarchy and waiting for approvals to come back down. The firm's market-based divisions can select pricing and product characteristics on the basis of better information that would reach the central office.

Decentralizing the firm reduces the divisional sharing of central office resources. Shared resources have some appeal because they are the source of economies of scale and scope. Some resources should remain in common or the benefits from operating the divisions jointly diminish to the point at which the firm should be broken up. However, shared resources create free rider problems as division managers seek to capture the services of common resources, while shifting their costs onto other divisions as much as possible. The result is competition between divisions for company resources and difficulties in accurately evaluating the performance of each individual division. The market-based organization increases opportunities for decentralization by allowing the company to shift functional specialties into the market divisions.

The question for a market-based organization is what activities to keep in the central office and what activities to move to the divisions. Decentralizing marketing and sales allows the company's divisions to get closer to their customers. Other activities, such as public relations and legal and government relations are traditionally the purview of the central office. Some aspects of finance are the responsibility of the corporation, which raises capital for the firm's divisions. Personnel and purchasing are well suited to decentralization because each division is best able to judge its needs and build relationships with suppliers. These advantages of decentralization often outweigh the returns to centralized purchasing and the market power that may be achieved by combining purchases.

Decentralization has its trade-offs, which is why many companies continue to cling to a functional organization. Keeping functions as separate departments increases central control by the corporate office, and top managers may be reluctant to give up power. Moreover, the heads of the functional areas may wish to preserve their positions within the organization. The members of functional areas are often trained specialists who enjoy associating with other professionals in their area of expertise. Keeping functional areas within a single unit can yield the benefits of specialization and division of labor. Finally, centralizing functions such as R&D and operations can increase cost efficiency through economies of scale.

These advantages of centralizing functions must be balanced against the benefits of market responsiveness because functional areas are the company's connections to its input markets. Finance connects the company to its sources of capital. Personnel not only manage internal human resources, it is often responsible for hiring in the labor market. Purchasing is the connection to the firm's suppliers of resources and manufactured inputs. The firm's R&D connects it to sources of innovation, including private research organizations, universities, or government labs.

Corporate strategy, especially what markets to enter and exit, is the responsibility of the CEO and the central office. The market-based divisions formulate and carry out competitive strategy in their respective markets. These strategies have important functional components. Thus, implementing strategy entails financial, labor, purchasing, technological, marketing, and sales policies. To the greatest extent possible, the functional activities should be carried out at the division level.

Marketing and Sales

Managers should examine whether the company should operate a central marketing department and maintain a unified sales force or divide these activities among the company's divisions. Marketing and sales tend to be performed by divisions when market responsiveness is important.

A central marketing department offers some advantages by unifying the company's branding strategies and coordinating the company's work with advertising agencies and advertising purchases. The company can obtain some scale economies in management and employment of marketing personnel.

The sales force can represent a significant portion of the company's total cost.[27] A unified sales force can be easier to manage because the company can apply uniform personnel training and compensation plans.[28] The company can determine the overall size of the sales force and coordinate sales calls and assignment of territories. With a unified sales force, the company's business clients do not receive multiple sales calls from different divisions of the company. Sales personnel are able to create bundles from any and all of the company's products and services.

When company divisions offer substantially different products and sell to significantly different types of customers, it becomes desirable to decentralize the company's marketing and sales functions. The company's divisions will likely have different branding requirements and engage in significantly different types of advertising. The company may wish to retain some centralized marketing that addresses corporate image making, such as GE's "We bring good things to life" ad campaign aimed less at consumers and perhaps more at investors, public policy makers, and corporate clients. However, with decentralized marketing the company's divisions can tailor expenditures and branding to their needs and monitor the cost-effectiveness of their marketing efforts.

Decentralization of the company's sales force offers advantages because the divisions can offer sales personnel different types of compensation depending on the types of goods and services the divisions provide. The company's divisions can adjust the size of the sales force and training to meet their specific requirements. The sales force assignments for the divisions can reflect the mix of new and older products, since new products tend to be sold through a company's own sales force while older products are often sold through distributors.[29] If the company's divisions serve different types of markets, the advantages of decentralization are likely to outweigh returns to a unified sales force.

Finance

The company's financial policies include how much capital to raise, the proportion of debt and equity, and the allocation of capital to competing projects available to the firm. The corporation generally centralizes raising capital and allocating its capital across the company's divisions. Thus, the corporate central office is a capital market intermediary, obtaining capital through the issuance of equities and debt, and providing capital to the company's units to fund projects based on their projected earnings streams.

The corporate board performs many of the intermediation tasks, acting in the interests of investors, setting management compensation; choosing dividends; evaluating tender offers; and determining spinoffs, mergers, and acquisitions. The corporate board provides functions that diffuse shareholders would have difficulty performing. With many shareholders, each shareholder has an incentive to "free ride" by relying on other investors to gather information about the firm's performance. The board has a fiduciary responsibility to monitor for shareholders as a group that overcomes the free rider problem. Moreover, the board can observe the details of management's plans without revealing them publicly.

The corporate board is a centralized intermediary for the multidivisional firm. To some extent, the corporate form obscures the performance of individual units in comparison with standalone businesses whose overall performance can be monitored directly. Investors rely on the company's annual and quarterly reports, along with industry data, to infer the performance of individual units. Because the corporation has shared resources and shared costs, breaking out each unit is an imperfect process.

Dividing the company's activities along market lines improves the information available to company managers about the company's market units. That information can then be made available to investors to give them a better picture of how well the company's market units perform. The financing function can be further decentralized by allowing the company's market units to pursue their own financing through the issuance of special equities and debt to fund projects at the division level.

Decentralization of the capital allocation process requires each unit to justify the returns of its investment projects relative to the market cost of capital to the firm. Thus, the competitive market cost of capital should serve as a guideline for the internal allocation of capital within the firm. The existence of clear capital market benchmarks allows the firm's management to decentralize the capital allocation process by letting division managers obtain capital within the firm based on market cost of capital. This process is imperfect in that the cost of capital to the firm is likely to differ from the cost of capital for each division if that division were to obtain funding as a "pure play". The individual product markets the firm serves entail differing amounts of risk, as does the strategy the firm pursues in those markets. Thus, it may be necessary to adjust the internal cost of capital to individual divisions to reflect the different riskiness of the ventures proposed within the firm. However, by using capital costs as a benchmark, the capital allocation process can be streamlined and increased autonomy granted to divisions. The company's central office can rely on the projected earnings of the projects division managers propose, in comparison with the appropriate cost of capital.

In determining their cost of capital correctly, companies need to look not only at the cost of debt, but also at the cost of money raised from equity. So much seems obvious, especially for managers that focus on shareholder value. However, for some firms, it appears that the cost of equity capital has not been used explicitly for strategic planning purposes. Thus, firms that appeared profitable in that their earnings were covering their operating costs and debt costs, were not providing their shareholders with a competitive return. These firms were actually destroying shareholder value. The correct measure of the value of the firm is the present discounted value of cash flow, discounted at the firm's cost of capital. The practical difficulty in applying this measure is determining cash flow from accounting data, which can obscure the results with rule of thumb for such things as internal cost allocation, capital stock depreciation, and amortization of expenditures, which bear little relation to the incremental returns to the company's activities.

Clearly, the full economic costs of capital need to be taken into account:

- The risk-free rate of interest net of inflation.
- The inflation premium.
- The risk premium.
- Transaction costs of raising capital.

In addition, there are short-run constraints on raising capital, so the opportunity costs of scarce capital resources come into play when allocating capital within the firm.

By bringing estimated capital costs within the organization to evaluate individual investment projects and divisional performance, company managers can increase the decentralization of the financing function. Thus, the expansion and entry decisions of division managers are linked to the opportunity costs of the company's investors. The function of raising capital from equity and debt remains centralized. However, by coordinating the use of capital within the firm through price signals, the company can decentralize the allocation of capital.

Management's allocation of capital within the firm represents a distinct internal capital market. For management to undertake this crucial task means that it has some advantage over the market in evaluating investment projects and in monitoring their progress. Managers generally will have higher quality information about the firm's prospects for success in particular projects than outsiders. The company has "organizational capital", that is, firm-specific information about the quality of the company's employees, technology, market and knowledge.

The details of the firm's tactics are known to its managers, including the rate of introduction of new products and processes, the details of marketing and sales plans, and minutia of operating costs. It makes strategic sense for the firm to keep such knowledge from its competitors. Thus, competitive information cannot be shared with potential investors. As a result, management allocation of capital to investment projects within the firm provides a necessary buffer between the firm and the market.

The selection of projects that compete for the company's capital is another management function. Management observes the full set of options, which are not observable to the markets, and long-run strategies that also need not be revealed to competitors. Thus, managers know "the road not taken" as well as the future course of the company. As a result, management is better placed to determine what projects are complementary and what projects are substitutes for existing and future projects.

Managers are able to monitor the performance of *ongoing* projects. Such monitoring requires significant transaction costs, since managers look at progress on a daily or even an hourly basis, including all of the factors that underlie success or failure, such as supervising individual employees, for example. Such detailed monitoring is beyond the capabilities of capital markets, which can only keep track of broader market developments and company earnings reports. Thus, management of internal investment provides some advantages over capital markets, and the firm cannot rely exclusively on market signals to decentralize internal investment fully.

However, the firm should decentralize the tasks of project approval and performance monitoring as much as possible. The means for achieving this decentralization is through performance incentives. The market-based organization creates the right conditions for the design of incentives because the company's divisions are responsible for most of their own costs and revenues, which means that net revenues for divisions provide a closer approximation to economic profit. The greater the extent of functional decentralization, the lower will be the shared costs and revenues. This reduces the need to use accounting allocations of joint costs and revenues to get a picture of division performance. Of course, the presence of joint costs and revenues provides the returns to consolidation that unites the divisions into a single firm. Thus there is a trade-off between the economies from sharing functions across divisions and the monitoring economies from decentralization of functions.

Personnel

The company's employment policies include determining the size of the labor force, the desired skills, the composition of permanent and temporary employees, and the level of compensation and benefits and mechanisms for monitoring and rewarding performance. Employment policy is crucial to implementing strategy because the company's personnel are usually its most important resources. The company's success is limited by the quality and performance of its personnel.

The company's managers can get the best performance out of employees through organizational design. The company needs to create tasks that will implement its strategy. Then, it must devise a hiring policy to obtain employees that are qualified to accomplish those tasks.

Once hired, employees should be matched carefully to the appropriate tasks and provided with incentives to act in the company's interest. Thus, managers need to carefully formulate hiring, internal allocation, and compensation policies.

Succeeding in these tasks requires the personnel function to be decentralized as much as possible. Efficiency gains may result from establishing a centralized human resources department to handle the administrative aspects of compensation and benefits or even from outsourcing some human resource functions. Even so, however, the company's interaction with its labor markets should be carried out at the division level.

The hiring function benefits from decentralization. Incomplete information is a general feature of labor markets, due in large part to the uniqueness of each employee and position offered. Because employees' labor services and employers' labor requirements are far from generic, hiring is a matching problem.

Employers often know little about the pool of available employees, both in terms of skills and wage requirements. Prospective employees often have vague ideas about overall employment opportunities, both in terms of working conditions and contract terms. While some labor market intermediaries exist, including both temporary help agencies and executive headhunters, companies generally establish their own supply networks with the aid of want ads, job fairs, school placement offices, and informal contacts.

Even when an individual company and a prospective employee manage to meet, information asymmetries remain. As employers sort through prospective employees, they attempt to gather information about each individual's ability, experience, and competing opportunities. Employees try to determine the job description, working conditions, future pay and promotional opportunities, the characteristics of fellow employees, and the company's current and long-term profitability. The process of getting acquainted continues long after the employment relationship begins.

Market-based divisions are best placed to determine what specialized labor skills they require. As technology evolves, customer demand changes, or competitors offer new products and services, the division will adjust its hiring strategies accordingly. Division managers have an incentive to learn about labor market conditions and the availability of different types of labor, and those closest to the task at hand should be the ones that recruit for a position. Thus, division managers will have the most accurate information needed to evaluate the company's labor demand and the market labor supply.

Management and Data Systems, an operating division of Lockheed Martin, hired a substantial portion of its workforce in less than one year using techniques that would be familiar to marketing and sales managers. The operating unit sought specialized personnel, particularly computer programmers and engineers. The company identified prospects in a variety of ways including in-house job fairs, employee referrals, and a toll-free hotline.[30] The in-house job fairs involved division managers in the selection process.[31] The company achieved success with employee referrals by compensating employees for hires and encouraging company employees to actively participate in the recruiting effort.[32] Such involvement illustrates the importance of decentralized hiring in screening employees. Management and Data Systems advertised its toll-free number on cable television and in newspapers, and tracked the responses to its ads.[33] The company kept track of potential employees using the same type of approach that companies use to monitor potential customers. In this manner, the division acquired highly specialized information about its job markets.

The company's market-based divisions are able to gather specialized information about job markets. Accordingly, the divisions can perform certain human resource management activities better than the central office. The company's central office can implement simple incentive plans such as the allocation of stock options that give employees a stake in the future of the company. Companies such as Starbucks Coffee and Lucent Technologies widely distribute such stock options. The central office has a role in determining some basic aspects of incentives, such as company compensation rules, but it should give divisions discretion in hiring and incentive design as well.

Purchasing

Purchasing is a vital function for companies, providing access to supplier markets just as marketing and sales provide access to customer markets. Purchasing refers to all supplier relations, including short-term buying, long-term contracts, and outsourcing. There is no general solution to the location of the purchasing functions within the company, although decentralization of purchasing has a number of distinct advantages.

Centralized purchasing yields economies of scale in purchasing systems, search for suppliers, and cost savings from increased buying power. This suggests that for generic inputs and materials at least, centralized purchasing may offer advantages. However, for generic goods and services, the purchasing function itself can be outsourced and the company can rely on wholesalers or other market intermediaries. If centralized handling of goods or processing orders offers substantial economies of scale, the company's purchasing systems can be consolidated, including warehousing, electronic data interchange, and order processing, as at Wal-Mart. Even with consolidated warehousing and data processing systems, however, the company can allow its market-based divisions autonomy in selecting suppliers.

Decentralized purchasing allows the company's divisions to tailor their purchasing specifications to division requirements and to adjust the amount purchased to sales and inventory variations. Decentralizing purchasing has the additional advantage that market-based divisions can forge closer supplier relationships, adapting contracts with suppliers to meet their needs.

It may be necessary to forgo some economies of scale in purchasing and buying power to provide divisions with the flexibility to switch suppliers or to use multiple suppliers for the same goods and services. Decentralizing the purchasing function for key inputs keeps division managers in touch with supply market conditions. They can monitor input prices and observe at close hand the division's arbitrage between supplier and customer markets. Centralized purchasing may make input purchasing appear as a generalized cost to the company, without making division managers aware of input price fluctuations.

Technology

The centralization trade-offs are particularly acute in technology. Centralization of R&D provides many benefits, including economies of scale and scope in the research function itself. The company can effectively monitor developments in basic science and engineering taking place outside the company. The company's research lab is an intermediary gathering information in the scientific and technological marketplace. The output of the company's research labs becomes a resource throughout the company that is shared across its many divisions. IBM or AT&T have traditionally maintained high-powered research labs that have produced scores of patents and even Nobel prizes.

However, the company's problem is to bring the basic research to market. Innovations in the company's central labs must translate into cost-saving process innovations or attractive product innovations. There is no guarantee that significant innovations produced in the company's central lab will attract the attention of division managers, who can then participate in the product development process. Conversely, the central lab may not work on problems that address the needs of the company's customers. Basic research is difficult to channel in any case because abstract problems may yield surprising payoffs and research topics with great appeal may turn into dead ends.

Centralization can stifle innovation by attempting to control the rate and direction of research. Moreover, centralized research can be isolated from the company's marketing, sales, and operations, not to mention from its customers. For this reason, companies sometimes resort to "skunk works", separate and sometimes secret divisions of the company charged with developing a new product or production process. An example of the value of decentralized R&D is IBM's Silverlake project in the late 1980s (see Exhibit 8.3).

Exhibit 8.3 IBM's Silverlake Project

A troubled computer development and manufacturing division of IBM located in Rochester, Minnesota, home of the Mayo Clinic, set out to build a new computer.[34] The division had encountered problems in its market, mid-range computers for business, with the entry of competitors including Digital Equipment Corporation (DEC), Hewlett-Packard, Wang Laboratories, Data General, Tandem, NCR, Nixdorf, and Fujitsu.[35] DEC in particular offered a line of computers that were compatible; that is, they were able to use the same programs and share information. In contrast IBM offered five lines of mid-range computers, all of them incompatible.[36] An effort to create a single advanced mid-range line, code-named Fort Knox, ended in failure as IBM tried to develop the machine at multiple research sites simultaneously.[37]

IBM Rochester's organizational structure was divided between its development lab and manufacturing operation. According to Roy A. Bauer, Emilio Collar, and Victor Tang, who were major players in the Silverlake project,

"The biggest problem about the structure, however, was that we were *part* of a business, not *the* business itself. We simply saw ourselves as functional entities — working parts of a larger whole. We designed computers. Or we engineered hardware. Or we manufactured machines. Because we didn't own the business ourselves — because we didn't see ourselves as a unit accountable to the market — we gave little thought to the strategic factors so critical to the success of a *business entity*. We paid scant attention, say, to our channels of distribution or to the independent software vendors who created so much of the applications software to be run on our machines. We didn't pay enough heed to our customers either. It was someone else's job — sales and marketing — to worry about that."[38]

Faced with choosing among literally thousands of possible product features, the project eventually turned to an analysis of the product features that its customers desired.

The focus on customer markets made the research and development project an eventual success. The project created a new computer, the AS/400, in two years, half the company's prior development time.[39] The computer became one of IBM's most successful products and was instrumental in IBM Rochester winning the Malcolm Baldrige National Quality Award.[40]

8.4 *Delegation of Decision Making*

The vertical structure of the company also includes the authority relationships in the organization. *Authority* refers to the rights of managers and supervisors to direct the activities of others within the organization. In turn, employees whose activities are directed by managers and supervisors have a responsibility to carry out their assigned tasks. *Responsibility* refers to the expectation that the employees will perform the tasks and be accountable for the quality of that performance. All organizations are *hierarchical* in that they consist of relationships in which managers exercise authority and employees have responsibilities.

An important aspect of the vertical structure of the organization is the number of layers of authority. There may be one or multiple layers of authority in the corporate office, with division managers reporting to the central office. Within the company's divisions, multiple layers of authority often exist. The number of layers of authority determines the lines of communication within an organization because decision-making involves assignments that travel down the organizational ladder and performance information that travels back up the organizational ladder.

An organization with many layers of authority is generally said to be *bureaucratic* because it has longer lines of communication that potentially reduce the speed of decision-making and adaptation to market developments. The cost of additional layers of authority is the potential for delay and inflexibility. The benefit of additional layers of authority, on the other hand, is that managers have fewer people reporting to them, increasing their ability to assign tasks and monitor performance. The number of layers of authority that the manager chooses thus reflects a trade-off between the organization's ability to make decisions quickly and the organization's ability to perform tasks.

Delegating Authority

Decentralization involves more than moving functional activities into the company's divisions. It also requires delegation of authority from the central office to division managers and, in turn, increased delegation of authority from division managers to other managers and employees. *Delegation of authority*, sometimes referred to as empowerment, requires giving employees the necessary information to complete their assignments and the necessary independence for effective decision making. The advantages of increased delegation are greater flexibility and market responsiveness of the organization and a potential reduction in layers of authority.

Classical management was concerned mainly with the establishment of a hierarchy and the allocation of responsibilities and authority within the hierarchy. Modern management shifts attention to the question of how individuals within the hierarchy can be motivated to respond to managerial directives and to further the interests of the firm. Chester Barnard pioneered the notion that managers derive their authority from the acceptance of that authority by employees.[41] He observed that "the individual is always the basic strategic factor in organization" and therefore "the subject of incentives is fundamental in formal organizations."[42] Barnard stressed the role of managerial leadership and concluded that "the most general strategic factor in human cooperation is executive capacity."[43]

Increased delegation of responsibility and authority down the organizational hierarchy allows the company's managers to widen their scope of supervision and flatten the organization by removing layers of management. The central office grants greater independence to managers of the company's market-based divisions, which do the same in turn for managers of operating divisions. Within operating divisions, the delegation process extends to individual employees. The tasks of employees generally are either oriented toward internal operations or external market relations.

Employees have many duties to the company, including remaining loyal to the company's interests, protecting confidential information, following instructions, providing information that affects the company's interests, avoiding negligent acts, and accounting to the company for receipts and expenditures.[44] The company also has duties toward its employees, including the duties of a principal. The company must also be loyal and honest toward its employees.[45] The company has a duty to compensate its employees, to reimburse employees for their reasonable expenses, and to hold employees "harmless for liability that may be incurred while the agent is within the scope of employment."[46] The employment relationship, then, is characterized by mutual trust and obligations.

Many of the employees in sales, purchasing, finance, and human resources represent the company's interests in market transactions.[47] These employees are market agents representing the interests of the company and acting as principal in transactions with third parties.[48] Purchasing employees negotiate contract terms with the company's suppliers. Sales personnel negotiate prices and contract terms with the company's customers. Human resource personnel negotiate employment contracts on the company's behalf. Many employees performing intermediate internal tasks serve the interests of internal customers.

Managers should recognize that many of the company's employees are market agents and as such require greater authority to respond to market conditions. Accordingly, they should give the employees greater responsibility to act in the company's interest. Employees can respond to information provided by customers or suppliers and tailor transaction terms accordingly. The costs of monitoring are lowered when agents have greater autonomy and managers can widen their span of control. Transaction costs are reduced when the employee does not need to have each transaction approved by a manager if the terms fall within specified guidelines.

Delegation of responsibility is an essential aspect of reducing the company's transaction costs. Delegating authority to individual sales personnel allows them to respond to customer needs. Southwest Airlines delegates significant responsibility to its personnel to address customer problems, incorporating this approach in the company's mission statement: "The mission of Southwest Airlines is dedication to the highest quality of Customer Service delivered with a sense of warmth, friendliness, individual pride, and Company Spirit."[49] Delegating authority to individual purchasing personnel allows them to take advantage of procurement opportunities and to respond quickly to supplier concerns.

Delegation creates the flexibility and responsiveness that are necessary to effective market making because employees can monitor market developments and respond appropriately. The company can obtain transaction cost savings from greater delegation of authority to sales and purchasing personnel. Along with better coordination of sales and purchasing activities, lowering transaction costs helps give the company a transaction advantage.

Delegation is the essential part of leadership and strategy implementation. Following a series of basic steps, managers

- Share goals and strategies.
- Communicate the required tasks.
- Grant independence.
- Inform.
- Monitor progress.
- Reward performance.

To share goals and strategies, managers inform employees *what they are trying to do and how they are trying to do it*. Knowing the company's goals helps enlist the support of employees and will be useful later on when initiative is required. Outlining the company's general strategies not only motivates employees but also stimulates creativity.

Managers communicate the required tasks by specifying *what they want employees to do and how they want them to do it*. Often, managers express disappointment that employees do not perform as desired when the managers never conveyed clearly what they wanted the employees to do.

Having specified the tasks at hand, managers need to get out of way. Giving employees the independence to carry out the tasks leaves room for initiative and creativity. New ways of doing things are the payoffs from delegation. Freedom of action is also an essential part of the trust that must accompany the relationship between the company and the employee.

If employees are to act independently, they must have the tools to get the job done. The required tools include not only the necessary authority and resources, but also the capacity to gather and process information. Employee innovation and efficient performance are conditioned by the quality of the information available. The company may need to reveal revenue, cost, or other data that traditionally has not been shared with employees. In some cases, the employees must be given the mandate to gather the required information themselves.

Granting autonomy does not eliminate the manager's responsibility; rather, it intensifies it. Managers need to substitute control with increased monitoring of progress toward the implementing strategy. Without waiting until the final completion of the tasks, some intermediate interaction and guidance can help. It may be simply answering questions, formulating progress reports, and reviewing the assigned task. Updates are costly and should not delay the project, but informal intermediate review is valuable because it enhances communication and reduces misunderstandings about the final task. Managers should recognize progress towards the goal.

Upon completion of the tasks, performance should be rewarded. Rewards can be financial and based on established incentive programs. Additional rewards can be devised for particularly successful projects. The manager must adjust rewards to reflect the employee's contribution to value, while recognizing the effects of serendipitous market forces.

Exhibit 8.4 Texaco

At Texaco's huge 100-year-old Kern River Oil Field near Bakersfield, California, management faced the possibility of abandoning oil fields that had become uneconomic. The value of the oil in the ground was set by the world price of oil. If the cost of production exceeded that price, the fields would have to be abandoned. Texaco discovered oil by improving its management; that is, it learned the value of delegation. The manager of the oil field, Stephen J. Hadden, reversed the long-standing command-and-control approach to oil production at Texaco:

> His first move was to tell workers and managers that sharing ideas, not protecting turf, should become their No. 1 priority. He created

(Continued)

Exhibit 8.4 *(Continued)*

62 teams of 9 members each, including foremen and production engineers, technicians, geologists, field workers and even outside contractors. Many were assigned to two or more teams. Each group was responsible for a certain number of wells.[50]

Along with delegation of authority to the work teams, the company began sharing information within the division and across the organization:

On the job, employees were given new powers to act on their own and the freedom to communicate with other departments without management approval. They also had unrestricted access to a new central computer that stored data on all aspects of operations, from each well's history to underground rock formations. Information that previously took days or weeks to obtain could now be retrieved in hours or minutes. Mr. Hadden also established a computer link with a Texaco research laboratory in Houston.[51]

Hadden described his strategy as "stay out of the way and give the people the resources to get the job done."[52] Morning meetings with the teams led to the development and application of a new method of enhanced oil recovery using steam injection. The result was a turnaround in the fortunes of the field, with increased worker productivity and more oil recovery: "Kern River has increased its recovery rate to 66 percent of all oil from 50 percent, and expects to push it up to 80 percent."[53] Alfred C. DeCrane, Jr., the chairman and chief executive of Texaco, told the *New York Times*, "Bakersfield is a microcosm of Texaco activities around the world."[54]

Teams

Managers also delegate authority by assigning responsibilities to *teams*, which are generally groups of employees from different functional areas of the company. Teams are given the responsibility for completing general tasks. Teams are the (hopefully) more productive cousins of committees, and thus they share the problems of committees but also some of their advantages. Committees are bodies that act under delegated authority of a larger group. The delegation of authority to a team can increase the productivity and flexibility of the larger group. Numerous companies, both large and small and in a wide variety of businesses including Boeing, Ford, General Electric, Monsanto, and Proctor & Gamble, rely on teams.

One of the keys to the success of IBM's Silverlake project was the creation of cross-functional teams consisting of engineers, programmers, marketers, and business strategists.[55] Cross-functional teams carry forward the process of decentralization by pushing functional areas further into the firm's market-based divisions, moving individual functional specialists to the level of specific projects.

Functional boundaries are lowered because individuals having different functional specialties and experience work shoulder to shoulder. The advantage of this approach is to make individual employees even more sensitive and responsive to market segments or even to individual customers the team serves. The customer service team brings employees even closer to customers since each team member is aware of who the customer is for the project at hand.

The value of teams is that they are composed of specialists, so teams take advantage of the gains from specialization and division of labor. Dividing and allocating tasks within the team enhances productivity as each member brings learning and experience to their area of responsibility. The traditional approach to obtaining the benefits of specialization and division of labor emphasized separation of personnel. The functional organization divides employees by specialty, with managers assigned to each functional area. Functional managers report to other managers who must coordinate the functional areas to achieve the overall task. Typically, employees in functional areas have little conception of the overall project, and operations and marketing thus may have little interaction. Within the functional areas, tasks are subdivided for the purpose of even greater specialization and division of labor. The assembly line, with each employee performing a mechanical task with endless repetition, carries this approach to an extreme.

Teams provide a means for employees of diverse functional backgrounds to interact directly, improving coordination and reducing the need for managerial intervention. Thus, the marketing manager and the engineering manager need not have meetings to coordinate the interaction between personnel in their respective departments. Marketing and engineering personnel that report to the same manager can be assigned to the same team and a single manager can supervise their interaction. For companies that rely heavily on teams, eliminating functional managers has advantages because it avoids the inevitable conflict with project managers that occurs in a matrix organization. Employees with different functional backgrounds can in principle report to the same manager.

By reducing the number of supervisors, teams allow increases in middle managers' spans of control. The number of middle managers can be reduced, further reducing the need for centralized control. The individuals within the teams and the teams themselves can report directly to middle management, thus shortening the lines of communication. Delegation of project management to teams requires increased sharing of information between management and employees, just as in the case of delegation of authority to individual employees. Teams thus reduce the role of management in data processing and, correspondingly, managers' strategy-making increases in importance.

Teams are not a new idea. Larger groups of people, whether legislatures, armies, or business organizations, have long formed smaller "working groups" or committees to focus on a specific project. The working group or committee frees up the rest of the larger group to work on other assignments. Moreover, the committee improves communication and reduces the time required to reach a consensus since fewer people are involved in the negotiation.

Teams work in the same way. Rather than involve the entire division or the company in each project, productivity can be increased by assigning a small team to each project. The division can handle multiple projects simultaneously by creating multiple teams. Teams become knowledgeable on their own project since team members can focus on the task at hand.

Because individuals can serve on more than one team simultaneously, the division can handle a multitude of projects by creating many teams. The company gains flexibility by creating teams with overlapping memberships. For example, suppose that a division has three employees. The division might create three teams of two people each, with each person serving on two teams. By assigning a different task to each of the three teams, the division handles three tasks, while each individual need learn only two tasks. Such an approach is preferable to assigning each task to different individuals if working in pairs raises productivity. It is also preferable to having all three tasks handled by the full group if three teams of two individuals are more productive.

Assigning of individuals to multiple teams increases productivity when individual members' skills and abilities complement those of other team members. It is also a way to deal with "slack" in the usage of a productive employee. Assigned to only one project, an individual's role may be very limited by the task at hand. For example, the input of an engineer in a production process may only occur at discrete intervals, resulting in employee "downtime" or idleness between projects. Through assignment to multiple projects, employees can schedule inputs so they can be productive in one project during periods of time when they are not needed on another project.

Teams can be reassigned to new tasks as they appear. Thus, the company is able to handle many projects of varying duration, creating a team that is matched to each project as it appears. As a project is concluded, the team is disbanded and its members are reassigned to new or existing projects. Such flexibility and rapid responsiveness is difficult to achieve with a rigid organizational structure. The standard response would be to assign components of the projects to functional areas (marketing, sales, production, design), with no assurance of coordination between the functional specialists and the possibility of delayed completion since the project can only be finished as fast as the slowest component. Assigning the project to a team composed of specialists from each functional area unites responsibility for the project, thereby increasing coordination among functional specialties. This approach should increase the speed of completion.

To achieve this flexibility in job assignments suggests that the company's division should not establish functional units so as to avoid the problems of the matrix organization, mentioned earlier. The matrix organization created dual lines of authority as project managers competed for control with functional managers. The solution is to eliminate functional managers so a group of employees composed of different specialties reports to the same general manager. The general manager then can draw from this functionally diverse pool of talent in creating teams. This need not imply that functional specialists are interchangeable in the team-assembly process. Rather, they are an internal storehouse of talent that is well understood by the company. The skill of managers in combining those individual skills into teams and matching those skills to projects enhances productivity.

The notion of a multifunctional talent pool within the company creates an internal market for skills and expertise, with individuals finding employment in various company teams. The traditional notion of an "internal labor market" meant rigid job descriptions, separation of functional areas, career ladders and many gradations in the corporate hierarchy. An internal skills market is the opposite, with flexible job descriptions, elimination of functional departments, and reduction in layers in the corporate hierarchy. Management must recognize that the opportunity cost of assigning an individual to a team is that it takes away time and energy that might be devoted to other projects. Thus maximizing productivity requires efficient allocation of skills within the company. There will be incentives for teams to lobby for access to talented members. The company must solve the difficult problem of allocating personnel to those teams where their incremental contribution is greatest.

By delegating authority to teams, managers can speed up the organization, avoiding the time-consuming communication with the corporate office to address operational details. At Monsanto's Pensacola, Florida, plant teams have taken on responsibilities for hiring, purchasing, job assignments, scheduling, production, and maintenance.[56] The *New York Times* reported a dramatic illustration of the benefits from Monsanto's delegated authority:

> Recently, when thunderstorms threatened, some workers decided to
> cut the plant off from the utility grid and switch key production lines

to Monsanto's own generators. The storms soon knocked out power lines — but the plant avoided a shutdown that would have clogged machinery and cost hundreds of thousands of dollars.[57]

The production team members in this case were closer to the problem at hand. Without the authority to act, workers seeking a go-ahead from management might have encountered delays in communication and decision-making that would have proved costly for the company.

The same is true for teams assigned to handle a customer account or a market segment. A team operating with delegated responsibility can make the rapid decisions necessary to address customer concerns. Rapid response translates into increased sales as customer demand for the company's services increases. Decentralization also enhances customer retention when customer complaints require a quick response.

Teams are not a panacea. To understand their usefulness, it is necessary to know their limitations. An old quip offers an insight on teamwork: What is a camel? The answer: a horse designed by a committee. Everyone is familiar with criticisms of committees: compromise solutions that please no one, endless meetings that fail to be productive, free riding on other members of the group, agenda manipulation by the committee chairperson to get a favored outcome. Committees substitute politics for productivity.

As with committees, teams may choose compromise solutions to problems that may be inferior to less popular solutions. Groups can be irrational, with logical inconsistencies arising out of voting or bargaining procedures. These potential problem areas suggest that each team should have a *single leader*, a manager to provide direction. Just as orchestras cannot interpret a piece of music without a conductor or a football team cannot choose a consistent strategy without a coach, so teams require leadership. As with committees, teams may waste time in discussions and negotiation. The self-directed work team may be useful when the project is well-defined, but it generally requires supervision. The group manager can take responsibility for decisions and choose the group's direction, thus reducing the need for costly negotiation.

Yet, advocates extol the benefits of *self-directed work teams*. How do teams make decisions if they are directing themselves? A closer look at the establishment and operation of teams provides the answer. The term is something of a misnomer. Teams are "self-directed" in the sense that management delegates authority to the team. That does not mean that teams are rudderless. Jack D. Orsburn, Linda Moran, Ed Musselwhite, and John H. Zenger explain that managers delegate tactical decisions to company teams:

Executives focus on strategic decisions, and managers and team facilitators clear the way for motivated workers to exceed ambitious teams standards they've set for themselves.[58]

Orsburn and his associates explain that teams go through five self-explanatory stages: start-up, state of confusion, leader-centered teams, tightly formed teams, and self-directed teams. They estimate the time period as six to nine months each for the first two stages and six to twelve months each for the next two stages, so that a minimum of two years goes by before the self-directed work team becomes fully established.

Such delays can be a problem for a company that hopes to increase its reaction time by continually creating new teams to deal with changing markets. One solution is to standardize the process of establishing, motivating, and coordinating teams, so the lessons employees learn are not specific to the team to which they happen to belong. The company should develop routines for establishing team leaders.

Orsburn and associates also observe that as the team emerges from the period of confusion, "most teams want one person *of their own choosing* to carry the flag."[59] They recommend that planners and managers encourage the rise of "strong team leaders," either appointing a supervisor, facilitator or team member at start-up or helping teams develop the selection criteria for an internally chosen leader.[60] The team leader can be a rotating position, and it can replace many traditional supervisory roles. With team leaders in place, company supervisors and middle managers take on different roles: acting as a liaison between teams and the organization, coordinating interaction between teams, and providing strategic direction to teams.

Team leadership presents some difficulties. Monsanto's experience was typical in that managers had trouble letting go of their authority while team members were reluctant to take on additional managerial tasks:

> Sometimes, no one wants to be team leader. Some teams have rotating leaders, others simply give the job to the most experienced person who does not firmly say no. The leader usually receives a small pay premium, but that is insufficient incentive in teams whose members have difficulty sharing responsibility.[61]

Managers, if anything, should understand that managing should be rewarded financially. Asking team members to carry on their regular tasks, while stepping forward as team leaders without additional compensation, is unlikely to succeed. With additional compensation, however, the team leader becomes simply a supervisor by another name. The internal leader quickly has an incentive to represent the viewpoint of upper management to the team members rather than representing the interests of the team members to upper management. The benefits of decentralization are attenuated as the team leader becomes another bureaucratic layer in the organization. By limiting the term of the team leaders, and rotating leadership within the team, the company can control the tendency to reinvent the hierarchy.

Even with internal leadership, group projects are notorious for free riding. Free riding occurs when individuals deliberately reduce their work and shift their responsibilities onto other team members. When members of a group are not rewarded individually, but only for the performance of the group, they have an incentive to work less and to rely excessively on the others in the group. Anticipating this, everyone in the group will shirk to avoid being taken advantage of by the other members. The result is that performance of the group suffers and everyone earns a lower reward. The problem is how to overcome these incentives to shirk. Group members are expected to police each other, making sure that no one shirks at the expense of the group. Self-policing creates additional work for group members, however, and can lead to interpersonal conflicts and resentment.

One way to resolve the problem of free riding is to base group rewards on some performance standard. The group rewards are earned only if it is able to vault over that standard. This approach to rewarding performance makes shirking costly since any individual member's shirking may be enough to make the group fall below the standard. As a consequence the contribution of each member of the group counts a great deal — the group is only as strong as its weakest member. Again, group managers are needed to monitor performance within each group and to take account of market conditions and other external effects on performance. Managers should take these factors into account when they administer the incentives offered to group members, just as they administer rewards for individual employees.

When many groups are engaged in the same task within the firm, the standard of performance can be set by tournaments, that is, competition between groups with rewards going to the best performers. Such an approach will encourage performance within groups, but it may discourage cooperation between groups. Again, managers are needed to coordinate group interaction.

Group leaders are important to the success of teams. Management should clearly specify the authority of the group leader and the mission of the team. Even if teams are run democratically, the team "chairperson" can still control the group's decision by manipulating its agenda. Within limits, choosing what is to be voted on and the order in which votes are taken can alter the outcome of votes. If the chairperson sets the agenda, then he or she can determine the group's final choice. Group members who understand that a different order of votes can produce a different outcome recognize and resent management agenda manipulation because only the votes are democratic, not the choice of the agenda. In contrast, employees may accept a well-defined authority relationship more easily than a fictional democracy. Moreover, the process of voting and discussion that accompany agenda setting and voting can be time-consuming, and consensus building can be a necessary but costly activity for a group leader. Thus, there is much to be said for managerial authority and leadership in running groups within a company.

Teams should not be a method of papering over employee problems, with more talented employees "carrying" less talented ones. Rather, assigning employees to teams should depend on where the employee brings the highest value added. Thus, the skills of the team members should be complementary to obtain the highest productivity. Rewards should balance group achievement and individual contributions. These rewards can include bonuses as well as future assignments and responsibilities.

Simply forming teams does not eliminate the need to motivate employees. Just as individuals can fail to act in the interests of the company so teams can fail as well. The team's members can simply be misguided or misinformed of company objectives, or they may cooperate in deviating from company aims. As with individual employees, aligning the interests of teams with the strategic objectives of the company requires a combination of economic incentives for team performance and management oversight.

Exhibit 8.5 General Electric Medical Systems

Just as individual employees must report to a single manager, so teams should not face dual or even multiple lines of authority. Managers at General Electric's medical systems division rediscovered this basic principle when it assigned two teams of engineers, one in the United States and one in Japan, to jointly develop software for ultrasound devices. The U.S. team in Waukesha, Wisconsin, consisted of 30 software design engineers, while the team in Hino, Japan involved 13 engineers skilled in ultrasound devices who were joint venture employees of Yokogawa Medical Systems.

For the first two years of the project, the U.S. and Japanese teams reported to managers in their home country:

> The local managers pushed the teams to emphasize features that would make their products popular mainly in their own markets. There was duplications — for example, each team developed graphics software — and they were frustrated by their inability to have critical questions answered by management and the members of the other team.[62]

Problems created by conflicting missions, or ill-defined missions, can be exacerbated when teams are involved.

(Continued)

> ## Exhibit 8.5 (*Continued*)
>
> For team members to work smoothly together, management must clarify the reporting relationships with the company:
>
> Last winter General Electric changed its structure so that both teams would report to a general manager in Waukesha who has the ear of top management. "Now we have a single place to go to get an answer," said Tracy Accardi, the leader of the Waukesha team. "We have one general manager who travels to Japan and focuses on global solutions."[63]
>
> After GE granted greater autonomy to the team and clarified the lines of authority, successful collaboration between the two groups eventually developed.[64]

8.5 Overview

Managers determine the vertical structure of the company based in large part on consideration of transaction costs. In evaluating the need to vertically integrate into production or distribution activities, managers compare the costs of administering these activities within the organization and the transaction costs of obtaining these services through market transactions. If it is costly to use the market in comparison with internal administration, managers expand the organization. If market transactions are relatively less costly, managers should choose to outsource many types of activities. By outsourcing activities, companies benefit from flexibility and competition between external providers of goods and services.

Managers improve the organization's execution of the company's strategy through decentralization. The organization lowers its transaction costs and administrative costs by shifting responsibility for functional tasks to the company's divisions. The organization further lowers its transaction costs through delegation of authority to the company's employees. The company enhances the effectiveness of the company's employees that act as representatives in the marketplace by granting them greater authority and requiring greater responsibility in carrying out transactions with customers and suppliers.

Questions for Discussion

1. You are the manager of a department of an insurance company. You must choose between establishing an information technology department to operate the company's computer system or contracting

with a service provider to operate the computer system. What are the factors that you must consider in making the decision?

2. Wal-Mart asks its vendors to simplify their pricing and delivery procedures. One of those vendors, Procter & Gamble, reduces inventories and paperwork by basing its product shipments on the rate of product withdrawal from the retail chain's warehouse.[65] Explain the benefits to Wal-Mart and Procter & Gamble from these actions. Do these actions give Procter & Gamble a transaction advantage over other vendors?

3. Media companies such as Disney and AOL-Time Warner are both creators of content and distributors. These companies also distribute content made by other companies and sell their content to other distributors. What are the advantages and disadvantages of making movies that are shown on your own television channel or cable television system?

4. Many types of manufacturers rely on wholesalers to distribute their products. How does using a wholesaler allow a manufacturer to lower its costs of sales and marketing?

5. Many types of retailers rely on wholesalers as a source of their products. How does using a wholesaler allow a retailer to lower its transaction costs of purchasing?

6. A European manufacturer wishes to expand its sales into Brazil. The manufacturer must choose between establishing a sales office in Brazil and relying on an export intermediary to handle its sales. What factors should the manufacturer consider in choosing between these two alternatives?

7. Computer manufacturers such as Dell rely on other companies to assemble some of its computers. What are some reasons why Dell might outsource assembly?

8. An entrepreneur has a new concept for a restaurant chain. The entrepreneur must decide whether to build the chain of restaurants or to offer franchises to independent owners. What are some of the factors that enter into the entrepreneur's choice between the two alternatives?

9. Ace Hardware is a cooperative owned by over 5,000 hardware stores. Each store is independently owned. Ace Hardware acts as a wholesaler providing both branded products and house brand products to the individual stores, as well as joint marketing.[66] What are some of the advantages and disadvantages of this ownership arrangement in comparison with independent stores served by wholesalers and in comparison with a company that owns a chain of hardware stores?

10. Consider Ace Hardware as described in the previous question. It operates in over 62 countries on six continents. What are the advantages and disadvantages of the company's ownership structure for international operations?

11. Ace Hardware stocks over 65,000 products. It advertises that personnel at its stores are "helpful hardware folks." Why is the emphasis on being helpful important in the hardware business? How might the company's ownership structure make its personnel more helpful in comparison with alternative ownership structures?

Endnotes

1. The company's strategy determines what markets the company will serve, which in turn guides the company's choice of divisions of the firm, its *horizontal structure*. The company's strategy also requires carrying out functional tasks. Those tasks that are carried out within the organization determine the company's *vertical structure*.
2. See the discussion in Chapter 11.
3. Ronald H. Coase, "The Nature of the Firm," *Economica* 4 (1937), p. 395.
4. Ibid., p. 396.
5. Franklin R. Root, *Entry Strategies for International Markets* (San Francisco: Lexington Books, 1994).
6. Keith Bradsher, "G.M. Plans to Spin Off Parts Division," *The New York Times*, August 4, 1998, p. C1.
7. http://www.solectron.com/about/index.html.
8. Joan Magretta, "Fast, Global and Entrepreneurial: Supply Chain Management, Hong Kong Style, An Interview with Victor Fung," *Harvard Business Review*, September–October 1998, pp. 102–114.
9. Ibid., p. 112.
10. Ibid., p. 111.
11. Ibid., p. 125.
12. Ibid., pp. 126–130.
13. Timothy J. Muris, David T. Scheffman, and Pablo T. Spiller, *Strategy, Structure and Antitrust in the Carbonated Sort-Drink Industry* (Westport, Conn.: Quorum Books, 1993), pp. 126–130.
14. Ibid., p. 128.
15. Ibid., p. 129.
16. http://www.bertelsmann.com/facts/facts.cfm?id=2018.
17. http://www.rtlgroup.com/corporate/.
18. The information about Disney in this paragraph is drawn from Seth Schiesel, "Where the Message Is the Medium," *New York Times*, July 2, 2001, Business Section, p. 1.
19. Shawn Tully, "The Modular Corporation," *Fortune*, February 8, 1993, pp. 106–113.
20. The account relies on company information about the joint venture at www.tbg.co.uk. On Anda, see also http://china-window.com/Shenzhen_w/business/company/sz2/ad/ad.html.
21. Allen R. Myerson, "A 20,000 Percent Bounce: Now That's Volatility," *New York Times*, August 23, 1998, p. 4; and Michael Parish, "Proposed

Transmission Companies Could Reward Investors," *New York Times*, August 23, 1998, p. 4.

22. Matt Murray, "Critical Mass: As Huge Companies Keep Growing, CEOs Struggle to Keep Pace," *The Wall Street Journal*, February 8, 2001, p. A1.

23. Tarun Khanna and Krishna Palepu, "Why Focused Strategies May Be Wrong for Emerging Markets," *Harvard Business Review*, July–August 1997. Khanna and Palepu also observe that the industrial groups recruit, train and allocate labor services in economies that do not have well-developed labor markets.

24. The hold-up problem is said to occur when buyers and sellers that share the surplus from exchange do not obtain the full returns to their investment. The economics literature on contracts has tended to focus on transaction-specific investment within bilateral relationships. See for example Oliver E. Williamson, *Markets and Hierarchies* (New York: Free Press, 1975); Oliver E. Williamson, *The Economic Institutions of Capitalism* (New York: Free Press, 1985); Benjamin Klein, R. G. Crawford, and Armen A. Alchian "Vertical Integration, Appropriable Rents, and the Competitive Contracting Process," *Journal of Law and Economics* 21 (October 1978), pp. 297–326; and Oliver D. Hart and John Moore "Property Rights and the Nature of the Firm," *Journal of Political Economy* 98 (December 1990), pp. 1119–1158. When contracts are incomplete or non-binding, renegotiation leads to contract hold-up, which creates incentives for under-investment as compared to the outcome that maximizes joint gains from trade for the buyer and seller. Legal contract rules, social norms, and market reputation are likely to mitigate potential hold-up problems.

25. David P. Hamilton, "Circuit Break: Gambling It Can Move Beyond PC, Intel Offers a New Microprocessor," *The Wall Street Journal*, May 29, 2001, p. A1.

26. Barnaby J. Feder, "New Microchip Design Is Introduced by Intel," *New York Times*, May 30, 2001, p. C6.

27. Gary L. Lilien and David Weinstein, "An International Comparison of the Determinants of Industrial Marketing Expenditures," *Journal of Marketing*, Winter, 1984, pp. 46–53.

28. On the design of sales force compensation, see Anne T. Coughlan and Subrata K. Sen, "Sales Force Compensation: Theory and Managerial Implications," *Marketing Science*, 8, Fall, 1989, pp. 324–342. On sales force decision-making see Andris A. Zoltners and Prabhakant Sinha, "Sales Force Decision Models: Insights from 25 Years of Implementation," *Interfaces*, June 2001, Vol. 31, Issue 3, Part 2.

29. Gary L. Lilien, "Advisor 2: Modeling the Marketing Mix for Industrial Products," *Marketing Science* 25 (February 1979), pp. 191–204.

30. Linda Micco, "Lockheed Martin Lures Techies with Innovative Recruiting," *HR News Online*, Society for Human Resource Management, March 31, 1997.

31. Ibid.

32. Ibid.

33. Ibid.

34. This story is told in Roy A. Bauer, Emilio Collar, and Victor Tang, *The Silverlake Project: Transformation at IBM* (New York: Oxford University Press, 1992).

35. Ibid., p. 19.

36. Ibid., p. 20.

37. Ibid., p. 21.

38. Ibid., p. 42.

39. Ibid., p. 6.

40. Ibid., pp. 6–7.

41. Chester Barnard, *The Functions of the Executive* (Cambridge, MA: Harvard University Press, 1938).

42. Ibid., p. 139.

43. Ibid., p. 282.

44. Robert N. Corley, Peter J. Shedd and Eric M. Holmes, *Principles of Business Law*, 13th ed. (Englewood Cliffs, NJ: Prentice Hall, 1986), pp. 306–307.

45. Ibid., pp. 302–303.

46. Ibid., pp. 302–303.

47. I emphasize the role of agents as market intermediaries in Daniel F. Spulber, *Market Microstructure: Intermediaries and the Theory of the Firm* (Cambridge: Cambridge University Press, 1999).

48. In law, a *principal* is a person who delegates authority to another to act in his interest, and an *agent* is a representative sent by the *principal* to interact with third parties. "Agency's intellectual distinctiveness is its focus on relationships in which one person, as a representative of another, has derived authority and a duty as a fiduciary to account for the use made of the representative position"; Deborah DeMott, "A Revised Prospectus for a Third Restatement of Agency," *U.C. Davis Law Review* 31 (Summer 1998), p. 1035.

49. http://iflyswa.com/about_swa/customer_servise_commintment/customer_service_commitment.html.

50. Agis Salpukas, "New Ideas for U.S. Oil: Teamwork in the Fields Helps Industry Rebound," *New York Times*, November 16, 1995, p. C1.

51. Ibid.

52. Ibid.

53. Ibid.

54. Ibid.

55. Bauer, Collar and Tang, *The Silverlake Project*, at p. 92.

56. Barnaby J. Feder, "At Monsanto, Teamwork Works," *New York Times*, June 25, 1991, p. C1.

57. Ibid.

58. Jack D. Orsburn, Linda Moran, Ed Musselwhite, and John H. Zenger, *Self-Directed Work Teams: The New American Challenge* (Burr Ridge, IL: Irwin, 1990).

59. Emphasis in original, ibid., p. 108.

60. Ibid., pp. 109–111.

61. Feder, "At Monsanto, Teamwork Works."

62. Aimee L. Stern, "Managing by Team Is Not Always as Easy as It Looks," *New York Times*, July 18, 1993.

63. Ibid.

64. Ibid.

65. Zachary Schiller, Greg Burns and Karen Lowry Miller, "Make It Simple, That's P&G's New Marketing Mantra, and It's Spreading," *Business Week*, September 9, 1996, updated June 14, 1997, http://www.business-week.com/1996/37/b34921.htm.

66. http://www.acehardware.com/root/faq.asp.

PART V

COMPETITIVE STRATEGY

CHAPTER 9

PRICE LEADERSHIP STRATEGY

Contents

A company has a *cost advantage* if cost efficiencies allow it to consistently outperform competitors and earn greater economic profits. By producing the same products and services at lower cost than competitors, companies can gain additional profits and still attract customers with lower prices. This chapter examines critical cost drivers and the determinants of cost advantage. The low-cost firm builds on its cost advantage by following a *price leadership strategy*.

The firm is able to price below its competitors based on its cost advantage. Lower-cost production allows the firm to offer no-frills products and services at prices below those of competitors. Lower-cost production also allows the firm to offer enhanced features of its products and services, while remaining profitable and pricing competitively.

Critical cost drivers include the firm's technical knowledge and investment in equipment and training. The firm's *technology* determines the relationship between its outputs and its inputs, including capital equipment, information technology, labor services, resources, and manufactured parts and components. The *cost efficiency* of the firm depends on how well the firm operates its technology and the mix of inputs purchased by the firm. Companies try to minimize costs by choosing the best mix of inputs such as capital equipment and labor services for the company's production technology. The best mix of inputs depends on the relative productivity of inputs and their costs. The cost of an input is the value of that input in its best alternative use (not the original purchase cost).

Size has potential advantages. The four *boundaries* of the firm are scale, scope, span, and speed, summarized in Table 9.1. The products

Table 9.1:   **Four potential cost economies associated with the boundaries of the firm.**

Types of Cost Economies	Definitions	Expansion of the Firm
Scale	Size of capacity	Growth
Scope	Product variety	Diversification
Span	Production sequence	Vertical integration
Speed	Rate of innovation	Accelerated innovation

and services the firm offers and the functional tasks that it carries out internally define the four boundaries. These boundaries determine four types of potential *cost economies*. *Economies of scale* are reductions in unit costs associated with higher levels of output production per unit of time. *Economies of scope* are reductions in per-product costs that result from producing a greater range of products and services. *Economies of span* are reductions in the costs of activities that result from increasing the extent of vertical integration. *Economies of speed* reflect reductions in the costs of R&D per unit of time associated with attempted increases in the company's rate of innovation.

Cost efficiencies are not enough to maintain a cost advantage. The company must continually reduce its costs to remain competitive. Accordingly companies should continually innovate to achieve and maintain a cost-leadership position in their industry.

Chapter 9: Take-Away Points

Managers should understand that a price leadership strategy is based on achieving a cost advantage. This requires attention to the total cost of the firm's activities as well as the stream of costs over time, rather than a focus on specific, short-term costs:

- The price leadership strategy allows the company to attract customers with greater value, to offer greater surplus to suppliers, and to capture additional value.
- Expansion of the firm's boundaries allows the company to achieve economies of scale, scope, span, and speed.
- Managers should compare the cost effectiveness of company activities with market benchmarks.
- Managers should be aware of the four phases of bringing technological change to market: research, invention, development, and innovation.
- Managers should apply external analysis and internal analysis to monitor the potential effects of technological change on process innovation.

- Companies must continually innovate to maintain a position of cost advantage.

9.1 *Cost Drivers*

Price leadership requires efficient operations. Keeping down costs does not mean reducing the costs of each activity individually but rather minimizing total costs by considering the interaction between companies' productive activities and their role as drivers of total costs. Moreover, total costs do not refer to quarterly costs, but rather to the stream of total costs over time. Companies cannot be penny wise and pound foolish, minimizing current costs while driving up future costs. It is the present value of the cost stream that matters.

Costs include more than explicit accounting costs. There are many hidden or implicit costs, particularly opportunity costs. The *opportunity cost* of an activity refers to the value of the best alternative opportunity that the company gives up. The opportunity cost of employing a productive input refers to the value of that input in its best alternative use. The opportunity cost of using an asset is best measured by its current market value, not by its original cost. For example, the cost of employing capital equipment is measured by its market value rather than book value. When market value cannot be used to measure some opportunity costs, it is sometimes necessary to impute or estimate costs. For example, if managers devote time to supervising an activity, the opportunity cost of those efforts are the returns to the best alternative use of their time.

The company pushes out its boundaries through growth, diversification, vertical integration, and accelerated innovation. Expanding the firm confers two advantages: The company avoids transaction costs in the marketplace and the company achieves potential economies from size. Expanding the firm also has disadvantages. Large organizations can have high management costs and bureaucratic inertia. Outsourcing gives the firm access to efficient and innovative suppliers. Companies resolve this dilemma by expanding those activities in which they have a comparative advantage and outsourcing the rest as much as possible.

Cost Efficiency

Companies strive to operate efficiently by choosing the appropriate technology and using it effectively. Efficient management of operations requires making the best use of the production technology, given guidance from engineering and operations management. The manager evaluates the technical and cost efficiency of company activities. Such an analysis is closely connected with decisions about the company's output level and product selection.

As a first step, the manager attempts to make sure that company operations satisfy *technical efficiency*, which refers to performance in terms of engineering, information technology, operations management, and other criteria. The company measures its *productivity* by measuring the rate of output per year (or other time period) relative to the input usage in that year, including labor services and capital equipment. For example, a manufacturing company tries to produce the highest rate of output per time period that is feasible given its inputs of capital equipment, labor, manufactured components, and natural resources. The company's information technology makes the best use of software, hardware, and networking elements. While the company seeks to achieve highest rate of production that is technically feasible given its input mix, it controls for desired product quality. Technical efficiency is a management problem as well as an engineering one. Implementing technical efficiency depends on employees working effectively.

In addition, the manager examines the *cost efficiency* of company operations. *Total costs* are measured by adding up the cost of all purchased inputs. Total cost also includes the opportunity cost of any other inputs that are not purchased. The company's total costs include the following:

- Cost of investment capital: interest costs, opportunity cost of equity capital.
- Labor costs: wages, salaries, bonuses, benefits for employees and managers.
- Capital equipment: machinery, information technology hardware, communications equipment.
- Purchase cost of inputs: manufactured parts and components, natural resources, utilities, rent on land and facilities, supplier services.
- Technology costs: licenses, royalties, software.
- Opportunity costs: market value or imputed value of company-owned assets and services.

Cost efficiency means choosing the right mix of capital equipment, labor, manufactured components, and natural resources. The desired *mix of inputs* is the one that achieves the company's output targets at the lowest cost. For example, depending on the relative cost of wages and capital equipment, the company might choose to rely on production methods that are relatively labor intensive or capital intensive. The manager compares production methods by comparing the productivity and purchase costs of labor, capital, and other inputs. The firm's relative cost performance depends both on the company's technical efficiency and its cost efficiency.

The cost efficiency of the company should be compared to competitive benchmarks, both for the costs of producing final products and the cost of functional operations. Managers should review data on productivity of labor, capital and other inputs and compare within the same industry, as well as to national and international productivity targets for similar activities outside the industry. Achieving cost efficiency is a management problem because the company must rely on employees making the right decisions about the company's input mix. Furthermore, employees must make the best use of information available about the cost of purchasing inputs and the productivity of those inputs.

Michael Porter emphasizes that *linkages* among activities pervade the value chain and hence the company's cost drivers are closely connected. The company must therefore coordinate and optimize its many connected activities. Moreover, vertical linkages exist between the company's many value activities and the value chains of suppliers and distribution channels. For example, Canon found it could almost eliminate the need to adjust its copiers after they were assembled by purchasing higher-precision parts from suppliers.[1] In this example, the company would reduce costs if the savings from reducing copier adjustment were greater than the premium paid for higher-precision parts.

Product design is another important cost driver. Cost efficiency involves more than producing a particular product or service at the lowest possible costs. By changing product features with attention to the costs of manufacturing and distribution, companies may be able to enhance product features while lowering costs. Thus, cost advantage and differentiation advantage (the subject of the next chapter) are closely connected. By considering the company's products as a bundle of services, the company can examine the most cost-effective way to deliver those services to its customers.

Cost efficiency not only includes controlling company processes in manufacturing and customer services; overhead expenditures, including management, can be a significant component of costs that require control. To the extent possible, costs should be assigned to activities that cause those costs through activity-based accounting. This accounting approach allows companies to observe more accurately the full costs of organizational activities as a guide to controlling costs and comparing them with market alternatives. However, arbitrary assignment of overhead costs will result in incorrect estimates of the costs of company activities, leading to incorrect strategic decisions. Overhead costs are those that are shared between activities and cannot be assigned to them arbitrarily.

Because practically any method of allocating joint and common costs is necessarily arbitrary, allocating those costs to individual activities can bias the costs of that activity up or down and lead to incorrect outsourcing decisions. By considering alternative combinations of activities, the company can effectively compare the performance of its organization with market alternatives.

Exhibit 9.1   CEMEX and Cost Efficiency

Cementos Mexicanos (CEMEX) is the largest cement company in the Americas. With a capacity of 56 million tons, it is also the third largest worldwide, behind Holderbank of Switzerland and LaFarge SA of France. CEMEX is headquartered in Monterrey, Mexico; the company has operations in more than 20 nations in Asia, Europe, and the Americas, and trades in more than 60 countries worldwide. CEMEX is a market leader in Mexico, Spain, Venezuela, Panama, and the Dominican Republic, and enjoys significant share in numerous other countries. The company is a major manufacturer and distributor of cement, specialty cement-based products, clinker (an intermediate cement product), ready-mix concrete, and gravel, among other products.

According to the company:

> If there is a secret to CEMEX's success, it is the company's ability to reap the benefits of integration and efficiency. CEMEX is constantly striving to drive back costs and maximize productivity through a relentless review and analysis of new and existing operations.

CEMEX managers use computers and the company's information technology system to continually monitor the company's worldwide operations. Managers check on "daily production, kiln conditions, crushing, bagging, and shipping output at any of CEMEX's operations around the world".

The company has entered into an extensive series of mergers and acquisitions, and now has operations in Mexico, Spain, Venezuela, Panama, Colombia, the Dominican Republic, Costa Rica, the Philippines, Indonesia, Egypt, and the United States. The company sends what it calls postmerger integration (PMI) teams of engineers, technicians, and managers to examine the operations, human resources, and capital expenditures of new acquisitions. The team prepares recommendations for improving productivity and reducing costs for newly acquired operations. For example, in the Philippines in 1998, "the PMI team helped Rizal Cement achieve cost savings of US$29 million for the first year of operations". In addition, the team seeks to integrate the newly acquired operations into the company's global network to improve overall productivity and reduce total costs.

(Continued)

Cement is costly to transport. One method of reducing total cost of delivered cement is to locate the company's many plants based on customer location. The company faces a trade-off between manufacturing economies of scale at the plant and the cost of transporting cement. Accordingly, the company monitors the total production and transportation costs in its network of manufacturing facilities.[2]

Learning and Knowledge

Keeping down the company's costs is difficult when managing complex operations. It should not be surprising that managers and the organization as a whole can improve performance through *learning* about the company's operation over time. Managers can discover ways to improve organizational routines and organization of manufacturing and other operations. Costs can be lowered by eliminating wasted effort and inefficient procedures.

Learning-by-doing refers to the lowering of costs due to repeated performance of manufacturing operations. Individual performance improves with repetition — just remember the first time you rode a bicycle! However, in the case of companies, repetition may not be enough. Cost savings sometimes attributed to learning-by-doing are often the result of company investment in capital equipment and employee training.

During World War II, the United States Emergency Shipbuilding Program that built the Liberty Ships achieved dramatic cost reductions. Peter Thompson looked again at the classic story of the Liberty Ships that has often been used to buttress the idea of learning-by-doing.[3] Using a newly declassified database, he found that productivity increases were due in large part to capital investment and reductions in product quality. Kazuhiro Mishina revisited the World War II building program for the B-17 heavy bomber, known as the Flying Fortress, and found that at Boeing's No. 2 Plant in Seattle, Washington, productivity changes improvements were due to greater scale and increased coordination by the production control department.[4]

Productivity gains might be the result of investment in human capital, research and development expenditures, information gathering, and organizational reforms that are independent of cumulative output. Accordingly, companies that seek to reduce costs simply by pumping up output may achieve just the opposite.

The company's critical cost drivers depend on its stock of knowledge about technology. This stock of knowledge is dependent on the training and abilities of the firm's employees, known as human capital. The company also has organizational knowledge in the form of patents, proprietary technological information and organizational routines. Companies develop their knowledge base by internal R&D, mergers and joint ventures, supplier alliances, partnerships, and transactions in the market for intellectual property.

9.2 *Economies of Scale and Scope*

Companies can achieve cost advantage by scale and scope. Even if two companies have similar technologies, the presence of scale and scope economies allows the company with greater output or more products to have a cost advantage over the other. Companies seek a virtuous cycle in which lower costs allow lower prices, leading to higher sales, allowing the firm to produce a higher output and achieve returns to scale and scope.

Opportunities for expansion depend on the constraints imposed by market demand and the firm's market share. The advantages of scale are limited by demand for a specific product or service. Scope economies are limited by the demand for product variety. Vertical integration economies are limited by matching input and output production within the firm, unless the vertically integrated company can supply excess inputs to other firms. Economies from speed of innovation are limited by the demand for new products and production methods.

Scale

Firms often define themselves in terms of size, usually quoting the annual value of sales, which provides some indication of total production per year since sales equal price times output. Firms frequently compare their success in terms of relative market share which provides a measure of size relative to competitors. While market share need not be an indicator of current or future profitability, firms strive to be the market leader in sales to achieve competitive advantage.

One source of advantage from sales growth can be the presence of technological economies of scale. The firm's technology exhibits economies of scale when average costs decrease as output increases. The firm realizes economies of scale if the marginal cost of the firm is less than the firm's unit cost; that is, the cost of expanding output by one unit is less than the firm's unit cost before the expansion. The firm's unit cost function decreases with output when the firm's technology has economies of scale, see Figure 9.1.

If economies of scale are present, then expansion of the firm's output lowers the firm's unit cost. The definition of economies of scale

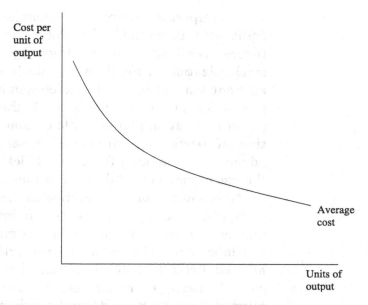

Figure 9.1: **The firm's technology has economies of scale when its average cost is decreasing in the firm's output.**

presumes that the firm is using the most efficient technology so that any gains from implementing technological innovations or from reducing waste have already been implemented. Of course, every firm can always find new areas for efficiency improvements. Economies of scale refer to efficiency improvements due only to scale, over and above those due to improving the firm's operating procedures. Identification of economies of scale does not require the firm to have a fixed technology. Rather, an important source of economies of scale is the choice of the most efficient production process for *each level of output*.

The advantages of economies of scale are well-known. Fifteenth-century merchants in the Mediterranean realized that larger vessels could carry more cargo at lower cost because the volume of the ship increased more than proportionately with its surface area and the size of the crew. The same scale economy is reflected in modern super-tankers and huge container ships. Volume-surface relationships have also driven expansion of firms in process industries from chemical manufacturing to brewing beer.

Adam Smith's famous story of a pin factory illustrates the advantages of the *specialization of function* and *division of labor*, which are important sources of scale economies. In a small pin factory, individual workers performed multiple tasks, such as rolling wire, cutting the wire, and attaching the heads of the pins. In a larger pin factory, workers could divide the tasks, specialize, and produce pins at a higher rate. The specialization of function and division of labor is evident in any company — employees are specialists in such tasks as engineering, design, accounting, finance, and sales.

An important source of economies of scale is the use of capital equipment. From the early days of the industrial revolution, manufacturers realized cost savings by replacing handicrafts with mechanization in textile mills. Mechanized agriculture increased the efficient scale of farms. Large economies of scale were realized in the production of electric power with the development of large central power generation plants. Scale economies also stem from standardization of parts and large-scale assembly, as Henry Ford used to advantage in making the Ford Model T. Increases in scale also allow the use of new materials such as plastics in manufacturing.

New sources of scale economies are continually being discovered. Computers are replacing routine information processing, calculating, and monitoring of equipment just as mechanization has replaced physical labor since the industrial revolution. Moreover, computers have allowed firms to make substantial cuts in administrative overhead costs. Increases in the information processing power of computers have allowed firms to make dramatic reductions in management personnel resulting in a fall in overhead costs. For example, firms such as electrical equipment giant Asea Brown Boveri were able to reduce central office staff by 90 percent. Efficiency improvements that reduce total overheads lower unit costs, but they have the added effect of reducing the firm's minimum efficient scale. In other words, a firm may be able to achieve all available scale economies at a smaller output than before. Thus, increased efficiency may lower the size of firms in some industries as firms are able to achieve the lowest possible costs at a smaller size.

Companies realized economies of scale in the operation of mainframe computers by centralizing data processing. These economies seemed to fade by the early 1980s with the advent of the personal computer, which brought tremendous computing power to the desktop. With interconnection of computer networks through the Internet, companies have again discovered overwhelming economies of scale from centralized computing. Internet service provider America Online operates vast server farms to handle traffic from its millions of customers as they connect to the Internet. Companies such as AT&T, IBM, and Sun Microsystems operate vast computer centers with many powerful servers, handling data storage, processing, and communications tasks for companies engaged in electronic commerce.

Economies of scale in its simplest form result from spreading out fixed costs across many units of output. If the firm's overhead costs, including accounting, legal, and other management functions, are relatively constant, expansion of output will lower unit overhead costs. The same holds for other types of fixed costs such as capital equipment. By using capital equipment to its full capacity, firms are able to reduce their unit costs. This explains the use of double or triple production shifts as a means of making full use of capital equipment.

Example 9.1 Spreading Overheads

A company has $100 of overhead costs that do not vary with output. If the company produces 50 units of output, overheads contribute $2 to unit costs. If the company produces 100 units of output, overheads contribute only $1 to unit costs.

The costs of acquiring and communicating information in large organizations can affect the optimal scale of the firm. The revolution in computers and telecommunications and the creation of information networks within large companies will create significant economies of scale. On the other hand, the information costs of monitoring employee performance and observing employee abilities can limit the efficient scale of the organization. To the extent that economies of scale transcend the production plant and are experienced across the organization, they will yield benefits for the market leader. The principal sources of economies of scale for market leaders must arise from efficiencies in communication, information processing, marketing, and research and development.

An advertising campaign is a fixed cost for the firm. Larger companies can spread those fixed costs across many units of output, allowing for substantial savings in marketing expenditures. This is the foundation of franchise businesses such as McDonald's, which can spread the costs of a national advertising campaign across all of its outlets. A smaller competitor making the same advertising expenditure would experience a greater increase in unit costs. Marketing expenditures are sunk costs so costly ad campaigns pose substantial risks for new entrants or for firms seeking to take on the market leader. This is why the market leader must continue to make marketing expenditures to maintain its competitive position.

Research and development, like marketing, is a crucial overhead cost in industries where the advantages of market leadership are particularly significant. The market leader can maintain a substantial R&D facility with less effect on unit costs than smaller rivals can. Then, a market leader that is cost-efficient can bring to market a one-two punch of low prices and continuous innovation. High R&D expenditures in themselves are no guarantee of success, as IBM found when it could not translate its billions of dollars in research expenditures and cutting-edge discoveries into marketable products in the early 1990s. However, if the leading firm's R&D efforts are productive, innovation can be achieved at lower unit costs than those of smaller rivals. This is not to say that the largest firms are always the most innovative. Smaller firms may design products that are a closer fit with customer requirements. Moreover, R&D by nature is an uncertain process and smaller firms may have the talent and the good fortune to make path-breaking discoveries before a larger rival.

The firm's total costs consist of the sum of fixed and variable costs, that is, costs that do not vary with output and costs that do vary with output. The firm's average or per-unit cost is obtained by dividing the total cost function by the number of units of output that are produced.

Example 9.2   Average Cost

The firm incurs a fixed cost of $200 that does not depend on its output level. If the firm produces 50 units of output it incurs variable operating costs of $150. The firm's average fixed costs are $200 divided by 50, which equals $4 per unit of output. The firm's average variable costs are $150 divided by 50, which equals $3 per unit of output. The firm's average costs equal $7 per unit of output.

The firm obtains *economies of scale* when average costs fall. One important source of economies of scale is the spreading of fixed costs over many units of output as noted previously.

Example 9.3   Economies of Scale

A company has fixed costs of $100. Suppose that producing 2 units of output entails variable operating costs of $8. Producing 5 units of output entails variable operating costs of $50. So, total costs of producing 2 units of output are $108 and total costs of producing 5 units of output are $150. The company's average or unit costs equal $54 when the firm produces 2 units of output and $30 when the firm produces 5 units of output. So, average costs fall as output increases.

Growing sales allows the company to expand output and to realize economies of scale. In this way, the market confers advantages in terms of cost economies that allow the firm to sustain its leadership position. Thus, achieving cost economies is a potential benefit of winning markets. However, striving for cost economies does not in itself guarantee success. Although costs from distribution efficiencies helped Wal-Mart's sales growth, Wal-Mart did not win the retail market solely as a result of its efficient distribution system. Wal-Mart's expansion strategy, pricing policies, and merchandising ability were primarily responsible for its growth in sales. The growth in sales in turn allowed Wal-Mart to take advantage of economies in distribution. American Airlines, on the other hand, a leader in sales in the 1980s and early 1990s, invested heavily in expansion of its fleet and in the development of airport hubs that potentially yield huge cost economies. American Airlines entered the 1990s with substantial annual losses from over-investment, forcing it to close down hubs, retire airplanes, and cancel orders for additional planes as smaller regional carriers with lower costs competed successfully for many routes. Thus winning markets can yield cost economies, whereas cost economies alone need not win markets.

Based on his examination of the origins and growth of large modern industrial enterprises in the United States, Great Britain, and Germany, Alfred D. Chandler (1990) argues that companies rarely continue to grow or to maintain their competitive position without cost reductions and efficient resource use, particularly through taking advantage of economies of scale and scope. He emphasizes that economies of scale and scope occur within the operating units of firms, whereas transaction costs within the firm generally are incurred through the exchange of goods and services between the firm's operating units. He observes that technological change, particularly in transportation and communication, made possible the creation and management of large-scale enterprises in production, such as DuPont, and distribution, such as Sears Roebuck. These enterprises took advantage of scale and scope within operating units.

Cost economies are not guaranteed by growth, however, but depend on whether the management can take advantage of technological opportunities. Large firms such as IBM, General Motors, and many others discovered that bureaucracy, inefficiency, and underemployed personnel can accompany growth. Some types of expansion have diminishing returns and it is up to management to discern how growth affects costs. Moreover, the presence of cost economies does not imply unlimited market opportunities. The firm cannot simply grow to achieve cost economies without recognizing the size of the markets it wishes to serve.

In the short run, companies that operate efficiently lower their average costs by growing production. In the long run, companies can reduce average costs by investment in cost-reducing facilities and capital equipment and other cost-reducing methods such as training and information systems.

Example 9.4   Cost-Reducing Investment

A company has the choice of two technologies. The first technology has a fixed cost of $100 with a per-unit variable operating cost of $6. The second technology has a fixed cost of $400 with a per-unit variable operating cost of $3. The company wishes to produce

(*Continued*)

200 units of output. Which technology should it choose? Since its average fixed costs are 50 cents, the first technology has an average cost of $6.50. Since its average fixed costs are $2 and average variable cost $3, the second technology has an average cost of $5. This means that the company should choose the second technology, even though it is initially more expensive, because operating cost savings are greater.

Economies of scale play an important role in international business strategy. Companies can realize economies of scale by boosting sales through export. If the company is experiencing increasing returns to scale at its domestic output, expanding output for export will lower average cost.

Economies of scale in production help to explain an important puzzle in international trade. Why do countries trade goods in the same industry? For example, there is intra-industry trade in automobiles: England exports cars to Germany and imports cars from Germany. Why does not each country manufacture its own cars? The answer is that consumers prefer to purchase a variety of cars. Both Britons and Germans buy British Jaguars and German BMWs. It makes sense to produce Jaguars in England and BMWs in Germany to achieve economies of scale, as long as those cost economies outweigh the cost of transporting the automobiles. In some cases, companies can achieve economies of scale and also avoid transportation costs and tariffs by producing cars near to their markets when there is sufficient demand. Thus, GM produces cars in Brazil and Toyota makes cars in the United States.

Exhibit 9.2   The Standard Oil Trust

The Standard Oil Alliance, a confederation of 40 companies, formed the Standard Oil Trust in 1882. According to the business historian Alfred D. Chandler, the Trust allowed a reorganization of production that resulted in economies of scale in producing kerosene. Under the trust, over a quarter of the world's production of kerosene was concentrated in three refineries, each with an average daily charging capacity of 6,500 barrels. Over two-thirds of the output of these U.S. refineries was exported. Chandler observes that the costs at the large refineries were below those of

(Continued)

> **Exhibit 9.2** (*Continued*)
>
> any other competitor. Although the Trust operated other refineries, the three large refineries helped the Trust lower its average cost of production of a gallon of kerosene by more than two-thirds:[5]
>
> Before 1882   1.5 cents per gallon
> 1884   0.54 cents per gallon
> 1885   0.45 cents per gallon

Scope

The firm's scope is determined by its range of product and service offerings. In addition to having the largest sales, many successful firms offer greater product variety than their rivals. In addition to the competitive advantages from offering customers more choices, cost advantages can result from variety as well. The firm achieves economies of scope if it can produce two or more products at a lower cost than if separate firms produced the products.

Economies of scope can result from spreading fixed costs over multiple products. Increasing the number of products will increase variable costs. For smaller output levels, sharing overheads yield cost economies, but as output increases this effect is overshadowed by higher variable costs.

> **Example 9.5 Economies of Scope**
>
> Consider a firm that produces both cars and trucks. The firm incurs a common fixed cost of $100 whether the firm produces only cars, only trucks, or both. If the firm produced no cars and only 4 trucks, its variable costs would be $64, so total cost would equal $164. If the firm produced no trucks but only 2 cars, its variable costs would be $16 so its total costs would be $116. The total costs of *separate* production of 4 trucks and 2 cars would be $164 + $116 = $280. Producing 4 trucks and 2 cars together has a variable cost of $144 for a total cost of $244. Economies of scope are realized since the cost of joint production is $244, which is less than the cost of separate production of $280. The cost saving of $36 reflects the avoidance of duplicate fixed costs of $100, which is offset by the additional variable cost.

Leading firms can achieve company-wide benefits of size by diversification into new products and expansion into new markets. The firm can spread the costs of communication, information processing, marketing, and research and development across its products and achieve a smaller unit cost if its total sales are greater than those of its rivals. Thus, economies of scope can be achieved based on the same forces that create economies of scale. A firm offering many products need not sell large quantities of each product to take advantage of economies of scope. For example, by producing multiple products, the firm is able to increase its overall output and enjoy the benefits of specialization of function and division of labor.

Computer-aided design and manufacturing and the development of the flexible factory imply that the firm need not sacrifice plant-level economies when it expands product variety. The product mix can be adjusted to satisfy individual customer requirements. For example, the Mitsubishi plant in Nagoya was able to produce different models on the same assembly line with minimal cost of adjustment when switching to different models by installing higher cost but flexible capital equipment.

The leading retailer Wal-Mart takes full advantage of economies of scope in distribution. Wal-Mart's superstores gain economies of scope by spreading store operating costs across products. Many economies of scope Wal-Mart captures are company-wide. Offering over 50,000 items allows Wal-Mart to spread the costs of its sophisticated computerized distribution system that has over 20 state-of-the-art regional distribution centers. The company employs a sophisticated satellite communications system with video conferencing and centralized inventory control that responds to customer demands and provides detailed sales information to its suppliers.

Service firms can also realize economies of scope. Of the more than 10,000 advertising agencies in the United States, only a few large agencies can realize the full advantages of scale economies. Advertising agencies are typically multi-product firms and achieve cost economies from offering a variety of services. Agencies provide content for various media categories such as television, radio, newspapers, magazines, and displays. The median cost savings from joint production for advertising agencies is substantial, exceeding 25 percent of total costs.[6] The majority of agencies achieve economies of scope by jointly providing combinations of individual media services.

The possibility of achieving economies of scope does not imply that expansion of product offering will lower unit costs. As with scale economies, limits on economies of scope can stem from the costs of organizing a multi-product firm. Moreover, many new products can fail to produce a market or can dilute the sales of the firm's existing products. General Motors incurred costs from offering a great variety of automobiles, including seven brands and over 60 models. While there may be scope economies within GM's divisions, such economies may be much harder to obtain across GM's divisions.

Manufacturing companies try to obtain economies of scope by sharing components across their products. For example, companies producing computers, mobile phones, and automobiles use common platforms across models to reduce costs of design, parts production, and final assembly. Nokia uses standard platforms for its mobile phones, varying the electronics and other features for different target markets. In automobiles, platforms refer to the chassis, including the wheels, steering, suspension system, and engine. DaimlerChrysler tried to reduce costs in its Chrysler and Mitsubishi divisions by sharing platforms to reduce the total number of platforms in the two divisions from an initial 29 to approximately half that number.

9.3 *Economies of Span and Speed*

Economies of scale and scope are achieved by producing higher output or a greater variety of products. Cost reductions can also be achieved by more fundamentally restructuring the company or remaking its technology through innovation. *Economies of span* are achieved through greater vertical integration of activities along the value chain. The company must continually reexamine the trade-off between increased coordination resulting from in-house production and increased flexibility from outsourcing production. *Economies of speed* refer to cost efficiencies that arise as a return to the scale of the company's R&D processes. Companies that engage in R&D must carry out and apply innovations at a faster rate than competitors.

Span

The *span* of the firm is a crucial aspect of the way in which firms define their activities. A characterization of the firm as a service company, manufacturer, wholesaler, retailer, or integrated manufacturer-distributor refers to the span of the firm. The firm's choice of its span is a vital component of its strategy. During much of the 20th century, there was a trend toward increased vertical integration of manufacturing. By the start of the 21st century, this trend was reversed in many industries. Information technology allowed firms to outsource a variety of activities. Outsourcing activities by manufacturing firms has led to a corresponding growth of the service sector of the economy as service companies have taken on outsourced functions.

The span of the firm is determined by the extent of primary and secondary production tasks it carries out. The company need not complete all the stages of production within its span. Rather, outsourcing any activity along the sequence is possible. Decisions regarding what activities should be carried out within the firm and what activities can be outsourced depend on comparing the effectiveness of carrying out each stage of production within the organization with market alternatives.

The company also must take into account the costs of coordinating production within the firm and the transaction costs of outsourcing services along the value chain. The firm's make-or-buy and distribute-or-sell decisions determine the span of the firm. These decisions establish the boundary between the firm and its upstream and downstream markets.

Example 9.6   Linkages within the Value Chain

Consider two different value chains for an equipment producer. The first involves costly product design of $7,000, and operations with extensive quality control costing $5,000. This combination results in lower service costs of $2,000. The second value chain involves more basic product design of $3,000, operations with cursory quality control of $4,000, and service costs due to product returns and maintenance expenditures of $8,000. Which value chain is the most effective? The firm cannot carry out cost minimization component by component because of linkages; it must consider total costs. Accordingly, the first value chain is the most effective because it has the lowest total cost.

Management's determination of the span of the firm's activities is closely tied to the quantity and variety of the firm's final output. Firms must decide what range of upstream and downstream activities they wish to undertake. Manufacturers choose which steps within the production sequence they wish to undertake. For example, manufacturers may produce all or some of the parts that make up the final product or they may purchase all of the parts and act as final assemblers. Manufacturers can design products internally or contract with independent product designers. Manufacturers can distribute their output themselves or they can sell the products through independent wholesalers and retailers. The span of the firm refers to the number and type of stages in the sequence of activities required to design, manufacture, and distribute a firm's product that are within the organization, (see Figure 9.2). Thus, the span of the firm refers to the vertical boundary of the firm. It should be emphasized that the notion of the span of the firm does not imply that production and distribution are a series of single, ordered steps. Rather, stages of production often involve a complex set of overlapping activities. For any set of activities, the firm must decide whether to carry out the activities themselves or whether others, either customers or suppliers, will carry them out.

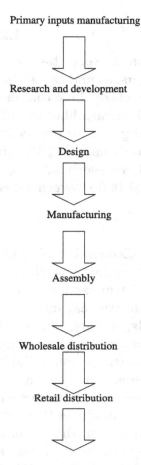

Primary inputs manufacturing

Research and development

Design

Manufacturing

Assembly

Wholesale distribution

Retail distribution

Figure 9.2:   The manager choose the span of the firm by selecting what activities to outsource and what activities to carry out in house.

Economies of span exist if the firm integrates into activities further upstream by taking on additional manufacturing and design tasks or if the firm integrates further downstream by adding additional distribution tasks. Economies of span result from sharing overhead and other common costs across stages of production.[7] Common inputs in a series of production processes often are cited as reasons for vertical integration. For example, the need to reheat steel is avoided by combining the blast furnace and rolling mill in a single enterprise.

Example 9.7   Economies of Span

A brokerage company carries out a sequence of three activities associated with completing securities trades for its customers. The brokerage (1) markets and sells securities, (2) processes orders for

(Continued)

> ### Example 9.7   (*Continued*)
>
> those securities in the financial markets, and (3) records purchases and sales. Each of the activities, if performed separately by a standalone company, would have a fixed overhead cost of $500. The variable costs of each activity are $40 for marketing and selling, $60 for processing, and $30 for recording purchases and sales. By combining the three activities, the company incurs only one fixed overhead cost, so its total costs are $630. The company saves $1,000 in overhead costs in relation to three separate companies.

Economies of span stem from many sources, including operating cost reductions, the ability to secure reliable supplies or distribution, and transaction cost savings. Alfred D. Chandler found cost savings from vertical integration in a number of industries including chemicals, metals, and machinery, but he notes that vertical integration occurred in many companies for other reasons such as coordination of operations and reliability of input supplies.[8] Larger firms can more easily achieve benefits from economies of span if there are scale economies at some vertical stages of production. Wal-Mart moves most of its products through its distribution system, making it a vertically integrated wholesaler and retailer.

Economies of span are of particular importance because of the company's merchant activities. Through their buying and selling activities, companies create distribution channels for their suppliers and marketing channels to reach their customers. The firm's vertical integration decision is much more than a decision about the stages of production. A retailer intermediates between its customers and its suppliers (wholesale distributors and manufacturers). A vertically integrated retailer such as Wal-Mart takes on important market-making activities by contracting directly with manufacturers, negotiating prices and product characteristics, monitoring sales, and communicating information about orders.

How should the vertically integrated company be organized? Which market level is the most important? Companies often think of themselves in terms of the level at which they started. Thus, a manufacturer that has expanded into wholesaling continues to perceive itself primarily as a manufacturer. Such perceptions should not be the determining factor in designing the organization.

Managers should determine the vertical stages at which the company can provide the greatest value added. Managers should compare the value added by the company to market alternatives and undertake an activity in-house only if its performance is superior to that of potential suppliers. The value added by a particular stage of production is the difference between the value received by the company's customers and the opportunity costs of its suppliers.

The manager should consider the spread between the price at which the company can buy the goods and services and the price at which the company can resell those goods and services, net of the cost of carrying out the critical transactions. The returns to a set of activities are evaluated based on the potential price spread between any two market levels from primary suppliers to final customer markets. The vertical extent of the organization should be targeted toward creating these critical buyer-seller transactions.

The choice between making or buying a critical input is a difficult but critical decision. Companies should use a market test to evaluate whether or not to outsource each of the firm's activities. A market test requires comparing the costs of carrying out an activity with those associated with outsourcing. A proper market test means not only that any individual activity is better carried out within the firm than by others, but also that any combination of activities is better carried out within the firm than by others. Checking every combination of activities corrects for errors that arise when individual activities have joint and common costs.

In evaluating the costs of carrying out an activity versus the costs of outsourcing, it is important to fully account for the associated *transaction costs*. For example, outsourcing an activity entails the costs of finding and negotiating with service providers. Performing that activity in-house entails the transaction costs of finding and negotiating with primary input providers. Transaction costs include all back-office processes associated with sales and purchasing, such as billing and invoicing. They also include the costs of setting and communicating prices or of gathering price information.[9]

When market transaction costs are high relative to organizational costs, managers should attempt to increase the span of the firm by producing products within the firm rather than from external suppliers — recall the discussion in Chapter 8. Conversely, when market transaction costs are low relative to organizational costs, managers should seek to reduce the span of the firm by outsourcing services and purchasing a greater share of inputs from suppliers.

Speed

The *speed* of R&D refers to the company's rate of innovation. The expected duration of an R&D project can be shortened by increased investment in equipment and additional personnel. Economies of scale and scope in R&D create *economies of speed*.

A firm achieves economies of speed if an increase in innovative activity lowers the average cost of innovation. This occurs if investment in R&D projects lowers the expected duration of R&D project per dollar invested or increases the success rate per dollar invested. The presence of economies of speed implies that efficiencies may be attainable through expansion of specific R&D projects. In addition, cost economies may be attainable through undertaking multiple R&D projects.[10]

Innovation plays a central role as a means of sustaining competitive advantage. There are three main types of innovation, as already noted. *Process innovation* refers to improvements in manufacturing and service provision (see the last section of this chapter). *Product innovation* refers to the development of new products and services (see Chapter 10). *Transaction innovation* refers to the creation of new types of market transactions that lower transaction costs and yield gains from trade through new combinations of buyers and sellers (see Chapter 11). Each of these types of innovations is closely associated with competitive advantage. Gaining and sustaining competitive advantage requires constant innovation to stay ahead of competitors: Cost advantage requires process innovation, differentiation advantage is based on product innovation, and transaction advantage depends on transaction innovation, as Table 9.2 illustrates.

What is innovation? *Bringing technological* change to market has four phases, as Figure 9.3 shows. First, *research* is the creation and discovery of basic scientific and technical knowledge. Second, *invention* is the process of converting and extending that knowledge to technology applications that can be disseminated, published, or patented. Third, *development* produces blueprints, industrial process specifications, and product designs that implement basic inventions. Finally, *innovation* brings these developments to market through the purchase and sale of designs and patents, construction of new and more efficient manufacturing plants, production of novel goods and services, and

Table 9.2:   Sustaining competitive advantage requires continual innovation.

Type of Competitive Advantage	Type of Innovation
Cost advantage	Process innovation
Differentiation advantage	Product innovation
Transaction advantage	Transaction innovation

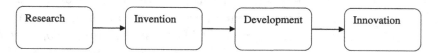

Figure 9.3:   The four phases of technological change.

creation of new types of transactions.[11] Innovation is the commercialization phase of the process of technological change.

As with the other boundaries of the firm — scale, scope, and span — the choice of innovative speed by the firm is a choice about which activities will be left to the market and which activities will be carried out by the firm. Should a company make or buy innovations? The question of whether or not to perform R&D is a special type of vertical integration question. Should the company do its own research, invention, development and innovation, or should the firm rely on others for these difficult but important tasks?

In part, the answer is the same as for other types of vertical integration. The company should consider doing its own R&D if there are economies of span, that is, economies of doing things in-house. In-house innovation efforts might also be desirable if there are high transaction costs of procuring innovations in the market as compared with the organization costs of in-house R&D. Finally, if R&D in the market is protected by patents and potential suppliers have no intention of selling, the company might be forced to carry out its own R&D. On the other hand, if combining production with in-house R&D offers limited joint economies, market transaction costs are low compared to organization costs of in-house R&D, and willing suppliers of technology are available, the company should consider purchasing or outsourcing innovation.

Despite these basic guidelines, the problem of whether or not to engage in innovation is a bit more subtle than the make-or-buy decision. Several factors suggest that managers should look *outside* the firm for innovations. The combined creativity of the scientific, technological, and business community far exceeds that of any one company. Managers who believe that the company should avoid anything that is "not invented here" will soon be left behind. No single company can expect to carry out sufficient basic research and inventive activities; it must rely on the vast collective output of the scientific community whether in universities, government and private research labs, and other companies. Also, companies may need to look outside the firm for many types of technology development because the applicability of basic science and inventive activities may not easy to observe until someone develops them into usable technology blueprints, licenses, equipment, and materials. Finally, the company may still not understand the impact of new technology until entrepreneurs create new products and processes and bring those innovations to market. At that point, the company may need to license the innovations or acquire the entrepreneurial start-ups. Many companies like Cisco Systems consistently use mergers and acquisitions as a method of acquiring innovations.

Freeman and Soete find that the increasing scientific content of technology explains the growing importance of corporate R&D. They emphasize the increasing specialization of scientific knowledge, the growing complexity of various types of technology such as electrical engineering or biotechnology, and the increasing scale of certain types of industrial and innovative processes as factors underlying the continuing shift from entrepreneurial R&D toward specialized R&D organizations and corporate R&D.[12]

Even though companies need to rely on others in the technology race, they sometimes still must have some amount of research, invention, development, and innovation in-house. The main reason is that companies often need to pursue these activities as a window on technology. Companies must engage in some parts of the innovation process to understand and keep pace with the marketplace. Company innovation is a window on the future of the industry; without it firms that observe innovations after competitors employ them will be too late to achieve a competitive advantage. Moreover, company innovation may be successful because the company may have the best understanding of its specific needs. This understanding allows the company to focus on just those applications of invention and then produce innovations that lead its industry. Some companies are able to maintain a creative edge over competitors by a combination of consistency of focus and large-scale R&D efforts.

Therefore, companies choose to both make and buy innovations. Managers should evaluate economies of scale and scope in the innovation process and structure the organization to increase the rate of innovation. Since changes in the firm's rate of innovation involve fundamental changes in the activities of the firm, innovation speed constitutes an important boundary of the firm. Firms choosing to increase their rate of innovation must hire scientific personnel, invest in specialized equipment, construct laboratories and testing facilities, purchase technical information, and obtain patents and licenses for technologies useful to their R&D efforts. When they have a comparative advantage in R&D, companies may choose to rely on in-house innovation. For example, Intel has maintained an innovative advantage in the computer industry during a period of intense technological change through its focus on the design and manufacture of microprocessors and large-scale R&D investment. Intel, the market leader in computer chips, has maintained its position by continual innovation. Its Pentium family chip was introduced even as its 486 family of chips achieved greater than expected success. Intel faced a number of important challenges, including the higher capacity Power-PC chip developed by the IBM-Apple-Motorola joint venture. In response, Intel's CEO Andy Grove sought to double the innovative speed of Intel by cutting the rate at which Intel introduces new chips from four years to two. Changes in the fundamental design of chips enabled Intel to increase its rate of innovation and even challenge the rate of computer chip development embodied in Moore's Law (Gordon E. Moore was cofounder of Intel): The

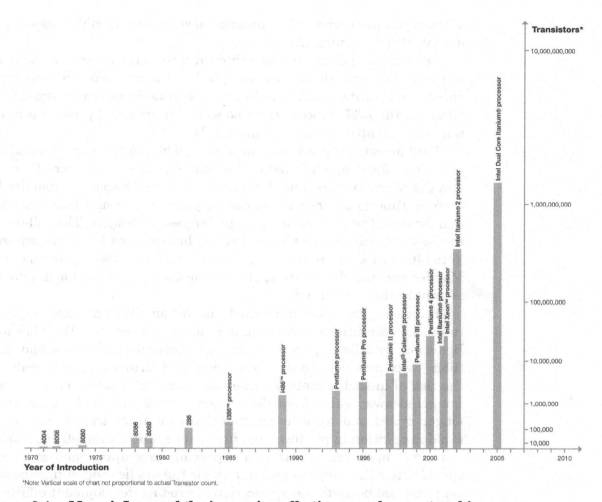

Figure 9.4: **Moore's Law and the increasing effectiveness of computer chips.**

Source: Intel Corporation, Moore's Law Poster microprocessor chart, Copyright © 2005 Intel Corporation, http://download.intel.com/pressroom/images/events/moores_law_40th/Microprocessor_Chart.jpg

number of transistors on a chip would double every 18 months.[13] Although the rate of capacity growth slowed somewhat, Intel continued to produce new generations of increasingly powerful chips including the Pentium I, II and III and 4, and the Itanium Chip. See Figure 9.4.

Increasing the speed of innovation generally entails changes in the firm's organizational structure and the design of incentives for innovation. Wheelwright and Clark emphasize competition through new product development and suggest that companies need to make organizational changes to speed up product development including integration of marketing, manufacturing and design.[14] Stalk and Hout of the Boston Consulting Group find that the "structure of an organization that facilitates rapid new product design and introduction is analogous to the structure of a fast-response factory."[15] James Brian Quinn points out that because "high risk in inherent in the very nature of the innovative process", companies need "a very flexible, responsive, decentralized organization structure."[16]

Economies of speed can be achieved if the firm has adequate scale in R&D. However, simply investing in R&D is not sufficient; the innovation process itself must yield returns to additional investment. In addition, the R&D process must be well managed and product innovation must be attuned to customer needs.

Ford invested the substantial sum of $6 billion to design a world car called the Mondeo, marketed in the United States as the Ford Contour and the Mercury Mystique. Ford hoped to achieve speed economies by coordinating its engineers and design personnel in multiple countries and focusing their efforts on a single large-scale project. The roll-out of the Mondeo was slower than expected, however, as Ford encountered difficulties in coordinating large design and engineering teams. The Mondeo succeeded in Europe, but the Contour and Mystique did not do as well in the United States.[17]

Increasing the speed of product innovation allows a faster response to market conditions and consumer tastes. Stores like The Gap and Benetton, by tailoring their inventory to current fashion and updating their styles with very short production and distribution cycles, reduced the risk of unsold inventory while increasing total sales. These chains competed away sales from department stores that had significantly longer cycles and were less responsive to market trends. Moreover, rapid innovation in products and processes reduces competitive risk by beating competitors to market and possibly preempting new entrants. Combining R&D projects also allows for the application of unexpected spin-offs and permits the application of product development efforts to basic discoveries. Increasing speed by investing in R&D reduces the risks associated with the pursuit of blind alleys. By innovating faster, inevitably sinking dry wells along the way, it is possible to correct missteps before it is too late, thus avoiding delays caused by half-hearted innovation efforts. IBM invested billions in basic research leading to important innovations in such areas as superconductors and the scanning electron microscope. In 1998, for example, IBM was the leading company in patents for the sixth consecutive year with 2,658 patents — 38 percent more than its nearest competitor and 40 percent more than its own performance in the previous year.

Although the innovation process is inherently uncertain, it is costly, requiring investments in laboratories, licensing crucial technology from others, and payments to technical personnel. Attempting to achieve a higher rate of innovation entails greater expenditures. The number of innovations the company's research efforts produce in a given period of time is an approximate measure of its rate of innovation, although innovations can vary greatly in terms of their market value and whether they are incremental or drastic in nature.

Example 9.8 Economies of Speed

A company invests $45 million in its R&D program and expects to produce 15 innovations per year in terms of patents, product roll-outs, or similar measures of performance. At 15 innovations, the company's average cost (per innovation) equals $3 million. If the company increased its R&D investment to $60 million, it would expect to produce 30 innovations per year. At 30 innovations, the average cost (per innovation) equals $2 million. The company would achieve economies of speed by increasing its investment in R&D.

The cost of R&D is closely related to the well-known S-curve. This relationship says that most R&D programs go through a slow initial period, followed by more rapid success and finally ending in a slowdown or maturation of technical progress. This corresponds to a period of rapidly rising R&D costs, followed by a period of slower R&D cost growth and ending with a period of somewhat faster R&D cost growth. The minimum efficient speed of the firm corresponds to the innovation level just below that at which total R&D costs start to have a higher rate of growth.

The acceleration of R&D costs described by the S-curve shows the limits reached by a given project. To conquer those limits requires fundamental technological change that puts the firm on a new S-curve. This means that the S-curve for a particular project is a short-run phenomenon. In the long-run, the firm attempts to create new S-curves by recognizing and taking advantage of fundamental technological change. Companies change their underlying technology or research methods to improve the efficiency of the research process.

9.4 *Process Innovation*

We now come to one of the most important questions of competitive advantage. How can a firm sustain its cost advantage? The practical answer is that the firm can remain a cost leader only by continually lowering its costs before competitors can. Because production costs in an industry often tend to fall over time, any specific company must run fast to stay ahead of the pack.

Few cost advantages can be sustained for long. Companies should not count on a specific strategic action to fend off competitors. Any specific production process eventually will be matched or surpassed by competitors. Any specific cost level will be undercut by competitors.

Improvements in operations and manufacturing that cut costs are known as *process innovations*. Without continual process innovations, a company will not remain a cost leader. Companies are forever installing new equipment, using new materials, or updating information technology. As a consequence, existing equipment will be obsolete, materials will be improved, and computer programs will be surpassed. This is what Joseph Schumpeter famously called "gales of creative destruction".

External Analysis

To maintain an overall cost advantage, companies must ride the wave of technological change. Thus, managers should apply external analysis and internal analysis to monitor the potential effects of technological change on process innovation.

Technological change is relentless. The four phases of technological change operate all the time. First, researchers continually generate new scientific and technical knowledge. Second, inventors constantly apply both existing knowledge and recent discoveries to novel technology applications. Third, established companies and entrepreneurs develop these basic inventions for commercial application. Fourth, innovative companies keep on bringing these developments to market in the form of new processes, products, and transactions.

Managers apply external analysis to monitor the four phases of technological change. What are the implications of scientific advances? What are the effects of recent discoveries? What are the basic inventions being developed? What are entrepreneurs and established innovative companies up to? Since technological change is by its very nature unexpected, early warnings become more valuable.

The process of technological change need not be smooth. There are hundreds of small discoveries and developments. There are fundamental changes in the way scientists and engineers think about a problem. Sometimes there are drastic breakthroughs that totally change the way people live their lives or business operates.

The industrial revolution that took place in Britain from about 1760 to 1830 involved many small and large process innovations. The mechanization of cotton textile manufacturing enabled British exports to compete with textiles from India and elsewhere in Asia. G. N. von Tunzelmann points out that the main inducement for innovators in the cotton industry was *time saving*, rather than simply saving on capital investment or land usage.[18]

Process innovations fundamentally changed product markets by altering the nature of competition. Henry Ford's moving assembly line lowered the cost of producing cars. Ford's use of interchangeable parts also played a role in reducing costs.[19] The price of a Model T fell from $850 in 1908 to $600 in 1913 to $360 in 1916 and its market share increased from 10 percent in 1909 to 60 percent in 1921.[20] Mass production of cars and the low prices offered by Ford and General Motors changed the nature of the auto industry from individually crafted luxury goods to mass-consumption products.

Scientific progress is accelerating. Although technological progress can be uneven, no industry is isolated from its effects. Companies in any industry must have an innovation strategy. For example, all industries are affected in some way by progress in computers and communications. Computer technology automates organizational operations and interactions with customers and suppliers for any type of company. Communications developments, including the Internet, impact market transactions and organizations across industries. Improvements in management practices apply to most companies. In addition, basic scientific progress, in such fields as chemistry, biology, and materials sciences, impacts many different industries.

External monitoring of technological change poses challenges because it takes many forms. Production technology can be embodied in manufacturing equipment and information systems. But production technology also includes routine methods of performing activities and managing operations within an organization. The company's costs further depend on its procurement costs and transaction costs. At any particular time, the company might need to improve manufacturing technology, organization of operations, or input procurement.

Internal Analysis

The external analysis of technological change must be supplemented by an internal analysis of three things: the company's technological knowledge, the company's ability to monitor external technological change, and the company's ability to generate innovations.

Managers need to take stock of the company's technological knowledge. They need to determine the characteristics of the company's manufacturing and operations processes in comparison with industry and state-of-the-art benchmarks. How advanced is the company's equipment? How well-trained are the company's personnel? If the company's processes are not up to date, should the company upgrade those activities or outsource to others who are better qualified?

Closely tied with the company's technological knowledge is its ability to monitor external developments. Does the company have the ability to understand the four phases of technological change? Are personnel sufficiently qualified to understand basic scientific developments or does the company rely on monitoring only final innovations when they are being introduced into the market? Should the company rely on others to provide technological information? Will managers be overwhelmed by the flow of information, whether generated internally or externally? Are the company's operations sufficiently flexible to apply the new information?

Having evaluated the company's stock of knowledge and its ability to monitor external change, managers should determine the company's ability to generate its own innovations. Can the company contribute at any of the four phases: research, invention, development, or innovation?

Technological leadership may be too costly for some firms given their organizational resources and competencies, and companies should not pursue technology leadership for its own sake. Instead, managers should determine whether the costs of technological leadership outweigh the expected benefits. If managers conclude that technological leadership cannot be attained in a profitable way given the firm's existing resources and competencies, a fundamental change of strategy may be called for.

Innovation Strategy

Managers need to include innovation in their strategic planning process. Although technological change is filled with uncertainty, innovation planning is essential. How does the company achieve price leadership? To achieve and maintain a cost advantage, the company must continually modify its in-house production to keep pace with technological change. Fundamental changes in technology used in manufacturing or in the provision of services alter the costs of both incumbent firms and potential entrants.

Companies that are the first to develop and implement process innovations become cost leaders. Yet even complex and proprietary production processes can be imitated. Moreover, unique production processes can be surpassed by technological change. The price leadership strategy requires continual management of change, both in terms of operations and sometimes at level of the entire business.

By effectively combining state-of-the-art equipment, best-practices operations, convenient locations and so on, companies may be able to operate efficiently. However, it is evident that similar costs are within the reach of competitors. Competitors have access to similar equipment in the market. Benchmarking best practices is essential to staying competitive but it is not necessarily a source of cost advantage since competitors are able to adopt the same best practices. Location of production and distribution facilities can lower costs, but competitors can locate nearby.

When companies are racing for a patent, there are both high costs and great returns to being the winner. If the company believes it has a good chance of winning and that rivals have little possibility of imitating or surpassing the patented invention, entering the race is worthwhile.

However, prizes in the R&D race do not always go to the first one across the finish line. Many inventions can be imitated or even bypassed by new technology. Imitation sometimes places followers at a disadvantage if the innovation leader builds a brand name and a customer base. But, imitation can yield advantages for followers if they avoid high R&D costs and learn from the mistakes of the early innovator. Followers that choose to adopt technology later can leapfrog to a more advanced technology. Late adopters can tailor new products to customer preferences surpassing early innovators whose products were not adopted. Compaq, Dell, and others surpassed IBM even though IBM introduced the first personal computer based on Intel processors. Nokia surpassed Motorola, even though Motorola pioneered mobile phones.

Digital imaging for cameras provides an example of technology leapfrogging. Eastman Kodak introduced an image-sensing chip capable of reproducing images digitally with a resolution twice that of 35-millimeter film, which it manufactured using charged couple device (CCD) technology. Two weeks later, a start-up company, Foveon, announced a chip with similar resolution but manufactured using lower-cost production technology, complementary metal-oxide semiconductor (CMOS). Foveon was founded by Carver Mead, co-inventor of very large scale integrated circuit (VLSI) design. Chips manufactured using that design have applications not only in digital cameras but also in next-generation cell phones.[21]

Cost efficiency is a dynamic process because technical knowledge continues to improve. Companies must apply innovations to keep ahead or just to keep up with their competitors. Michael Hammer and James Champy found that application of computer technology to customer service or procurement required a fundamental redesign of company activities that they termed business process reengineering.[22]

Redesigning processes involves changing the company's basic operations technology rather than tinkering at the edges. The use of modular components reduces assembly costs for manufacturers while placing greater responsibility on suppliers to combine parts into modules. At a Canadian plant putting together GM's Chevrolet Tracker, a simple two-worker process that involves slipping a single component into the roof of the car in 30 seconds replaces a complicated process that took "six minutes and as many as 12 employees to install — roof liners, wiring, dome lights, coat hooks, rear washer hoses and grip assists."[23]

GM worked on redesigning its auto plants to facilitate delivery and use of components made by suppliers. GM's modular Blue Macaw plant in Gravatai, Brazil, relied on suppliers to assemble components — GM simply snapped the components together like a large version of a model car kit. The plant was laid out in a T shape to allow 16 suppliers to work inside GM's plant.[24] Having on-site suppliers reduced the costs of coordination.

R&D projects should be subjected to net present value analysis just as other types of investment projects are. Managers need to evaluate the expected costs of engaging in R&D projects. They should consider the expected competitive advantages of obtaining new technology by factoring in the likelihood of success, the expected duration of the project, and the anticipated value of the innovation. Then, the manager can evaluate the net expected benefit of the R&D project.

The R&D process itself can be a source of cost advantage. As already emphasized, companies that engage in significant levels of R&D may achieve economies of speed, lowering their average costs of innovation. Corporations may be able to sustain a substantial R&D effort by sharing the costs across business units. If a company has highly talented research personnel and experience in invention, it may be able to outperform its competitors in R&D. To achieve price leadership, however, the company must be able to translate basic research into cost savings.

Innovation by many firms tends to be technology-driven so that product innovations are "supply-push" rather than "demand-pull." Companies that have been subject to government regulation or have been otherwise sheltered from competition tend to focus on basic research and less on development. For example, before the Bell System was dismantled in 1984, AT&T's Bell Labs produced significant innovations. As a long-distance company, AT&T continued its tradition of high quality research. After AT&T spun off its Western Electric equipment and manufacturing unit and Bell Labs to form Lucent, the market value of the independent company rapidly increased, suggesting that the newly formed company was more efficient at bringing its basic innovations to market than it has been as a unit of a larger company.

Many large American, European, and Japanese companies have created independent research units, known as "skunk works" to address the flexibility and specialization required for innovation.[25] The key question about skunk works is whether they will eventually be integrated back into the company that created them. If they remain independent, will they be spun off as a separate company or are there benefits to retaining ownership of their technology? If the start-up is integrated into the organization, will its creativity be stifled by the same bureaucracy that it sought to avoid? Ideally, the influence goes the other way, with the creativity and emphasis on marketing by the start-up changing the parent organization.

The manager should compare in-house R&D with market alternatives. Companies should choose to bypass costly R&D if they can obtain new technologies by licensing the innovation after others develop it or by purchasing components from more innovative suppliers. Companies without a comparative advantage in R&D firms should devote less effort and investment to new product or process development, leaving these tasks to customers, suppliers, or competing firms.

The company can outsource production to the most qualified suppliers. Outsourcing can yield a cost advantage if suppliers are highly efficient and if the company's contracts with its suppliers allow it to reduce costs below those of competitors. Outsourcing production does not necessarily yield a cost advantage, however, if competitors are being served by the same suppliers. A company can achieve a cost advantage by outsourcing if suppliers are quicker to adopt new production processes and if the company's competitors pursue inefficient in-house production and are slow to switch to more efficient external suppliers.

Outsourcing also can yield advantages if the company has exclusive supplier contracts or if it is able to obtain price concessions from suppliers by high-volume purchases or long-term contracts. The company also can obtain cost advantages from outsourcing if it is able to lower the cost of its suppliers through close working relationships, technology sharing, improved communication procedures, or product redesign.

9.5 *Overview*

Achieving and maintaining a cost advantage over competitors requires cost efficiency and innovation. Cost advantage is only sustainable through development and application of process innovations at a more rapid rate than competitors.

Companies can obtain a cost advantage through economies of scale, scope, span, and speed. The ability of the company to achieve economies of scale depends on the level of its sales. The extent of economies of scope is limited by the range of the company's product offerings. Economies of span are realized to the extent that vertical integration lowers total costs. For vertical integration to be desirable, organizational governance costs must be less than the transaction costs of outsourcing production. Finally, economies of speed depend on the extent and effectiveness of the company's R&D activities.

Cost economies are not guaranteed by the size of the firm. Rather, they depend on the ability of management to induce employees to perform effectively both individually and collectively. To the extent permitted by market demand and organizational effectiveness, the firm pushes out its boundaries through growth, diversification, vertical integration, and accelerated innovation. Although cost economies are an important aspect of organizational design, they are not necessarily the main determinants of the firm's boundaries. Cost economies can be easily offset by the inflexibility and information costs of a larger organization. Moreover, internal economies must be compared with market alternatives. Outsourcing production and services often increases the organization's flexibility and market responsiveness. Moreover, sacrificing cost economies through decentralization of the organization's functions can often increase the creativity and performance of the organization.

Price leadership requires continual innovation. Companies use external analysis and internal analysis to formulate an innovation strategy, comparing the advantages of carrying out their own R&D with the benefits of obtaining process innovations in the technology market. Since few firms can go it alone, price leadership can require a mix of in-house R&D and technology procurement.

The price leadership strategy may not be feasible given the firm's resources and competencies in comparison with those of competitors. The manager must compare the pursuit of cost advantage with the alternatives of differentiation advantage and transaction advantage, then seek the type of competitive advantage that yields the greatest returns.

Questions for Discussion

1. Give specific industry examples of economies of scale, scope, span and speed.
2. A company producing tractors currently sells its products domestically. The managers of the company would like to increase sales by offering the company's products for sale in other countries. What are the potential cost effects of the company's international strategy?
3. Why are production costs of making a product such as steel different across countries? Compare the United States, France, Mexico, Korea, and China.
4. What are the benefits of training an employee? What are the direct costs of training an employee? What are the opportunity costs? Suppose that the company has spent substantial amounts training an employee and that technology changes, thus requiring more training. Should the company consider the amount already spent in training that employee in deciding whether or not to retrain the employee?
5. A company that manufactures cars makes 25 different models. Is there a cost to variety? What are the costs to the company of introducing one more model of car? What are the cost economies from producing so many models? Why do car companies try to use common platforms and components in different models?
6. Two companies that manufacture plastics plan to merge. How might their unit costs be reduced by the merger?
7. Advances in computers and telecommunications make it less costly to communicate within organizations and between organizations. How do such advances change the shape of organizations? Why do these types of technological changes seem to reduce the extent of

vertical integration? Why might these types of technological changes lead to a greater amount of outsourcing?

8. An appliance manufacturer wishes to incorporate computer chips in its coffee makers to improve their performance and to provide additional features. Should the company develop its own computer chips or rely on an outside supplier? What factors should managers consider in making this decision?

9. A pharmaceutical company wishes to develop new types of drugs based on developments in biotechnology. Should the company develop its own drugs or rely on contracts with independent laboratories? What factors should managers consider in making this decision?

10. A manufacturing company has invested billions of dollars in developing and establishing its production facility. At the time of its completion the production facility has lower per-unit costs than any other production facility in the industry. Is the level of the company's expenditure sufficient to guarantee that the company will maintain its cost advantage over its competitors?

11. A manager must choose between two R&D projects. One project is customer-focused with a known payoff within two years. One project involves basic science with an uncertain possibility of leading to a commercial innovation in five years. How should the manager go about comparing the two projects?

Endnotes

1. Michael Porter, *Competitive Advantage: Creating and Sustaining Superior Performance* (New York: Free Press, 1985), p. 76.
2. Company information from the company's website, cemex.com.
3. Peter Thompson, "How Much Did the Liberty Shipbuilders Learn? New Evidence for an Old Case Study," *Journal of Political Economy* 109 (February 2001), pp. 103–137.
4. Kazuhiro Mishina, "Learning by New Experiences: Revisiting the Flying Fortress Learning Curve," in *Learning By Doing in Markets, Firms, and Countries* by Naomi Lamoreaux, Daniel M.G. Raff, and Peter Temin, eds. (Chicago: Chicago University Press, 1999).
5. The account of Standard Oil is based on information from Alfred D. Chandler, "Fin de Siècle: Industrial Transformation," in *Fin de Siècle and its Legacy*, Mikulas Teich and Roy Porter, eds. (New York: Cambridge University Press, 1990).
6. Alvin J. Silk and Ernst R. Berndt, "Costs, Institutional Mobility Barriers, and Market Structure: Advertising Agencies as Multiproduct Firms," *Journal of Economics & Management Strategy* 3 (Fall 1994), pp. 437–480.
7. Elsewhere, to refer to the cost savings from combining stages in a production process within a single firm, I introduce the term *economies of*

sequence; Daniel F. Spulber, *Regulation and Markets* (Cambridge, MA: MIT Press, 1989).

8. Chandler, "Fin de Siècle," p. 37.

9. The importance of transaction costs in evaluating the make-or-buy decision was first made by Ronald M. Coase, "The Nature of the Firm," *Economica* 4 (1937), 386–405. Harold Demsetz observes that performing activities in-house also involves purchasing decisions. See Harold Demsetz, "The Theory of the Firm Revisited," in Oliver E. Williamson and Sidney G. Winter, eds. *The Nature of the Firm* (Oxford: Oxford University Press, 1991), pp. 159–178.

10. The widely used "S-curve" introduced by McKinsey & Company relates R&D effort in terms of investment to the performance of innovation. Foster (1986) describes the S-curve as the infancy, explosion, and gradual maturation of technological progress across firms but also speaks of the returns to investment of an individual development laboratory. Thus, for a specific project, the inverse of the S-curve is nothing more than a cost function for R&D with an initial range of performance levels for which there are increasing returns followed by a range of performance levels for which there are decreasing returns. See Richard D. Foster, *Innovation: The Attacker's Advantage* (New York: Summit Books, 1986).

11. See Fritz Machlup, *The Production and Distribution of Knowledge in the United States* (Princeton: Princeton University Press, 1962). See also Chris Freeman and Luc Soete, *The Economics of Industrial Innovation*, 3rd ed. (Cambridge, MA: MIT Press, 1997), pp. 6–8.

12. Freeman and Soete, *The Economics of Industrial Innovation*, p. 16.

13. Gordon E. Moore's original paper was "Cramming More Components Onto Integrated Circuits," *Electronics* 38, no. 8 (April 19, 1965). See also http://www.intel.com/research/silicon/mooreslaw.htm.

14. Steven C. Wheelwright and Kim B. Clark, *Revolutionizing Product Development* (New York: Free Press, 1992).

15. George Stalk, Jr. and T.M. Hout, *Competing Against Time* (New York: Free Press, 1990), p. 120.

16. James Brian Quinn, *Intelligent Enterprise* (New York: Free Press, 1992), pp. 293–294.

17. Daniel Howes, "Ford is Aware There Are Limits to Globalization in the Auto Industry," *The Detroit News*, Tuesday, October 19, 1999, www.detnews.com/AUTOS/howes/howes991019.htm

18. See G. N. von Tunzelmann, "Time-Saving Technical Change: The Cotton Industry in the English Industrial Revolution," *Explorations in Economic History* 32 (1995), pp. 1–27.

19. See J. T. Womack, D.T. Jones and D. Roos, *The Machine That Changed the World* (New York: Rawson Associates, 1990).

20. Freeman and Soete, *The Economics of Industrial Innovation*.

21. The discussion of Foveon is based on John Markoff, "Low-Price, Highly Ambitious Digital Chip," *New York Times*, September 11, 2000, p. C1.

22. Michael Hammer and James Champy, *Reengineering the Corporation: A Manifesto for Business Revolution* (New York: Harper Business, 1993).

23. Dave Phillips, "New Approaches Threaten Big Labor," *The Detroit News*, December 21, 1998.

24. Barbara McClellan, "Brazilian Revolution," *Ward's Auto World* 36 (September 2000), pp. 69–74.

25. Quinn, *Intelligent Enterprise*, p. 294.

CHAPTER 10

PRODUCT DIFFERENTIATION STRATEGY

Contents

A company has a *differentiation advantage* if enough customers prefer that company's products over those of competitors. The firm follows a differentiation strategy by producing products that provide added value relative to its competitors. The firm gains additional revenues that exceed the higher costs of the enhanced features. The company attracts customers who discern product differences and are willing to pay more for its products than for those of competitors. Increased customer willingness to pay allows the company to raise prices while still offering greater customer value. This chapter presents the product differentiation strategy and examines the basic drivers of product differentiation.

The product differentiation strategy requires the development of a differentiation advantage. The company *positions* its products in brand space in comparison with existing and anticipated brands of competitors. In addition, the company provides customers with information that affects their purchasing decisions and the way they evaluate the product after purchase. The company or its partners supply *complements* that provide enough benefits to customers to distinguish the company's products. Transaction costs and customer convenience are important aspects of customer benefits that are examined in the next chapter.

What are the sources of a differentiation advantage? Companies seek to differentiate themselves through many activities that go into providing products and services. The firm's internal processes such as design, manufacturing, quality control, and personnel training are potential sources of differentiation. Differentiation also comes through interaction with critical suppliers, distributors, and partners.

Differentiation advantage is temporary, as with all competitive advantages. A company's products may be unique, but creative competitors can copy or imitate product features. Even if uniqueness can be maintained, other companies can offer new goods and services that deliver even greater customer benefits for critical market segments. Competing products and services may come from out of the blue with features that are completely different from those of the goods they challenge. The greater the benefits from a company's products, the greater is the incentive for competitors to counter with their own offerings. Technological change constantly creates new opportunities for product development. Sustaining a differentiation advantage requires continual product innovation.

Chapter 10: Take-Away Points

Managers should develop and emphasize the differences between the company's products and those of competitors that offer the greatest competitive advantage:

- Differentiation advantage adds value by increasing customer willingness to pay for the firm's products relative to those of competitors.
- Differentiation advantage allows the company to attract customers with higher prices, offer greater surplus to suppliers, and capture additional value.
- It is not enough to be unique; product differentiation must add value by delivering greater customer benefits relative to costs than competitors' products.
- Managers must carefully balance their efforts in research, invention, development, and innovation.
- Managers should apply external analysis and internal analysis to monitor the potential effects of technological change on product innovation.
- Companies must rely on continual product innovation to achieve and maintain a differentiation advantage.

10.1 *Customer Benefits*

Being different is not enough. Differentiation cannot be restricted to the physical or technical features of the company's product or service. What counts are customer preferences, so that the product differentiation strategy must target customer value. The company's products must deliver greater *customer benefits*, so that customers will have a greater willingness to pay for the products.

Companies choose the features of their products strategically. Managers need to ask how the company's products serve the needs of their customers in comparison with the products offered by competitors. Managers should consider the benefits customers receive, the costs of production, and competitors' actual and potential offerings. By enhancing product features relative to competitors, a company increases customer willingness to pay for its product relative to competitors products, which increases total value and yields a differentiation advantage.

Customer Preferences and Product Differentiation

Economists are fond of saying: "You cannot argue tastes."[1] By this, they mean that a rational dispute should end when it becomes a matter of personal preferences since these are not subject to persuasion. That presumably explains why logical advice is rarely followed in practice — arguing tastes is highly popular! What are book, movie, and restaurant reviews if not arguments about taste? People debate the relative merits of works of art and music, of different types of food, or of places to go on vacation.

Fashion trends for art, architecture, clothing, or cars show that people do influence each other. Moreover, people are influenced by the choices of others, from blockbuster movies, to best-selling novels, to political bandwagons. Individual tastes are apt to change as people develop an affinity or dislike as they repeatedly consume a product. Readers of children's books remember Dr. Seuss's character who found that he liked green eggs and ham after being persuaded to try them by his friend Sam-I-Am. The tastes of populations can change as immigrants bring different interests and traditions, or as age distributions change over time.

Ultimately, what matters is the entire customer experience from search to purchase to consumption. The customer's choice of products depends on the customer's income and preferences, which in turn depend on personal characteristics, experience, cultural background, social interaction, and other factors. The customer's overall experience reflects his or her interaction with the marketplace and consumption of the product.

Customers have preferences over the range of possible product characteristics. This range of possible variations of product features is called the *brand space*. Different brands are different combinations of product features. A car has a complicated set of features from engine specifications to body design that distinguish a Ford Taurus from a Volkswagen Beetle.

Product differentiation can be based on both subjective and objective factors. *Subjective* factors are those that are entirely a matter of taste, and people need not agree on the desirability of the product features.[2] *Objective* factors are those features for which customers are in nearly universal agreement on desirability. These features are usually called product quality.

Despite this distinction, product differentiation is ultimately subjective; that is, it all comes down to a matter of taste. Because products have many features, people may not agree on the right *mix* of features, even if they agree on the appeal of individual objective features. The art and science of product differentiation is choosing product features that attract the targeted customers, even if it is all a matter of taste.

How do individual customers feel about specific features of the company's products? Some prefer to drink very sweet sodas while others prefer their beverages dry. Many want to eat crunchy cookies while other prefer soft and chewy ones. Some like red cars and others like blue ones. These are called subjective features.

Those features that everyone agrees upon, the objective features, are often based on quality. Everyone prefers a battery that lasts longer, a tire whose treads resist wear, an appliance that rarely breaks down. There is general agreement that cars should have better gas mileage, that refrigerators should have greater energy efficiency, or that home insulation should have less heat conductivity. People generally prefer computers with faster processing speed and greater memory capacity.

Yet, although everyone agrees that more quality is a good thing, people differ on their willingness to pay for higher quality. Some people buying a wireless phone or handheld organizer have a higher willingness to pay for battery life than others. So, even though everyone agrees that more battery life is a good thing, some value it more than others. Although quality is an objective feature, then, the level of willingness to pay for quality is not something everyone would agree upon.

Customer differences in willingness to pay for product features, whether subjective or objective, are a basis for segmenting markets. Companies take advantage of differing willingness to pay for quality by offering a product line, or several product lines, in which higher quality goods cost more. Airlines offer coach, business, or first class seats for which quality improvement goes together with a higher ticket price.

In practice, most products or services are bundles of many features. Because product features come in bundles, customer choices inevitably involve a trade-off. Which bundle of features best addresses the customer's need, even if no product provides exactly the desired set of features? Because choosing products is usually a multidimensional problem for consumers, it comes down to a matter of taste again, even if some or all of the product features are objective.

Consider the case of cookies that differ in terms of two subjective features: consistency (from soft to crunchy) and sweetness (from less sweet to sweeter). Figure 10.1 shows the wide range of possibilities and

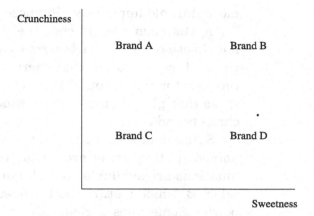

Figure 10.1: **Four brands of breakfast cereal with two taste features, any of which might appeal to a consumer.**

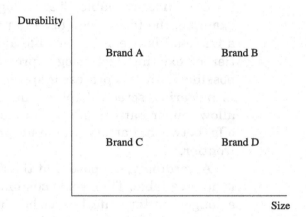

Figure 10.2: **Four brands of appliances with a quality feature (durability) and a taste feature (size). All other things equal including price, customers should prefer brands A and B.**

suggests how four brands can combine the features in different ways. Brands A and B have the same crunchiness as each other but different sweetness, while brands A and C have the same sweetness as each other but different crunchiness. Different consumers might purchase any of the four brands.

Next, consider products with characteristics that are combinations of quality and taste features. For example, a household appliance such as a refrigerator has quality features such as durability and taste features such as size. Everyone would want a more durable refrigerator, but people will differ on desired size depending on their need for storage space and the roominess of their kitchen. Figure 10.2 shows four brands that are different combinations of these two features. All other things being equal, such as price, customers would purchase only brands A and B because, size requirements aside, all customers prefer more durable appliances. Suppose now that brand B is not available. Then, customers would choose either brands A or D; they still would not choose brand C. Customers would face a trade-off between durability and size. Those customers that particularly wanted a large refrigerator would purchase brand D with less durability, whereas those that placed greater emphasis on durability than size would purchase brand A.

Suppose now that both features are quality features, as battery life and computing power are for laptop computers. If all four brand combinations are available, as in Figure 10.3, customers would all wish to select B, which features both longer battery life and greater computing power. Sometimes a clear winner such as brand B is available and emerges in the market. This type of advantage allows a company to dominate other brands by consistently enhancing product features.

However, a technological advance that enhances multiple features is often not available. Technologies inevitably face trade-offs. For example, the technology used to produce laptop computers encounters a trade-off between battery life and computing power. Along the frontier of existing technology, providing a greater battery life is only possible with less processor speed. Even if technological change occurs in batteries, screen displays, data storage, and microprocessors that allow longer battery life with greater processor speed, the new trade-offs between battery life and processor speed would open a new frontier.

Accordingly, suppose that the clear winner, brand B in Figure 10.3, is not available. Then some customers would prefer brand A, which has a longer battery life but is heavier to carry, while other customers would prefer brand D, which is easy to carry but has a short battery life. This example illustrates that even though everyone agrees on each

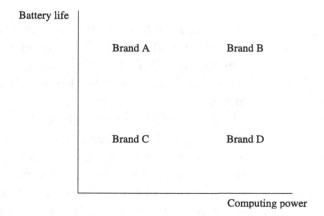

Figure 10.3: Trade-off between battery life and computing power for laptop computers. All consumers would prefer B. If B is not available, consumers would choose either A or D.

of the quality features taken one at a time, the choice of brands will still be a matter of taste. Generally, there is no one brand that is the best in all quality dimensions. The available brands offer a trade-off in quality features. The best mix of these features is a matter of taste, and people can differ in their most preferred brand.

Even if the product has only one feature, which is a quality feature that everyone agrees upon, the level of willingness to pay for that feature differs across consumers. Therefore, even objective measures of quality such as durability or battery life are a matter of taste.

Computers differ in amount of memory, all other things such as the speed of computation being equal. Even though everyone agrees that more memory is a good thing, people will differ in their willingness to pay for computer memory. Although memory is a quality feature, the willingness to pay for that memory is a matter of taste. Providing computer memory is costly for computer manufacturers. Accordingly, some brands will offer a computer with a lot of memory at a high price whereas other brands will offer a computer with less memory at a lower price. The consumer will buy more memory up to the point where the additional expenditure required for more memory exceeds the benefits from additional memory.

Product Lines

Companies address the trade-off between quality and price by offering product lines with different combinations of features at different prices. Consumers purchase the product within a particular line based on their willingness to pay for additional product features.

Willingness to pay depends on subjective features, too. Suppose that cola drinks are identical except in terms of sweetness. An individual consumer may prefer a cola that is very sweet whereas another prefers a cola that is less sweet. The consumer that prefers a cola of a certain level of sweetness will have the highest willingness to pay for the level of sweetness that is just right and would be willing to pay somewhat less for a cola that is either sweeter or less sweet.

Customers cannot always be counted on to reveal their preferences when answering surveys or participating in marketing focus groups. They do reveal a lot about their preferences through their actual choices in the marketplace. When a customer chooses between two different price-quality bundles or between two products with distinct sets of features, they reveal something about their willingness to pay.

For example, service stations offer three types of gasoline at the pump — regular, super and premium, — with each costing a bit more than the preceding grade. Customers 'reveal' that they are willing to pay more for higher levels of octane through their choice of what gasoline to pump into their car. Generally, a quality discount is present.

As the customer moves up the product line, the product or service costs more but the average cost of the quality feature decreases.

Customers voluntarily reveal their willingness to pay for quality by their choice in a product line. Choosing between low- and high-end products reveals a low or high willingness to pay. Customers' revelation of their willingness to pay allows the company to price to each segment, charging more for higher quality but providing an incentive to purchase greater quality by discount — essentially offering a better deal as quality rises. This approach increases profitability while giving customers greater selection and creating higher net value for each individual customer.

The company may chose to serve the low, middle, and high ends of the market with targeted business units or even separate divisions. Marriott offers economy hotels (Fairfield Inns), moderately priced suites (Residence Inns), moderately priced hotels (Courtyard Hotels), full-service suites (Marriott Suites), and full-service hotels (Marriott Hotels and Resorts).[3]

Example 10.1   Self-Selection and the Product Line

Suppose that there is one company and two consumers. The company offers two types of suitcases: standard and premium. Everyone agrees that premium suitcases are more durable than standard ones. The price of a standard suitcase is $75 and the price of a premium suitcase is $150. The first consumer has a willingness to pay of $100 for a standard suitcase and a willingness to pay of $200 for a premium suitcase. The second consumer values a standard suitcase at $140 and a premium suitcase at $210. What suitcases will the two consumers purchase?

The first consumer gets a surplus of $100 – $75 = $25 for the standard suitcase and a surplus of $200 – $150 = $50 for the premium suitcase. So, the first consumer will choose the premium suitcase. The second consumer gets a surplus of $140 – $75 = $65 for the standard suitcase and a surplus of $210 – $150 = $60 for the premium suitcase. So, the second consumer will choose the standard suitcase. Even though the second consumer has greater willingness to pay for both types of suitcases than the first consumer, the second consumer has a lower willingness to pay for the quality increment. The example is summarized in Table 10.1.

As the example illustrates, product differentiation is a matter of subjective customer preferences rather than an objective measure of

Table 10.1: Willingness to pay for product quality.

	Willingness to Pay for Standard Model	Customer Value from Buying Standard Model	Willingness to Pay for Premium Model	Customer Value from Buying Premium Model
Consumer 1	$100	$100 – $75 = $25	$200	$200 – $150 = $50
Consumer 2	$140	$140 – $75 = $65	$210	$210 – $150 = $60

product features. Looking at technological product features will ultimately be misleading, causing the firm to lump together customers with widely different tastes and to provide different products to customers with similar tastes. Companies must look outward to their customers to discover what their markets look like. The product differentiation strategy requires providing customers with more value than competitors.

Some managers mistakenly define their company's business strictly in terms of the characteristics of the products they produce. As a result, these managers become product-oriented rather than customer-oriented, defining their business on the basis of the features of their own products. Managers should focus instead on the services that customers receive.

Theodore Levitt identified *marketing myopia* as an important source of management failure.[4] He observes that the railroads lost customers because they saw themselves in the railroad business rather than the transportation business, and therefore they did not offer new transportation services (cars, trucks, airplanes). Hollywood long viewed television as a threat rather than a competitive opportunity to expand entertainment. Levitt notes the troubles the oil industry encountered because it viewed its product as gasoline, not more broadly as energy, fuel, or transportation services. He warns that managers often become enamored with product features, production technology, and research and development, while neglecting the needs of the customer.

The myopia problem is due to a management concern with physical features or other aspects of the company's products rather than the benefits that they provide to the final customers. Often, the list of product characteristics creates a narrow perspective. Auto companies emphasize attributes of their cars, such as speed, horsepower, gas mileage, size, or color. The customer is cognizant of these things, but is fundamentally trying to solve a transportation problem — getting to work, picking up the kids, or getting groceries. For some, the car provides additional services including conspicuous consumption. Automakers often emphasize the relative performance of the car compared to others in its "class", a narrow category of cars of the same size.

Customers often consider a much wider set of substitutes, including other types of vehicles or other forms of transportation, such as commuter rail lines.

Software makers stress the features of a program and differences with previous versions, including screen icons, tool bars, and multitasking features. The user thinks in terms of carrying out certain tasks, such as writing a letter or a novel. WordPerfect Office offered 50 automatic "QuickTasks" such as send message, create newsletter, and create address list. This feature attempts to go beyond technical descriptions of features such as "drag-and-drop" for moving text and pictures to address the *services* that the software provides.

Restaurant chains describe their menus and the decor of the restaurants. Yet, customers visit the restaurant wishing to dine, to entertain others, or to celebrate an occasion. The alternatives to eating out can include take-out or convenience foods. Entertainment alternatives include movies, theaters, and museums. By taking a narrow view of the restaurant service, managers can miss potential competitors providing dining or entertainment services.

A manager at a major telecommunications company once suggested that the company's pay phone business did not need to worry about competition from wireless phones. The reason he offered was based on technology differences: Pay phones used land lines and wireless phones did not. Yet from the customer's point of view, the two devices were simply alternative ways of communicating, that is, substitute products. The wireless phone offered the customer the ability to call untethered while walking, driving, or riding on a train, not to mention the payment convenience and the cleanliness of a personal phone.

In defining its markets, the manager should focus instead on the customer's point of view. What services does the customer receive? How does the customer perceive the product? What needs does the product or service satisfy or what problems does it solve? What is the customer's willingness to pay for the firm's product? What are the competitive substitutes currently available or potentially available to the consumer?

Defining markets does not begin with the company's technology, traditions, or product features. Rather, it begins with the benefits that customers receive. A company defines its markets based on the characteristics of the company's actual and potential customers and the services they seek. Customer preferences are the foundation for the company's strategic decisions about pricing, marketing, product design, and innovation. Without a clear understanding of what customers desire, the company cannot identify its current and future competitors.

Exhibit 10.1   Fax Machines

To illustrate the difference between marketing myopia and market definition based on customer benefits, consider a company that produces fax machines. How does the company define its business? How should the company view its market? What type of product differentiation strategy should the company pursue? The company has several alternatives: by product, by service features, or by customer benefits.

Suppose that the company producing fax machines views its competitors as being limited to other companies that produce fax machines, such as Cannon and Ricoh. The company is likely to establish a fax machine division that develops expertise in producing fax machines. Managers will keep up with the latest developments in fax technology, achieve economies of scale in production, and diversify by developing a full line of fax machines. The market segments can be defined in terms of the quality and features of low-, middle, and high-end fax machines. This narrow market definition is strategically dangerous since the company sees only other fax machine makers as its competitors.

Taking a broader view, the company's managers might define their market in terms of *service features*, especially document transmission. Then, the competitive alternatives available to customers widen considerably to include not only fax machines, but also computer fax, e-mail, and express mail. Sales of computers with modems far exceed sales of fax machines. Specialized software allows a person with a computer to send a document stored in a computer file over the phone lines to a fax machine. In addition, computers can receive faxes. Thus, the company's fax machines compete with computer hardware and software that can be applied to sending and receiving faxes. By widening their perspective, managers could anticipate and even take part in the trend, producing or reselling fax modems and computer fax software. Accordingly, the company would create a document transmission division that would present its customers with various solutions to the problem of sending and receiving documents.

Widening their reach still further, the managers could identify the *customer preferences and needs*. The fax machine provides communication services. Customers use it in transmitting and preparing documents, exchanging messages, and sending bills and invoices. Its competitors include not only computer fax, e-mail, voice mail and express mail, but also the telephone companies,

(Continued)

Exhibit 10.1 (*Continued*)

paging companies and Internet service providers. Messages generated by fax machine, e-mail, or voice mail can be directed to a single messaging center that is operated by a telephone company, an Internet service provider, or some other type of company. The end-user accesses messages from the message center using a telephone, wireless equipment, a PC, or some other device.

A number of product innovations make unified messaging possible. For example, Lucent Technologies Octel Messaging Division is a leading provider of advanced voice messaging with over 110 million mailboxes in 90 countries worldwide. Headquartered in Milpitas, California, the Octel Messaging Division serves more than 100,000 business, government, and educational customers worldwide, and more than 140 wireless and wireline service providers in 25 countries. Octel offers a number of innovative products, including a family of servers for business customers that allows voice messaging, fax messaging, and call processing on platforms ranging from a PC-based model to the Octel 350 that supports up to 30,000 users. The company offers software such as the Octel Unified Messenger that adds features to Microsoft's Exchange software, setting up a single mailbox for voice, e-mail, and fax that users can access and manage from a multimedia PC. The Intuity Audix Multimedia Messaging System uses industry standard hardware components for businesses to integrate voice, fax, and e-mail. The Intuity Interchange is a messaging network platform that uses server technology to create a hub-and-spoke, multimedia messaging network.[5]

Unified messaging means that companies in what previously were different modes of communication compete with each other, creating some degree of convergence in voice messaging, paging, faxing, and e-mailing. In this example, not only do the substitute services compete in the provision of messaging, but companies offering combinations of the substitute services are competitors. Phone companies, paging companies, and Internet service providers become competitors in messaging.

The managers of the fax machine company overcome marketing myopia by understanding that product innovations create new communications services. These communications services are substitute products for the basic fax machine, even though the means of providing competing services differ considerably. By recognizing that it is in the communications business rather than in the fax machine business, the company can anticipate who will be its competitors.

(*Continued*)

Exhibit 10.1 (*Continued*)

The company can enhance its fax machine to handle different types of communication. Moreover, the company can create new products and services that address the communications needs of its customers. The fax machine company can set up a wider-ranging communications division that offers alternative products to its customers, and thus it can avoid being stuck supplying an obsolete specialized product.

10.2 *Product Differentiation*

Companies seek to develop strategies that result in differentiation advantage. Just as cost leadership means low cost relative to competitors, so differentiation advantage means that customers prefer the company's product relative to competitors. Differentiation strategies include the choice of product attributes, the company's position in brand space, pricing methods, and the provision of information to customers and the availability of complements.

Choosing Attributes

The extent to which the firms earn returns from differentiation depends on the extent to which customers value product differences, the cost of providing differentiated products and the extent of differentiation from competitors. Particular product attributes are potential avenues of differentiation only if customers care about them. Customers must also be able to discern the difference between competing products.

Customers must be willing to pay for the differentiated product. The additional benefits they receive must be greater than the cost of providing the enhanced product. Product differentiation only adds value if the additional customer benefit obtained from enhancing the product is greater than the additional cost of enhancing the product. A company can only obtain a competitive advantage by creating more value than its competitors. So a differentiation advantage means that after the product innovation, customer value net of the costs to suppliers and the costs of using the company's assets increases.

The choice of product attributes for differentiation depends on the industry. Retailers have a number of critical attributes. Retail stores differ in terms of *location*, *hours of operation*, and *personal service*. Internet retailers have the advantage of avoiding the real estate costs of a bricks and mortar store, presence on the customer's computer,

24 hours a day/7 days a week operation, and personalized Web pages. Traditional stores have the advantage of interaction with sales personnel, experiencing the merchandise, and immediate availability. Some companies with both stores and Internet presence make the mistake of keeping their two operations separate. Companies found that they could benefit by combining stores and websites to take advantage of the best combination.

Retail stores differ in terms of the *number* of products they carry, often measured in terms of stock-keeping units (SKUs): A supermarket carries a wider assortment of SKUs than a convenience store such as 7-Eleven; a category killer such as Toys-R-Us has a vastly wider assortment than an independent toy store or specialty toy store such as Zany Brainy. Stores differ in terms of the *selection* of products and brands they have available, an upscale clothing store such as Nordstrom is distinguished from a discount store such as Kohl's. Retailers differ in terms of product availability, also called *immediacy*. Blockbuster distinguishes itself from other video rental stores by offering a guarantee of availability. Providing immediacy requires carrying a larger inventory of some items, reordering more frequently, and operating more sophisticated information systems that track both inventory and shifts in customer demand.

Some of these methods of differentiation are illustrated by the differences between Amazon.com and Barnes & Noble.com. The two companies have many similarities. Both offer somewhat similar Internet websites, both allow customers to order for home delivery, both have wireless and online access, both offer essentially the same access to books and recorded music, and both provide editorial content such as book recommendations. Moreover, both companies operate warehouse distribution systems. The companies differ along other dimensions. Amazon exists only on the Internet whereas Barnes & Noble is affiliated with a parent company that operates a chain of superstores featuring a wide selection, sales personnel, and a coffee shop. In contrast to Barnes & Noble, Amazon built online brand recognition by moving early into Internet book retailing, and then extended its brand to a wide array of products. Amazon branched out into auctions, electronics, toys and video games, tools and hardware, lawn and patio products, kitchen products, and even automobiles. Although both companies offer access to extensive catalogs of in-print books and access to networks of suppliers of out-of-print books, the two Internet retailers dispute which one has the largest selection of books. Amazon refers to itself as "Earth's Biggest Selection." Barnes & Noble replies that "With 800,000 titles on its shelves, it now offers the largest selection of in-stock titles of any bookseller online," and notes that with "30 miles of shelving, it is the largest collection of books under one roof in the history of bookselling." In October 1999, seeking to differentiate its business methods, Amazon.com filed suit against Barnes & Noble.com alleging patent infringement of its trademark 1-Click technology that allows customers to order with one click of the mouse button.

Many financial intermediaries such as banks and brokerages differentiate themselves in terms of product selection: Citigroup offers a broad range of financial services including insurance, banking, and securities brokerage while Charles Schwab advertises its mutual fund supermarket. Financial firms offer various complementary services, such as stock market information and financial planning advice. Some securities brokerages provide more complete information services whereas discount brokerages provide generic information. Another important source of differentiation for financial institutions is customer trust. Banks advertise that they are insured by the Federal Deposit Insurance Corporation (FDIC) so as to distinguish themselves by offering trustworthiness beyond the commonly guaranteed level.

Manufacturers selling products to consumers seek to distinguish their offerings by subjective characteristics such as design, packaging, or brand image and quality or performance measures such as durability, energy efficiency, or ease of maintenance. Manufacturers selling to other businesses tend to emphasize technological prowess and reliability of service. ExxonMobil's Chemical division emphasizes its expertise in developing high pressure low density polyethylene (LDPE), one of the largest segments of the world plastics market. In distinguishing its technology licensing program, ExxonMobil Chemical highlights its access to two world-scale technical centers; focus on design safety and flexibility; extensive process, product, and applications training; access to future product and process development; and tailoring of polyethylene product grades.[6]

Manufacturers can distinguish their products through the choice of distribution channels, selling through specialized dealers or upscale or discount stores, or striving to make their product available everywhere. Petroleum companies have tended to vertically integrate by owning gasoline stations, differentiated not only by location but also by brand name fuel and other products, convenience stores, and credit cards.

Position

Companies seek a differentiation advantage by maneuvering for *position* in brand space. This means that companies choose product features strategically in response to actions or anticipated actions of competitors. Companies must choose products and services that meet the needs of customers more effectively than competitors' products and services.

The basics of product positioning can be illustrated with a simple brand space. The brand space describes product features. Features can

Basic C   Brand B   Brand A   Brand C   Complex

Figure 10.4:   Brand space and brand locations for a software product.

be any type of feature or mix of features. Recall that with multiple types of quality, customers differ on their most preferred mix. Each location in the brand space is an address where a company might locate a potential brand. For example, Figure 10.4 shows a brand space for software, ranging from basic to complex. At the basic end, the software has few applications but great ease of use. At the complex end, the software has many applications but is more difficult to use.

Every customer is likely to have a most preferred potential brand. Some customers will prefer basic software, others complex software, and still others will desire somewhere in between. The population of potential customers can be seen as being located along the brand space, just as customers might be physically located in a geographic region. The strategies described for the case in which the brand space is a line also apply to more difficult situations with multidimensional product features.

Suppose that a company in the market already offers brand A, as in Figure 10.4. Where should a competitor locate in the brand space? There are three possible solutions. The first solution is to get basic and appeal to ease of use. If the competitor offers brand B, then it will capture customers seeking the most basic software and some customers located between brand B and brand A. The second solution is imitation. By locating very near the existing brand, customers will find it hard to distinguish between the two products, excluding consideration of price and brand image, and the two companies will split the entire market. The third solution is to get complex and appeal to those who are more interested in applications than ease of use. By locating at C, the competitor's brand captures all of the customers interested in having more complex functions, that is, all the customers above C, and also some of the customers located between A and C.

Managers should evaluate all three of these strategies; each entails different revenues, different customers, and different costs. Any of the three strategies may be attractive depending in large part on the way that customers evaluate software features and how many customers favor each type of brand. The effectiveness of the strategies is also affected by whether the existing company can reposition its brand.

The choice of positioning strategy depends on the extent of competition and the market demand for variety. The imitation strategy is risky: It involves the two brands fighting for the middle of the market because the basic and complex ends are not necessarily well-served. Fighting for the middle with a one-size-fits-all strategy opens the door to future entrants coming in at both ends of the market. New competitors will offer very basic or very complex software and steal those market segments from the established companies. A strategy of excessive differentiation is also risky. If the two companies run off to the ends of the market, with one company tending to the basic end of the market and the other to the complex end, new entrants can attack the middle of the market by offering a brand with broad appeal.

Finally, note that a company can offer multiple brands. If a company offers brands A and B, a competitor will have an easier time differentiating itself by serving the more complex end of the market by offering brand C. Alternatively, if the company offers brands B and C, a competitor could then be attracted by the basic end of the market.

The product differentiation strategy is subject to attack from both the high end and the low end. On the high end, competitors offer still better products with greater customer value. On the low end, competitors offer generic or basic products or services. A basic product that is low quality but offered at a sufficiently lower price will attract consumers with lower willingness to pay for product quality. Such a strategy is useful for entrants seeking to offer new products and services that can compete for the low end segment of an established firm's market.

Example 10.2 Divide and Conquer

Suppose that there are two consumers and two software companies. Suppose that one company offers an advanced software product at a price of $200. The first consumer has a willingness to pay of $260 for the advanced software and the second consumer has a willingness to pay of $220 for the advanced software. The competing company offers a basic software product at a price of $100. The first consumer values the basic software product at $150 and the second consumer values the basic software product at $130. What type of software will the two consumers purchase?

The first consumer chooses to buy the advanced software because that consumer obtains surplus equal to $260 – $200 = $60 from buying the advanced software and surplus of $150 – $100 = $50 from buying the basic software. The second consumer chooses to buy the basic software because that consumer obtains surplus equal to $220 – $200 = $20 from buying the advanced software and surplus of $130 – $100 = $30 from buying the basic software. The competitor's strategy of offering basic software attacks the low end of the market and successfully segments the market. Would the outcome change if the entrant lowered the price of basic software to $80?

To counteract the divide-and-conquer strategy, companies attempt to offer a product line that addresses the needs of many market segments. Offering a product line has many advantages. The company extends the benefits of its brand name to a wider array of products. Each customer has more choice of product features, and the company can tailor pricing more closely to the segment. Total sales increase because customers obtain products that more closely match their preferences. A number of problems are also associated with offering a product line. The company encounters increased production costs of offering varieties. It will have lower scale economies, although they may be partly offset by economies of scope across the product line. There are transaction costs of marketing and distributing many different products. Providing variety can entail higher inventories and greater use of costly shelf space.

Pricing Methods

Pricing methods provide a means of differentiating the product. Sears offered "everyday low prices" to differentiate its stores from competitors. Many discounters offer price stability as an inducement to customers who may not know the exact price of specific items when they plan a shopping trip, but who will choose a store that has traditionally offered low prices and that keeps its prices relatively stable, avoiding the rapid swings that come with a cycle of sales and markups. Procter & Gamble switched to a policy of steady low prices with less reliance on promotions to compete more effectively with national brands while counteracting lower priced generic goods.

Many companies use discounts, promotions, and sales to stimulate interest in their product and service offerings. Department stores feature frequent markdowns for every occasion. These price fluctuations draw attention to the company and are a good marketing device. They separate price-conscious shoppers, who are willing to visit the store to take advantage of the sale, from other customers who buy on impulse whether or not there is a sale and from customers whose time is sufficiently valuable that they cannot alter their plans to take advantage of a sale.

There is no obvious prescription. A steady pricing policy reduces customer transaction costs. Buyers turn to a company because they are able to rely on past experience as a guide. Companies can then win the market with a low-pricing policy, counting on repeat sales in the future to drive down unit costs and allow continued low prices. Alternatively, a flexible pricing policy draws attention to the company and allows for a rapid response to market events that suggest raising or lowering prices.

The choice between these pricing approaches depends on strategic considerations. When facing low-priced competitors, the company must find ways to beat the low prices or to enhance customer value through product and service differences. When facing companies with rapidly fluctuating prices, a pricing policy that insures customers against price risk has advantages. The problem with a steady pricing policy is that customers will buy from the steady pricing firm when the rival company's prices are high and not when the rival company's prices are marked down for a sale. The solution to this is to provide guarantees, such as airline frequent flyer programs that reward customer loyalty by assurances that expenditures will be lower over the long term.

Prices must be easily understood. Companies that confuse potential customers with complex options and discounts on list prices that are difficult to calculate can be outmaneuvered by prices that are simple. In this way, straightforward pricing provides a service to customers by reducing the costs of comparison shopping. Customers will even pay a premium for convenience. For this reason, complex marketing and promotional schemes can be defeated by presenting customers with a basic price plan.

Where competitors' prices are excessively high, the company can follow an everyday low price strategy. Where others' prices are complex and difficult for customers to understand, the company should present simple pricing rules that may sacrifice some short-term returns for long-term building of customer confidence. If rivals disguise their prices, requiring shopping, bargaining, or other time-consuming methods to discover the price, the company should provide easily accessible pricing information. Where competitors constantly change their prices, creating risk and uncertainty for customers, the company can offer fixed prices or pricing guaranties, such as most favored nations clauses. Where rivals bundle unwanted goods and services, the company can unbundle, pricing components separately and offering customers the flexibility to create their own options packages.

Information

The *brand* is central to a product differentiation strategy. A brand name conveys a vast amount of information to customers about the product. The brand name summarizes quickly the many different attributes of the product, conjuring up the complex characteristics of a wine or the specifications of an automobile. The brand name further communicates the company's reputation for quality or customer service. Successful brands, often enhanced with trademarks and colorful logos, are widely recognized and instantly understood by customers.

Customer perceptions of the company's product are its *brand image*. Brand image depends on marketing, media information, promotions, word-of-mouth, and the customer's experience with the product. The attributes of the product will affect customer perception of the brand. Different brands offered by competitors do not guarantee product differentiation because customers must perceive a difference. The reliable performance of a brand in relation to customer expectations creates *brand equity*. The company that owns the brand receives returns to performance though continued sales.

Companies must invest in product quality and customer service to maintain their brand equity. Poor performance can quickly destroy brand equity. Tylenol turned around a product tampering problem with improvements in packaging. Bridgestone sought to restore customer confidence in its Firestone brand by offering replacement tires in a massive recall of defective tires that had been sold on Ford sport-utility vehicles. Ford Motor Company also faced criticism associated with the Firestone tires. To try to maintain confidence in its brand, Jacques Nasser assured customers in an advertisement, "You will have my personal guarantee that no one at Ford will rest until every recalled tire is replaced."[7]

A company's brand name may be on more than one product. By extending the brand over multiple products, companies differentiate a product line from those of competitors. Joseph Campbell started his company in 1869, but it was not until 1895 that the company marketed a ready-to-serve beefsteak tomato soup. By 1902, the company offered 21 varieties of soups with the *Campbell Soup* brand. Although the company owns *Pepperidge Farm*, *Godiva* chocolatiers, *Prego* spaghetti sauces, and *Pace* foods, it is perhaps best known for its soup.[8] Honda employs an international brand extension strategy. It enters developing countries by first selling Honda motorcycles, thus establishing its brand image in the country with a more affordable product before it begins to sell cars.[9]

International sales of products contribute incremental revenue, but customer preferences can differ across countries. Some types of products such as *Coke* and *Pepsi* are sold in many countries without much variation. Other products, such as books and magazines, depend on local interests and language. Providing standardized products has the advantages of economies of scale in production and distribution. Providing tailored products has the advantage of increasing demand because the product variations address customers' needs. International companies must consider this trade-off in designing their products.

In matters of taste and customer choice, information matters a great deal. Information here denotes both factual communication and image advertising. The consumer's willingness to pay depends in part on the extent of information available about a product's features. Because information affects customer willingness to pay, information can be a key source of differentiation advantage.

Information can affect buyer perceptions of the product both before purchase and during consumption. To the extent that tastes can be influenced by persuasion, information plays a role in altering customer preferences. Company messages can enhance the customer's experience when using the product. Fashionable logos on clothing or cars can contribute to customer benefits.

Consumers receive information from many sources, including friends and family, advertising, websites, and independent reviews such as *Consumer Reports*. Consumers are deluged by product information about goods they do not intend buy and yet they encounter high costs of gathering information about those goods they do intend to buy. The cost to consumers of finding and digesting information about the attributes and availability of products reduces the overall value they receive.

By providing detailed information about product features, companies can distinguish their products and services. It does not do any good to offer a better product if consumers are not aware of its features and distribution channels. Informative in-store displays and trained sales personnel that provide detailed information distinguish products from those of competitors. Broadcast, print, direct mail, and other forms of advertising contain information about product features to help distinguish between brands and increase customer willingness to pay.

Manufacturers and retailers have discovered that the Internet is a powerful medium for customers to obtain detailed information on demand. Most company information on the Internet takes the form of detailed product catalogs. Some company websites respond to customers in various ways. Amazon greets customers at its online store by name with suggestions of new products and services due to software that conducts a statistical analysis of the customer's past buying patterns. Clothing retailer J. Crew's website guides customers through a set of questions to select the best pair of chino slacks: his or hers, plain or pleated, a dozen types of fit, and multiple colors. Customers can compare detailed photos, fabrics, and prices. Companies transmit more information through these websites by increasing the degree of interaction with customers.

10.3 *Sources of Differentiation Advantage*

The manager's external analysis and internal analysis suggest sources of differentiation advantage. If the company decides to follow a product differentiation strategy, the company pursues unique market relationships and develops unique organizational strengths.

Distinctive market relationships are a source of differentiation advantage. The quality of components, parts, and other inputs are critical elements of manufactured goods. Companies can derive a differentiation advantage by partnering with critical suppliers. A company's suppliers and subcontractors affect the quality of service the company provides.

Supplier alliances provide a means of improving product quality that might not be attainable with the firm's own resources and competencies. Retailer Marks and Spencer combined its knowledge of customer preferences and food preparation technology with its suppliers' knowledge of food products and manufacturing by jointly developing product specifications. Motorola's Paging Products Group developed new manufacturing equipment with its suppliers narrowing the millions of possible variations in product features and reducing design to manufacturing cycle times. Phillips Consumer Electronics Company held joint product quality improvement meetings with its component suppliers, involving personnel from purchasing, materials, operations, and information systems.[10]

Jordan D. Lewis suggests that companies cooperate with their suppliers to create greater value by defining customer benefits, identifying value creation opportunities, sharing technology road maps, and sharing development risks. He adds that it is necessary to involve the supplier early on in the design process when the supplier's product has a substantial effect on customer value. Close cooperation in design is needed when the supplier is a key innovator.[11]

A company can distinguish its products by providing *complements* itself or by enlisting the help of partners. Recall from Chapter 4 that complements are goods and services that increase a customer's demand for the firm's products. The availability of compatible software applications affects demand for computers and operating systems. Demand for players of compact disks or for digital video disks (DVD) is enhanced by a greater variety of recorded offerings. The demand for cameras depends on pricing and availability of compatible film. The availability and quality of complements enhance product differentiation.

Distinctive market relationships help a product differentiation strategy based on complements. Partners provide complementary goods that a company might not be able to supply given its internal competencies. By partnering with suppliers of complementary goods, companies help to ensure that those goods will be better tailored to the needs of its customers. Exchange of information with developers of complementary goods is particularly valuable when compatibility depends upon technically complex features. Microsoft shares information about its Windows operating system with developers of applications to make sure that the applications run effectively with Windows and provides extensive services and training for developers.

A company's distributors and the customer services they offer affect customer benefits and brand perceptions. The quality of customer service at retail outlets is an important part of the customer's purchasing decision. The quality of repairs and maintenance that automobile dealers provide after purchase affects the benefits automobile owners receive and plays a role in purchasing decisions. The services provided by value-added resellers increase customer benefits from computer hardware and software.

Consider next the firm's organizational strengths. The company's internal analysis attempts to discover resources and competencies that can be used to distinguish its products for others. Basic scientific and technical knowledge, patents, and other intellectual property are critical in creating distinctive product designs. The company's engineers, product developers, and other creative personnel offer enhanced product designs. The company's brand image can be further developed to identify its products to consumers. The company can share information with customers that helps them to distinguish unique product features.

Michael Porter observes that "differentiation grows out of the firm's value chain." The firm's internal processes provide many different means to distinguish the firm, including inbound logistics, operations, outbound logistics, marketing and sales, and customer service. Management can design the firm's infrastructure to improve brand image and performance in customer service. Human resource management can provide superior training in these activities to improve product quality and to enhance customer experiences. The company can invest in technology development to develop unique product features and advanced servicing techniques.[12]

Exhibit 10.2   Sony Playstation 2

Sony's Playstation captured a convincing lead in video game equipment, selling over 70 million units. As Sony readied its next generation, Playstation 2, it faced formidable competitors: Sega Dreamcast, Nintendo GameCube, and Microsoft Xbox. What was Sony's product differentiation strategy for the new machine?

The four companies offered different brand names with different track records in the field — Microsoft, for example, was a relative newcomer in video games. Sony planned an extensive publicity campaign. The four companies also had different market-entry timing. Sony could be characterized as a first mover even though Sega was already in the market with a considerably smaller installed base. Sony Playstation 2 would enter the market a year before Nintendo and Microsoft, and the company planned an extensive publicity campaign.

Sony had the largest installed base of players. As market leader, the company had the largest number of games in its format, including Crash Bandicoot, Final Fantasy VIII, Tomb Raider,

(Continued)

Exhibit 10.2 (*Continued*)

Metal Gear Solid, Ape Escape, Resident Evil, Tony Hawk Pro Skater, Syphon Filter, Gran Turismo 2, Madden NFL 2000, and NHL 2000. To take advantage of its market leadership and the large number of Playstation games in use, Sony's Playstation 2 machine would be backward-compatible; that is, it would be able to run existing games. The competitors offered systems with new formats, requiring the purchase of new games. Based on its installed base and large number of successful games, Playstation 2 would most likely have the largest number of new games available for its format, a critical consideration for video game enthusiasts.

All four companies pursued a strategy of proprietary formats, with no machine capable of running a game made for another format. Each company relied on independent software developers. According to Sony, over 200 partners would support Playstation with complementary products, including 89 publishers with license agreements to develop content and 45 developers of programming tools and middleware. Sony, Sega, and Nintendo required licenses and charged fees to its software developers, while Microsoft chose to pursue an open systems strategy that did not require licenses to write software for their machine.

Sony presented Playstation 2 as a complete entertainment center that integrated music and video with games. The machine would run both CD-Rom and Digital Video Disc (DVD) formats. Although Microsoft offered DVD movie capability, Nintendo and Sega did not. The machines differed in computing speed: Sega offered 200 MHz, Nintendo 405 MHz, Sony 295 MHz, and Microsoft 733 MHz. Thus, Sony was near the low end of central processing unit (CPU) power, with three quarters of Nintendo's clockspeed and less than half of Microsoft's clockspeed. Sony's machine offered a 147 MHz Sony Graphics synthesizer, whereas Sega had an NEC PowerVR video chip, Nintendo had a 203 MHz ATI chip, and Microsoft's Xbox a 300 MHz X-Chip made by nVidia. Sony's machine ranked third in memory with 38 MB, compared to Sega (28), Nintendo (40), and Microsoft (64). The computing speed, graphics chip, and memory of the competitors would permit more life-like graphics and more complex games.

Sony's product differentiation strategy seemed to depend more on complements — video games — rather than on the attributes of its product. At its introduction, Playstation 2 expected to offer more games in its format and less computing speed and memory capacity than two of its competitors. Would Sony's product differentiation strategy be enough to beat the video-game challengers?[13]

10.4 *Product Innovation*

How can a firm sustain its differentiation advantage? As with cost advantages, specific differentiation advantages are very difficult to sustain. Competitors are likely to imitate or surpass the features of any successful product. The only way to achieve and maintain a differentiation advantage is to continue to improve the company's products or to develop new ones. A product differentiation strategy requires innovation.

Product innovation is the development and introduction of new types of goods and services. Companies create many product innovations by applying inventions obtained from scientific and technical research. Some product innovations are new combinations of existing features. Other product innovations involve branding, fashion, and image. Just as manufacturing technology can become obsolete, so products can be swept aside by "gales of creative destruction."

External Analysis

The company performs an external survey to identify customer preferences and the features of competing products. In addition, the company examines the technological developments that have an impact on product features. The company's product innovation strategy attempts to discern opportunities to match inventions with customer needs.

Companies may seek to purchase or license inventions that can be useful in designing products. As noted in the case of cost advantage, technological change is so far-reaching that no company can go it alone. Product innovations necessarily involve combining external technological change with the company's market and technical knowledge.

As in the case of cost advantage, managers apply external analysis to monitor the four phases of technological change. The survey considers the implications of scientific advances for the design of new types of products. Managers also examine the potential of recent discoveries to enhance the features of existing products or to allow creation of novel products. Next, they consider how basic inventions being developed translate into marketable products. Finally, managers attempt to forecast the types of goods and services entrepreneurs and established innovative companies are planning to offer.

The external analysis monitors changing customer preferences under evolving market conditions. Fundamental changes in transportation, communication, lifestyles, and working conditions create opportunities to offer new types of products that address the changing needs of the company's customers or to identify new prospects.

The introduction of new products necessarily means that the firm will compete in the future with products that differ from its own current offerings. This is why it is crucial to understand that substitute products will not have the same characteristics or features as existing products but may deliver greater customer satisfaction. Managers performing an external analysis must cast their net widely to understand how products that look very different from existing products can present significant competitive challenges. The external analysis must focus on innovative solutions to customer problems.

Internal Analysis

Preparing for product innovation requires managers to perform an internal analysis. As noted in the discussion of process innovation in the previous chapter, managers consider the company's technological knowledge, ability to monitor external technological change, and capacity to generate innovations. The manager seeks to determine the extent to which the company's product innovations will rely on internal resources and the need to work with external suppliers and partners.

The manager analyzes the features of the company's existing products in comparison with competitor offerings. The manager seeks to determine whether the company's products have additional life or whether they are subject to replacement by products with enhanced features or entirely different types of products.

Because product differentiation depends on customer benefits, product innovation should be directed at providing those improvements that benefit customers. Arbitrary applications of cutting-edge technology may satisfy company engineers and technology fans, but fail the market test. By addressing customer benefits rather than improvements on existing products, companies effectively anticipate competition from substitute products and services with completely different features and distinct methods of distribution.

Exhibit 10.3   Transmeta and the Crusoe Chip

Intel dominates the market for microprocessors not with a single chip but through continual product innovation. The company maintains its differentiation advantage by continually introducing new generations of processors, each representing huge leaps in computer speed. Intel packs millions of transistors onto a single chip. The vast increase in computational power with each generation allows Intel to build chips with significant new functions while remaining backward-compatible with earlier generations of

(Continued)

Exhibit 10.3 (*Continued*)

x86 chips. Intel supports its differentiation advantage through massive investment in fabrication plants that provide high quality production and reliable supplies. Intel reinforces its differentiation advantage by its advertising campaigns and by placing its Intel Inside logo on the many types of computers that contain its chips. Intel's differentiation advantage appears invulnerable and the company has sustained its position against a host of competitors.

At the beginning of 2000, the Transmeta Corporation emerged from the self-imposed secrecy it had maintained since its founding almost five years before. The company announced a line of microprocessors named after the fictional stranded survivor, Robinson Crusoe. Through its product innovation in processors, Transmeta sought a differentiation advantage over Intel and other chip makers. Like Intel's chips, Transmeta's chips would be backward-compatible, that is, they would run software written for earlier generations of processors including those made by Intel. However, the Crusoe chips would do so by relying on code-morphing software that would translate applications software into simpler instructions for the Crusoe chip. By shifting tasks to software, Transmeta could use fewer transistors, thus greatly simplifying the design of their chip. While Intel's chips relied on putting more transistors on a chip, Transmeta's combination of software and processor design reduced the need for so many transistors on the chip. Moreover, the chip could fine-tune its operations on the fly.

The advantage of the Crusoe chip design would be its low power consumption. A chip with lower power usage would be ideal for portable applications including Internet-compatible wireless phones and subnotebook computers. At a time of rapid growth of mobile computing, many customers would be willing to trade off some computational speed for longer battery life.

The Transmeta chip offered energy efficiency, consuming less than one watt of power on average. The chip would supply 600MHz computer speed and still allow laptops to run for eight hours on a single battery charge instead of less than three or four hours for a conventional chip. As additional advantages, the chip offered smaller size and reduced operating temperatures.

Transmeta chose not to challenge Intel directly on the larger desktop market but instead to maximize its differentiation advantage in wireless phones and mobile computers. Sony, Fujitsu, Hitachi, IBM, and NEC designed notebooks using the Transmeta.

(*Continued*)

Exhibit 10.3 (*Continued*)

Dave Ditzel, CEO of Transmeta Corporation, said that "Hitachi's new device lets you put the Internet in your hand and surf the Web with the stroke of a pen." According to Ditzel: "Transmeta's Crusoe microprocessor allows Hitachi to put all the performance of a PC into a handheld computer with up to seven hours of battery life."[14] The chip was included in Sony's ultra-light 2.2 lb notebook named PictureBook that features an integrated digital camera. Gateway and America Online chose the Crusoe chip for their jointly developed Internet Appliances.[15]

Intel responded with a press conference announcing the future introduction of its Pentium III processor for portable computers that would have comparable low power usage but less processor speed (500 MHz when plugged in to an electrical outlet and 300 MHz on battery power). Intel did not mention Transmeta's name in its press conference. Transmeta's David Ditzel attended Intel's press conference and observed "Intel is clearly playing catch-up ... and Transmeta is not going to stand still."[16]

Transmeta's Crusoe chip represented a product innovation in processors. The firm was founded in 1995 by David Ditzel, who had been a designer of Sun's scalable processor architecture (SPARC). Ditzel and others formed the company with the intent of creating a new chip design based on fundamental advances in basic research.

An entrepreneurial start-up, Transmeta pursued a perceived market opportunity in processors that use low amounts of power. Although the founders were already highly skilled, the firm had to develop competencies that would allow them to substitute software for hardwired transistors. Ditzel observed, "We realized there were less than a dozen people in the world capable of doing this ... We had one simple plan: hire all of them." Among its programmers, Transmeta hired Linus Torvalds, the creator of the Linux operating system. The company obtained the necessary resources and competencies by hiring the best-qualified people in the world and assembling an effective team. To secure a differentiation advantage, the company would have to continue turning out innovative chips.

Innovation Strategy

Product innovation is an integral part of any company's product differentiation strategy. To keep a differentiation advantage, a company must successfully introduce a stream of product innovations. The product innovations must stay ahead of the competition to yield incremental revenues. Although the innovation process is inherently unpredictable, managers must guide the innovation process.

One important aspect of the innovation process is the *rate* of innovation. Companies try to determine how often product changes should be introduced. How significant should changes in the company's products be? Given the underlying rate of invention, managers determine how frequently to introduce product changes, which in turn determines the extent to which the company's product changes. Frequent product changes have the advantage of keeping up with technology and staying ahead of competitors. However, frequent product changes increase the costs of production, distribution, and marketing. The company must take account of the trade-off between these benefits and costs of product innovation.

Consumer purchasing patterns are an important determinant of the rate of product innovation. The company's current products compete with its own installed base. A company risks saturating the market so that consumers may not wish to upgrade. Moreover, a company's current products compete with its own future products. If a product is steadily improved, customers may delay their purchase to wait for improved versions in the future. To avoid harming itself, the company must choose the right rate of product innovation. Each product introduction should sufficiently improve the product's features so customers are willing to switch to the new model. Ford and General Motors understood this process well for many decades, offering highly touted annual model changes and trade-in allowances. Intel mastered this process through many generations of computer chips. Microsoft also grasped the desirable rate of innovations, guiding consumers through a long sequence of upgrades of its DOS and later Windows operating system.

Some product introductions are minor and calculated to differentiate the product at a low cost. Such changes can be as simple as small changes in ingredients in soup or redesigned packaging. Such product variations can keep a brand image fresh and stimulate consumer interest. However, such changes can be easily emulated by competitors. Relying on many incremental changes can involve costly marketing and design changes that might be better saved for major product innovations. After several updates of a product design, a more complete overhaul is often necessary, as often occurs with annual automobile model changes.

Some product innovations are so dramatic that they fundamentally change customer perceptions. These product innovations create entirely new product categories and new directions of product evolution. An example of such a path-breaking product innovation is the Apple II personal computer introduced by Steve Jobs and Stephen Wozniac after founding Apple Computer in 1977. The product was an important spark of the information age, helping to create the vast market for personal computers, not to mention software and peripheral equipment such as printers.

Another example of a path-breaking product was the Sony Walkman portable cassette player introduced in 1979, which began a long line of personal entertainment devices including the portable compact disc player, the Watchman portable television, the mini-disc player, and the MP3 device for replaying music downloaded from the Internet. Although the transistor radio had shown that consumers wanted portable entertainment, the Walkman's stereo headphones made listening to music on the go a strictly personal and high-quality experience. The feeling of a movie soundtrack to one's life so inspired author William Gibson that he coined the term *cyberspace* to describe data processing with the intimacy of a Walkman.[17] The story of the Sony Walkman shows that dramatic innovations may involve well-known technology. The portable cassette recorder was an existing product. A Sony engineer removed the recording mechanism and speaker and added some light headphones that he found next door in another Sony laboratory. The Walkman sold over 185 million units with over 600 hundred model variations.[18]

In addition to the rate of product innovation, managers attempt to determine the *direction* of product innovation, that is, what types of improvements or new features the product will have. The range of possible product improvements depends on underlying technological change and available inventions. Customer preferences are the best guide for selecting among potential product improvements. In planning its innovative investment and direction of innovation, the firm must anticipate the rate and direction of competitor innovation.

Managers must carefully balance their efforts in research, invention, development, and innovation. Companies should compare the returns to these diverse activities. Simply focusing on basic research and inventions is worthwhile if the company can bring the results to market through licensing technology to others. In turn, some companies obtain research and inventions from academia, government laboratories, and the technology market, commercializing the discoveries through development and innovation. Companies that have made major research discoveries and significant innovations may wish to focus substantial additional efforts in development and innovation to reap the commercial benefits of their discoveries.

Product innovation is an important activity at General Electric, which was founded by the great inventor Thomas Edison and his partners. Under Jack Welch, General Electric directed its scientists and engineers to focus on short-term development and innovation activities. Projects included improving the efficiency of turbines and fine-tuning dishwashers and washing machines. As chairman and CEO Jeffrey R. Immelt took the helm at GE, he redirected GE's research budget toward more long-term research and invention activities. The company's three central research labs, at Niskayuna, New York, Bangalore, India and Shanghai, China work on basic technologies. Individual businesses such as Power Systems, Medical Systems, and Appliances are concerned with product development and innovation. Global Research Senior VP Scott Donnelly observed that scientists at GE combine long-term research and interaction with the company's marketing: "This is the best of both worlds."[19] According to Donnelly, "The better a company's technology is, the more products it eventually sells."[20]

A company's efforts in research, invention, development, and innovation depend on its scientific and technical knowledge and abilities relative to suppliers of technology. Managers should compare the costs of developing internal technological abilities relative to the costs of market procurement. They should also take into account the costs of managing innovation within the organization versus the transaction costs of sourcing innovations in technology markets. When knowledge is costly to develop, purchasing the technology from companies that specialize in developing and licensing innovations can have advantages. Another important consideration is the ability of the firm to protect its intellectual property when developing its own innovations. If the technology is subject to copying or imitation, the returns to investment in R&D are likely to be lower than with sufficient protection for intellectual property.

Creating or identifying critical inventions is not enough. Innovation requires moving from technological developments to products that satisfy customer needs by executing product design, manufacturing, marketing, and distribution. The complexity of these tasks explains why some companies experienced difficulties in moving from invention to innovation. The classic example is Xerox, whose Palo Alto Research Center (PARC) produced many fundamental inventions. PARC developed the laser printer, the thin transistors used in making flat-panel displays, and the standard for Ethernet linking of computers in local networks. PARC created the graphical user interface with onscreen computer display of icons to click on, which Apple operating system and Microsoft's Windows software would feature. PARC built the Alto personal computer, the first computer to use a mouse and a graphical user interface. Xerox failed to benefit fully from these revolutionary inventions because other companies turned them into product innovations. For example, Hewlett-Packard offered popular low-priced printers and copiers. By the year 2000, 30 years after founding PARC, Xerox contemplated selling the research unit.[21] The example of Xerox illustrates how a company failed to capitalize on its inventions although many other companies benefited by creating product innovations based on them.

A more subtle problem in guiding product innovation is deciding on the right generation of technology. Moving too late opens the door to competitors who move first. A successful product can build a brand reputation that creates a first-mover advantage. The product may garner sales from early adopters, customers who enjoy trying new technology. However, moving too soon need not guarantee success. Early movers can miss the improvements of the next generation. An imperfect initial product can harm the company's reputation. The company could have reduced its costs by learning more about the technology and delaying its product introduction. The company blazing a trail sometimes finds itself doing all the work for others — America Online and Microsoft, for example, are known for pursuing a "fast follower strategy," improving on the innovations of others with substantial investment in R&D and marketing as well as acquiring successful start-ups.

Timing product development and introduction is perhaps the most difficult part of product innovation strategy. The trade-offs between early and late innovation underlie differentiation advantage. The timing problem must be addressed repeatedly since maintaining a differentiation advantage requires continual innovation.

The costs of battling for a first-mover advantage are illustrated by the early competition for devices to provide wireless access to the Internet. According to *The Wall Street Journal* columnist Walt Mossberg:

> "Imagine that the Internet could be viewed only on tiny, two-inch screens with green backgrounds capable of displaying fewer than 10 lines of blocky black text, and no color or graphics. On top of that, visualize these screens filled with an endless series of confusing menus that must be navigated to get where you want to go online. And then consider that only a few websites could be viewed with any semblance of clarity or organization. Now, imagine that your e-mail and Web information flowed onto these puny screens at frustrating speeds far slower than even the slowest dial-up computer modems sold today. Finally, imagine that, in order to compose an e-mail, or an instant message, or to fill out an order or construct a search, you had to peck out your words on a keyboard so clumsy that at least three taps were required for each character."

Mossberg continued: "That is the typical experience in the year 2000 of using the so-called wireless Web ... it's awful, and not ready for prime time." Of the many models out of the gate early, each had pros and cons. The RIM Blackberry was good for e-mail, the Sprint cell phone with built-in personal digital assistant and Web browser was best for phone users, while the Handspring VisorPhone offered advantages for dialing using contact lists. Mossberg's advice to consumers was to wait a year for better devices.[22]

The costs of delayed innovation, on the other hand, are illustrated by Lucent's experience in switching equipment. Lucent delayed the development of a new generation of optical networking equipment. The equipment, known as OC-192, uses a single wavelength of light to transmit up to 10 gigabits of digital information per second.[23] Instead, Lucent worked on the next generation of technology known as OC-768, which would have a 40 gigabit per second transmission capacity, as well as other advanced technology. The result was that Nortel, which embraced the earlier technology, developed a huge lead. According to William T. O'Shea of Lucent, "At that point in time, that stuff was just getting started, and we and Nortel were just about neck-and-neck in optical. Nortel now is several times our size."[24]

Innovative product differentiation creates competitive benefits by attracting customers away from competitors or by pre-empting entrants. However, product innovation has many costs. Excessive model changes compete with the company's past models. Moreover, customers may not adopt products with extensive changes, even if the changes deliver customer value in and of themselves, if they encounter switching costs. Customers may be unwilling to upgrade to a new software technology if the learning costs exceed the benefits of additional features. The customer weighs the value of existing products against the price of the new product, transaction costs of obtaining the new product, and switching costs of adapting to the new product. The company must recognize the customers' total cost: price, transaction costs, and switching costs.

The company engaged in product innovation faces hidden costs as well. Organizational modifications may be needed to produce and sell a new product, including training of sales and customer service personnel. The company's distributors also encounter adjustment costs in terms of sales and customer support. The company must compare the competitive advantages from product differentiation with the costs of implementing product change. The company should implement the change only if the additional benefits to customers are greater than the additional costs of producing and delivering the enhanced product features.

America Online (AOL), before its merger with Time-Warner, achieved a leadership position in providing Internet services by emphasizing ease of use for customers rather than the most advanced technology. With over 25 million customers, accounting for almost 40 percent of the time spent online in the United States, the company identified its mission as: "To build a global medium as central to people's lives as the telephone or television ... and even more valuable." How did the company maintain its differentiation advantage in comparison with Yahoo, Microsoft, and other Internet service providers? It pursued a strategy of continual innovation, repeatedly issuing new versions of its software, while continuing to emphasize its ease-of-use differentiation strategy. However, to maintain an advantage, AOL also expanded its *distribution* channels through its "AOL Anywhere" strategy.

The AOL Anywhere strategy included access not only through the Web on the personal computer, but also through television, Internet-ready phones, and handheld computers. Thus, to maintain a differentiation advantage, the company offered customers not only ease of use, but perhaps just as important, ease of access.

Exhibit 10.4   Super Jumbo Aircraft

The rivalry between Boeing and Airbus exemplifies competitive product innovation. A new entrant into the super-jumbo category, Airbus spent well over $10 billion to develop its A380 aircraft.[25] The double-decker Airbus plane carries from 481 to 656 passengers in five passenger configurations on long-range routes.[26]

In contrast to Airbus's strategy of a completely new design, Boeing planned to produce a revamped stretch version of its older 747 aircraft for a development cost of about $4 billion.[27] The Boeing 747 dates back to 1970, with incremental improvements spread over 15 different model variations. To increase the range of the plane, the wing would be widened near the fuselage to boost the amount of fuel storage space, and a trailing-edge wedge added to increase the lift of the wings.[28] Boeing also planned to add a variety of more powerful, fuel-efficient engines for customers to choose from. The updated aircraft, which Boeing called the 747–400, would transport between 416 and 524 passengers with an extended range and is available in four models. In addition to the basic 747–400, Boeing also announced plans to offer a high capacity domestic version that would carry up to 568 passengers, a combination aircraft with passengers in the front and cargo in the back, and an all-cargo freighter.[29]

Boeing's strategy for product innovation was that of incremental change in contrast to Airbus's strategy of offering an entirely new design. Boeing's reasoning was that the market for 747 and larger airplanes over the next two decades is small, little more than 1,000 planes, of which only about 330 will be super-jumbos carrying over 500 passengers. Boeing expected demand for super-jumbos to be driven by growth of traffic on intercontinental routes and the need to deal with airport capacity constraints by carrying more passengers per flight.[30] The expected size of the market, anticipated needs of the airlines, and competition from Airbus led Boeing to pursue cost leadership as a response to product innovation.

Within months, Boeing reversed course. It had received no orders for the stretched 747 while Airbus already had 66 orders for its super-jumbo. Accordingly, Boeing discontinued the project.

(Continued)

Exhibit 10.4 (Continued)

Boeing's new response to Airbus consisted of an almost-supersonic plane called the Sonic Cruiser or the 20XX. The plane would not compete on size, since it carried between 175 and 250 passengers, but on speed, taking an hour off a transcontinental flight. The estimated cost of developing the Sonic Cruiser — about $9 billion to $10 billion — was similar to the cost to Airbus of developing their super-jumbo.[31] Would Boeing and Airbus maintain their innovation strategies?

10.5 *Overview*

Product differentiation is one of the three sources of competitive advantage, along with cost advantage and transaction advantage. Companies seek a differentiation advantage by selecting product features that increase customer willingness to pay in particular targeted market segments. Differentiation is not necessarily achieved through improvements in quality or advanced technical features. Instead, differentiation depends on creating additional customer benefits relative to market alternatives.

Companies attempt to position their products relative to their competitors, both reacting to and anticipating competitor differentiation strategies. In some cases, firms emphasize minor differences in product attributes as they fight for the broad market. In other cases, firms try to distance their products from those of their rivals to protect against direct competition and capture targeted market segments. In other cases, companies attempt to significantly outperform competitors in an effort to capture most of the market. With less product differentiation, companies compete on price and pursue cost leadership as a source of competitive advantage. With greater product differentiation, companies compete on product attributes and related customer services, leading to increased costs, higher prices, and greater marketing efforts. Companies pursue differentiation advantage through advertising, brand management, variation of product attributes, dissemination of information, pricing policies, and provision of complementary products.

Specific differentiation advantages tend to be temporary. Companies must rely on continual product innovation to achieve and maintain a differentiation advantage. A company whose products have become generic products can enhance product features through innovation or pursue cost advantage instead. Faced with challenges to both cost leadership and differentiation strategies, companies should consider strategies to achieve a transaction advantage. These are examined in the next chapter.

Questions for Discussion

1. Companies that try to sell products in different countries encounter differences in customer demand. Customer preferences differ across countries for such products as food, entertainment and clothing. Why are some products marketed in the same way across countries (for example luxury handbags) whereas other products are marketed differently (for example biscuits)?

2. Marketing myopia happens when managers consider competition only from similar products and ignore important competition from substitute products. How does Internet usage compete with other forms of entertainment such as watching television, going to the movies, and reading books?

3. A mobile phone with more features such as a larger screen or built-in organizer may be more bulky to carry. How should managers make choices about the combination of product features?

4. Increasing the features of the mobile phone also increases the cost of production. How should managers evaluate the trade-off between greater customer benefit and greater cost?

5. Consumers love product variety, whether in clothing styles or automobile options. Companies incur costs in providing variety. How should managers determine how much variety is enough?

6. A clothing store advertises that it has a large selection and lower prices, but customers must drive farther to reach the store. How do transaction costs affect the customer's benefit from the clothing store?

7. Consider two stores with similar merchandise. A department store offers individual customer service but charges higher prices while a discount store offers almost no service but has lower prices. Do these stores compete with each other? Are there some customers who choose between shopping at the two stores? Are there customers who visit both stores?

8. An electronics store sells both video cassette recorders (VCRs) and digital video disk players (DVDs). As customer demand shifts away from VCRs toward DVDs, how will the store's inventory, marketing, and pricing policies change for the two products? Should the store shift gradually to another format or all at once?

9. Home theater systems offer a bundle of DVD player, speakers, woofer, amplifier, and cables. Why would customers prefer a bundled system over individual components?

10. How long can companies expect to maintain a differentiation advantage without changing their product? Consider the longevity of such products as Oreo Cookies. How often do products change in electronics or automobiles? How often do clothing companies change their designs?

Endnotes

1. George J. Stigler and Gary S. Becker, "De Gustibus Non Est Disputandum," *American Economic Review* 67, no. 2 (March 1977), pp. 76–90, argue that tastes are stable over time and similar among people, with different choices affected by income and experience. See also the empirical analysis in Edwin G. West and Michael McKee, "De Gustibus Est Disputandum: The Phenomenon of 'Merit Wants' Revisited," *American Economic Review* 73, no. 5 (December 1983), pp. 1110–1121.

2. Economists refer to products that differ in terms of a feature that everyone agrees upon as "vertically" differentiated, and products that differ in terms of features that are a matter of taste, as "horizontally" differentiated.

3. The company split in 1993 between Host Marriott, which purchases the properties, and Marriott International, which operates properties and franchises hotels and suites, under the various company brand names. See Gary Hoover, Alta Campbell, and Patrick J. Spain, eds., *Hoover's Handbook of American Business* (Austin, TX: The Reference Press, 1995).

4. See Theodore Levitt, "Marketing Myopia," *Harvard Business Review* 53 (September–October 1975).

5. The information in this paragraph and the product descriptions are drawn from http://www.octel.com/whoweare, Lucent Technologies, 1999.

6. See http://www.exxon.mobil.com/chemical/licensing/pe_licensing/index.html

7. Advertisement, *New York Times*, September 1, 2000, p. A15.

8. Historical information about Campbell is from "A Condensed History," *New York Times*, July 30, 2000, p. 9.

9. "Face Value: The Car Man Who Says No," *The Economist*, August 5, 2000, p. 64.

10. These examples are due to Jordan D. Lewis, *The Connected Corporation: How Leading Companies Win Through Customer-Supplier Alliances* (New York: Free Press, 1995). See particularly pp. 86–105.

11. Ibid., note 10.

12. Michael Porter, *Competitive Advantage: Creating and Sustaining Superior Performance* (New York: Free Press, 1985), chapter 4; see especially pp. 120–122.

13. Technical information on the systems comes from John Markoff, "Microsoft's Game Plan," *New York Times*, September 2, 2000, p. C1.

14. Company press release at http://www.transmeta.com/press/PRhitachi 120700.html

15. Company press releases at http://www.transmeta.com/press/pcexpo.html

16. The information in this paragraph is from John Markoff, "Responding to Small Rival, Intel Sees Mobile-Chip Edge," *New York Times*, October 11, 2000, p. C4.

17. Bruce Headlam, "Walkman Sounded Bell for Cyberspace," *New York Times*, July 29, 1999, p. D7. William Gibson introduced the term *cyberspace*

in his 1984 novel *Neuromancer* (New York: Ace Books, reissued edition, 1995).

18. Phil Patton, "Humming Off Key for Two Decades," *New York Times*, July 29, 1999, p. D1.

19. The information in this paragraph is based on Rachel Emma Silverman, "GE Goes Back to the Future," *The Wall Street Journal*, May 7, 2002, p. B1.

20. Donnelly is paraphrased in Silverman, ibid.

21. Reed Abelson, Claudia H. Deutsch, John Markoff, and Andrew Ross Sorkin, "The Fading Copier King," *New York Times*, October 19, 2000, p. C1.

22. Walt Mossberg, "Walt Does Wireless," *The Wall Street Journal*, September 29, 2000, Weekend Journal, p. W1.

23. Seth Schiesel, "How Lucent Stumbled: Research Surpasses Marketing," *New York Times*, October 16, 2000, p. C1.

24. William T. O'Shea, Lucent's executive vice president for corporate strategy and business development, is quoted in Schiesel, "How Lucent Stumbled."

25. Airbus is 80 percent owned by the European Aeronautic Defense and Space Company and 20 percent owned by the British BAE Systems. The European Aeronautic Defense and Space Company N.V. (EADS) was founded on July 10, 2000, the result of a merger between Aerospatiale Matra S.A. of France, Construcciones Aeronáuticas S.A. of Spain, and DaimlerChrysler Aerospace AG of Germany. According to the two CEOs of EADS, Philippe Camus and Rainer Hertrich, "The new integrated structure is as important for a transparent and efficient industrial organization as for its new projects like the super Airbus A3XX," http://www.eads-nv.com/eads/en/index.htm

26. John Tagliabue, "Airbus Clears Plans to Build Long-Range Jumbo Jet," *New York Times*, December 20, 2000, p. W1.

27. "Thank You, Singapore," The Economist, September 30, 2000, p. 63.

28. Jeff Cole, "Wing Commander: At Boeing, an Old Hand Provides New Tricks in Battle with Airbus," *The Wall Street Journal*, January 10, 2001, p. A1.

29. The company home page for 747-400, http://boeing.com/commercial/747family/index.html

30. "Boeing Current Market Outlook," http://boeing.com/commercial/cmo/4da05.html

31. Laurence Zuckerman, "Boeing's Planned Jetliner to Be Almost Supersonic," *New York Times*, March 30, 2001, p. C1.

CHAPTER 11

TRANSACTION COORDINATION STRATEGY

Contents

The transaction coordination strategy involves more than jockeying for position in product markets. A company has a *transaction advantage* if it consistently creates more value than its competitors by developing innovative transactions. A company obtains a transaction advantage by lowering transaction costs for its customers and suppliers or by making new combinations of buyers and sellers. What distinguishes a transaction advantage from other types of competitive advantage is that the firm coordinates its purchasing and sales activities to realize market opportunities. This chapter explores the sources of transaction advantage.

Recall that the value created by the firm represents customer willingness to pay minus supplier costs and the costs of using the company's assets.[1] The company creates greater value than competitors by carrying out transactions at lower cost or by discovering better combinations of customers and suppliers. There are several important ways to pursue the transaction coordination strategy. As *intermediaries*, companies lower transaction costs in many ways such as designing better contracts or creatively applying technologies such as point-of-sale terminals or websites.[2] As *market makers*, companies create and operate markets by aggregating demand and aggregating supply, adjusting prices, and balancing supply and demand. *Entrepreneurs* create new combinations of customers and suppliers in various ways such as opening different marketing channels or expanding domestic and international markets.

Transaction advantage is temporary, as is the case with cost and differentiation advantages. To develop and maintain a transaction advantage, managers must continually seek new methods of interacting with customers and suppliers. Managers go beyond standard analysis of manufacturing costs or product differentiation by developing distribution channels and supply chains and coordinating these activities. Transaction advantage arises from innovations in business methods and market design.

Chapter 11: Take-Away Points

The transaction coordination strategy involves intermediating between buyers and sellers, establishing and operating markets, and creating new combinations of buyers and sellers. Managers develop innovative transaction methods to pursue a transaction advantage:

- Transaction advantage adds value by lowering transaction costs or by creating new combinations of buyers and sellers.
- Managers should apply the transaction triangle framework to focus on coordination of transactions with its customers and suppliers.
- Managers should recognize that their company can sometimes be bypassed when their customers and suppliers transact directly with each other.
- Acting as intermediaries, companies create value by providing convenience to customers and suppliers.
- Acting as market makers, companies create value by establishing and operating markets efficiently.
- Acting as entrepreneurs, companies create value by bringing together new combinations of buyers and sellers.
- Managers should apply external analysis and internal analysis to monitor the potential effects of technological change on transaction innovation.
- Companies must engage in continual innovation to improve transaction methods to achieve and maintain a transaction advantage.
- Transaction advantage depends on developing and coordinating relationships with customers and suppliers.

11.1 *Transaction Costs*

Transactions costs, while subtle, are often the most important types of costs. Customers encounter costs of searching and shopping, learning about product features, finding out prices, negotiating the terms of exchange, placing orders, keeping track of payments and receipts, and arranging for delivery. It is often said that time is money, meaning it is valuable. Because many transaction costs are time-consuming, customer *convenience* affects the customer's value of the firm's products.

Companies incur costs of marketing, sales, purchasing, hiring, and financing. Sales personnel incur the costs of time spent searching for customers, informing customers about prices and products, and conducting the sales transaction. Transaction costs include the costs that companies incur in negotiating contracts, writing contracts, and monitoring the performance of contract partners. Transaction costs also include the cost of back-office processes within the organization, including such activities as billing and invoicing. Transaction costs incurred by the company's suppliers reduce their supplier value. By lowering transaction costs of their customers and suppliers, firms create greater total value.

Convenience

Economic transactions take time.[3] Buying products and services has two costs: the price paid and the indirect time costs of finding the product, making the purchase, using the product, and obtaining related services. What counts is the total cost to the customer, that is, the purchase price plus transaction costs. Companies that deliver products with the greatest convenience gain a transaction advantage.

No matter what customers purchase, whether food, clothing, entertainment, or gifts, they usually must search for the best store. Visiting merchants takes even more time. Customers spend time traveling to stores; finding out prices, product features, and what is in stock; and locating an available salesperson, not to mention waiting in the checkout line. Almost anyone who has bought an automobile knows the excruciating feeling of waiting for hours at an auto dealer to haggle over the price of a car. Consumers spend substantial amounts of time in any given week shopping for housing, food, clothing, insurance, and other necessities.

Managers should explore ways to increase customer convenience. Often this involves reducing the time buyers need to complete a purchase. By improving store layouts and increasing the speed at the checkout counter, stores cut customer transaction costs and deliver greater value. The design of transactions also affects their costs. Complicated contracts increase the time and effort customers spend on transactions. The need to choose between complex option packages also decreases customer convenience. Managers should understand the trade-offs between the benefits of increasing customer choices and the cost of the time customers spend making choices.

Going online sometimes cuts the costs of search dramatically for certain types of products. Consider the time costs of contacting several dozen insurance companies to find the policy that best suits your needs and budget. Finding and meeting a number of insurance agents would be so time-consuming that you would be tempted to cut your search short and possibly settle for second best. Shopping online with such services as Answer Financial, Quick Quote, Quotesmith, Insweb, or InsureMarket makes it possible for customers to successfully complete such a search in a matter of minutes. When buying a car, the endless driving around to different dealers and interminable visits to the much-dreaded auto showroom can be discouraging. Shopping through online car-buying services such as Kelley Blue Book, Edmunds, Auto-by-Tel, MSNAutos, or Autoweb reduces transaction costs by letting customers locate dealers or explore option packages more quickly.

Buying a house is likely to be even more time-consuming. Traveling around with a real-estate agent can be exhausting, and the brain begins to reel after viewing only a handful of houses. On top of that, real-estate agents sometimes steer prospects toward certain neighborhoods or control the order in which houses are viewed. With electronic real-estate listings such as Realtor.com, Home Advisor, Home Fair.com, or Rent.net, home buyers regain some control over the search process. Using these types of services, a home buyer can set search parameters such as price range, location, and house types and instantly generate a list of available houses. The virtual search could be done at the buyer's convenience rather than on the realtor's timetable and could be repeated as frequently as desired to see if any new houses had appeared on the market. This allows buyers to narrow their search to the most likely prospects, cutting down the time spent viewing houses.

However, many customers have found that shopping online creates other types of inconveniences. It can be difficult to get customer assistance or to return products to online retailers. Shopping online also entails delivery delays that can be avoided by going to traditional stores. Moreover, many customers prefer personal service and the ability to inspect goods offered by bricks-and-mortar retailers. Thus, many companies have discovered the advantages of multi-channel marketing approaches, allowing customers the best of both worlds.

Whereas traditional merchants are open during so-called business hours, limited hours of operation will soon be an anachronism for many types of businesses. The online store stands ready to provide sales and service 24 hours a day, 7 days a week. The Gap advertised that its online store is "always open". Electronic banking has changed the notion of "banker's hours"; now Citigroup tells us that "the Citi never sleeps". Advances in computer and communications systems and commerce on the Internet provide services to fit the many and varied needs of customers around the clock. Not so long ago, the phone company urged us to let our fingers do the walking through the Yellow Pages. The vast set of online directories and search engines can significantly reduce the time costs of collecting information about products and suppliers.

The hero of Jules Verne's early science fiction classic *Around the World in Eighty Days*, Phileas Fogg, defied conventional wisdom and won a wager by traveling around the world in record time. Even in the jet age, people complain about travel times. While there will always be the need for human contact, many kinds of travel can be left to the virtual world, which can be traversed in nanoseconds. Online stores can be visited instantly, no matter how distant in geographic space. Museums abroad or tourist information about other countries are a couple of clicks away. As the Internet became international, the electronic traveler could move around the virtual world at speeds defying science fiction.

Whether at work, at home, or on the road, electronic communications generate substantial time savings. E-mail avoids the endless games of telephone tag that result when communications are out of sync and saves the considerable time needed to prepare traditional paper-based correspondence. More advanced Internet and intranet communication using shared documents and other methods allows workers to contribute to a common document without time-consuming meetings. Just like video cassette recorders let people adjust television viewing to fit their schedules, shared documents give employees greater flexibility in workplace scheduling, saving time at work. Companies such as Xerox advertise the time-saving benefits of document management. Although the promise of telecommuting is far from being realized, the time savings from reduced commuting time would greatly enhance productivity and improve the quality of life. Employees would avoid traffic jams and follow more flexible work schedules.

Automation of Transactions and Transaction Costs

One way that companies lower transaction costs is by using technology to automate buying, selling, and contracting. Companies lower the transaction costs for their customers and suppliers by automating the exchange process, applying computer technology to complex transactions, and employing enhanced communication methods. The application of information technology (IT) to back-office processes further lowers transaction costs. Internet commerce reduces certain types of transaction costs for a company's suppliers and customers. Small changes in transactions have far-reaching effects on competition because transaction costs are a significant part of overall costs.

Automation of transactions allows the substitution of capital for labor services in the production of transactions. Thus, a buyer can place an order from an online catalog without dealing directly with sales personnel. The capital used to automate transactions consists of computer hardware and software and data transmission networks. Such capital substitutes at the margin for costly labor services that are applied to routine commercial tasks, including the time that employees spend communicating with customers and suppliers regarding prices, product availability, ordering, billing, and shipping, and the costs of managing those employees. Thus, advances in transactions technology allow firms to reduce their transaction costs by choosing the optimal capital-labor mix in their commercial activities. Costs are lowered in comparison with alternative transaction technologies that rely primarily on sales personnel to process transactions.

Application of computation to transaction data is another innovation in transaction technology that potentially lowers transaction costs. By applying information technology to process data from transactions, companies can lower the transaction costs of complex transactions, such as real-time price adjustments on auction websites such as eBay. Companies can also match large numbers of buyers and sellers on electronic stock exchanges such as electronic communications networks (ECNs). Computation of transactions data further allows companies to link external transaction systems with their internal computer systems, thus increasing the frequency, rapidity, and accuracy of communication and allowing links to production and inventory management systems within each organization. Companies have made increasing use of so-called enterprise software to manage their operations and back-office systems. Companies can reduce their transaction costs by using the information from market transactions to update inventories and production plans and reduce the costs of engaging in purchasing and sales.

Communication of transaction information allows buyers and sellers to transact with the firm at remote locations and at different times. Thus, the buyers and sellers in an auction on eBay need not be present at the same location and can participate in the auction at different times. This reduces the transaction costs by avoiding the costs of travel and the costs of holding meetings, whether those costs would be borne by the firm or its customers and suppliers. Thus, technological change in information processing and communications results in innovations in the technology of producing commercial transactions.

The introduction of the Universal Product Code (UPC) and the use of bar code scanners fundamentally changed the nature of retail transactions. The innovation diffused at high speed. Only two years after the first product carrying the UPC symbol, a package of Wrigley's chewing gum, crossed a scanner at Marsh's Supermarket in Troy, Ohio, in June of 1974, over three-quarters of supermarket products carried the symbol.[4] The use of UPC codes spread quickly to other consumer goods industries such as clothing, household products, and toys and to commercial and industrial products. The use of bar codes not only speeds the customer through the checkout counter, reducing wait times and increasing productivity of cashiers, it also reduces the cost of generating information used to track sales patterns and to update inventories. The code also allows retailers to adjust prices easily. Moreover, the information generated at the checkout counter has changed the relationship with wholesalers and manufacturers; it has become a part of a system of electronic data interchange and Internet commerce that has automated ordering and billing systems.

By lowering transaction costs of retailing in these ways, UPC codes and scanner technology fundamentally changed the nature of both retailing and wholesaling. By the end of the 1990s, over 80 countries had organizations that issue product codes.[5] These innovations, together with other advances in information technology, contributed to the rise of the large discount retailers such as Wal-Mart and Target and category killers such as Best Buy and Circuit City in electronics and appliances and Home Depot in hardware and building materials.

There is a broad range of inter-business transactions that includes wholesale trade as well as company purchases of services, resources, technology, manufactured parts and components, and capital equipment. Companies also engage in many types of financial transactions, such as insurance, commercial credit, bonds, securities, and other financial assets. Companies can automate inter-business transactions by substituting computer data processing and Internet communications for costly labor services.

Automation of transactions significantly lowers inter-business transaction costs. Traditionally, inter-business transactions begin with a buyer looking for inputs or a supplier seeking buyers for its goods and services. Buyers and suppliers search for each other through advertising, trade shows, brokers, and dealers. Suppliers send out sales agents. Buyers then negotiate with potential sellers concerning product specifications and prices, and perhaps conclude a spot transaction or form a long-term contract. After the agreement has been reached, the transaction still involves ordering, billing, arrangements for transportation, confirmation of payments, and acceptance of delivery.

Innovations in transaction technology have significantly reduced the transaction costs of purchasing for companies. Before the transaction, Internet technology may lower the cost of searching for suppliers or buyers and making price and product comparisons. Search costs can be significant relative to the value of the product, particularly for small purchases. According to a manager in the chemical industry: "When you're dealing with one or two drum quantities, the cost of comparison shopping can be more than the value of the product."[6] Sales personnel acting as sales representatives have traditionally carried out such mundane tasks as tracking product availability and pricing and supplying such information to customers. By automating these information services, e-commerce relieves sales personnel of these tasks, allowing them to concentrate on account management and marketing strategy.[7]

During the transaction, innovations in transaction technology can reduce the cost of communicating with counterparts in other companies regarding transaction details. Transactions over computer networks avoid many of the associated costs of interpersonal economic exchange, including the costs of travel, time spent on communication, physical space for meetings, and processing paper documents. Companies avoid the need to translate computer files into paper documents, a process that generally involves errors, delay, and costly clerical personnel. Companies automate this process by mediating transactions through websites and electronic data interchange (EDI).[8]

After the transaction, innovations in transaction technology allow companies to lower costs of communication, to monitor contractual performance, or to confirm delivery. In addition, companies can apply information generated by the transaction to update their inventory, production, and accounting records by automatically linking their transactions to software used for managing all aspects of the firm including sales, purchasing, and operations.

The potential cost savings in this area are substantial. Processing a purchase order manually, including paperwork, data entry, phone calls, faxes, and approval requests, can be quite expensive, so online transactions might easily reduce costs by a factor of 5 or 10 or more. Anecdotal evidence indicates that such cost reductions are possible. British Telecom estimated that by moving external procurement functions to electronic commerce, it reduced its costs from \$113 to \$8 per transaction.[9] MasterCard estimated that the cost of processing purchase orders fell from \$125 to \$40, with the time involved cut from 4 days to 1.25 days.[10] Lehman Brothers found that a financial transaction costs \$1.27 for a teller, \$0.27 for an ATM, and \$0.01 for an online transaction.[11] Online brokerage fees fell below \$5 in comparison whereas traditional discount brokerage fees exceeded \$50, suggesting a decrease in costs in back-office operations and brokerage transactions with financial exchanges. Even if such estimated savings were greater than average or varied across industries, their aggregate impact would be significant. Thus, advances in transaction technology not only reduced transaction costs but also widened the scope of possible transactions.

How does automation of transactions impact productivity? Nobel-prize-winning economist Robert Solow identified the problem that has come to be known as the Solow paradox: "You can see the computer age everywhere but in the productivity statistics."[12] Even though computers are widely used business tools, do they really help lower costs for companies? The Solow paradox raises some doubts about the effects of advances in information technology (IT) on productivity improvements. Companies spend a lot of money on buying computers and IT services, including employee training. For computers to enhance productivity at a company, the benefits from employing IT must exceed its costs. Managers carefully evaluate the incremental benefits of IT spending. Just getting the most advanced computers, software, and networking equipment does not guarantee increased profitability. Instead, managers should tailor IT expenditures to business applications.

The overall effect on productivity of advances in computers and communications has been decidedly mixed.[13] This presents a puzzle, given the high level of investment that companies have made to apply these new technologies. It may be that the costs of adjusting to new technologies may offset some of the productivity benefits. Robert J. Gordon showed that in the latter half of the 1990s, information technology failed to measure up to the standard set by the industrial revolution; productivity has exploded in the manufacture of computers, semiconductors, and other types of durables but productivity growth has decelerated in the rest of the economy. Gordon argued that "the speed at which diminishing returns have taken hold makes it likely that the greatest benefits of computers lie a decade or more in the past, not in the future." Moreover, according to Dale W. Jorgenson and Kevin J. Stiroh, "the empirical record provides little support for the 'new economy' picture of spillovers cascading from information technology producers onto users of this technology."[14] In contrast, a Federal Reserve Bank study found that the use of information technology, including not just computer hardware and software but also communications equipment, contributed substantially to productivity growth in the late 1990s.[15] These mixed results present a challenge to managers as they deploy and modify company IT systems.

Just as IT does not guarantee productivity improvements in the firm's organizational processes, automation of transactions and market reorganization do not guarantee profitability in e-commerce. The rapid entry and collapse of many e-commerce companies known as the dotcom shakeout demonstrated that company business plans needed more than creative IT applications to generate profits. Predictions of vast efficiency gains from business-to-business e-commerce were based on conjectures about how many inter-company transactions would shift to the Internet. Expectations about cost gains from e-commerce depended on possible efficiencies from automation of transactions and the potential economic advantages of intermediation and centralized exchanges.

One significant problem managers face is measuring the benefits and costs of IT. Because e-commerce is in its initial stages, it is difficult to evaluate the productivity impacts of IT investment. Evaluating the effects of IT on transaction costs is also difficult to measure in the initial stages of market reorganization. Measurement of productivity growth in services presents special problems. Productivity measures are geared to manufacturing, while e-commerce targets merchant activities. Jack Triplett and Barry Bosworth observed that economic changes attributable to e-commerce cross traditional production boundaries. As an example, they compare the purchase of a book from a traditional retailer with the purchase of a book from an online retailer.[16] Comparing the prices in the two settings ignores various differences, such as failing to include the costs of travel and time involved in visiting the traditional retailer while explicitly counting the costs of shipping and handling for the online purchase.

In e-commerce, measurement of productivity gains should accurately reflect the total net benefits that result from shifting bilateral inter-company transactions to intermediated exchange. Managers should carefully evaluate the benefits from reductions in marketing, sales, and procurement costs. In procurement, benefits include not only paying lower prices for goods the company purchases, but also possible reductions in the time and expense involved in procurement activities. In marketing and sales, benefits include not only higher customer demand and increased revenues but also potential reductions in the time and expense of marketing and sales activities. In evaluating the effectiveness of e-commerce systems, companies need to measure their own transaction costs.

Perhaps because they are difficult to measure, transaction costs have traditionally been treated as overhead or unavoidable back-office expenses. By correctly applying automation to transactions, companies can reengineer organizational processes and reduce management and employee costs. By employing innovative e-commerce technology, companies can reduce marketing, sales, and procurement costs while improving convenience for their suppliers and customers.

11.2 *Intermediaries*

An *intermediary* matches buyers and sellers. An intermediary such as a retailer can match many customers with the products of many suppliers. An intermediary such as a real-estate broker needs to match a particular home seller with a particular buyer because each house is different. An intermediary must lower the costs of matching not only in comparison with other intermediaries, but below what buyers and sellers could achieve by dealing directly with each other. Intermediaries lower transaction costs by providing customers and suppliers with accessible distribution channels, distinctive sourcing techniques, superior contract designs, and better information.

Intermediaries achieve a transaction advantage by delivering *greater convenience* to customers and suppliers. Intermediaries create value by improving the matchmaking process. Intermediaries also create value through services that enhance the value of the match between customers and suppliers. Intermediaries include retailers, wholesalers, financial firms and business representatives. Specialized intermediaries include realtors, banks, insurance agents, securities brokers, travel agents, and theatrical and sports agents. Intermediaries can be the counterparty to buy and sell transactions, such as a retailer. Intermediaries can also be basic matchmakers, such as realtors, who earn a commission but do not take part in the transaction.

The Transaction Triangle

Many business methods are available for reducing transaction costs, and managers need a consistent framework to evaluate and develop new business methods. Managers often think of customers and suppliers in a disconnected fashion. Customers are the responsibility of marketing and sales, while suppliers are the responsibility of the company's procurement units. However, the company's buying and selling transactions often are closely related. By conceiving of the business as an intermediary between its customers and suppliers, managers can gain a better strategic understanding of value creation.

This section introduces the *transaction triangle framework*, illustrated by Figure 11.1. The transaction triangle framework highlights the role of the firm as a coordinator of transactions with its customers and suppliers. The transaction triangle helps managers recognize *potential competition* from the firm's customers and suppliers. The firm loses business opportunities if its customers and suppliers choose to *bypass* it and engage in direct exchange with each other.

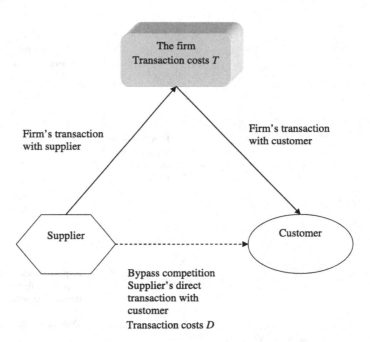

Figure 11.1:   The transaction triangle.

The firm must offer lower overall transaction costs in comparison with the buyer and seller dealing directly with each other. The transaction costs of the intermediated exchange, which equal T, must be less than the transaction costs of direct exchange, which equal D.

To avoid being bypassed by its customers and suppliers, the firm must develop transactions that create at least as much value as direct exchange between its suppliers and customers. The firm must offer lower overall transaction costs in comparison with the buyer and seller dealing directly with each other. The overall service the firm provides, including quality of service and convenience, must be better than customers and suppliers could do themselves. *The firm has a transaction advantage if the transaction costs of the intermediated exchange are less than the transaction costs of direct exchange.*

For example, suppose that the *transaction costs of direct exchange* between a customer and a supplier are $D = \$50$. If the firm handles the transaction as an intermediary between its customers and suppliers, the *total transaction costs of intermediated exchange* (including those of the firm itself) are $T = \$20$. The firm has a *transaction advantage* of $\$30$ as compared with direct exchange because it creates convenience for its customers and suppliers. Without such a transaction advantage the firm faces bypass competition from its potential customers and suppliers, as Figures 11.2 and 11.3 indicate.

To have a transaction advantage over bypass competition, retailers must offer their customers a better combination of prices and availability of goods, quality of service, and convenience than customers would be able to obtain by dealing directly with manufacturers or wholesale

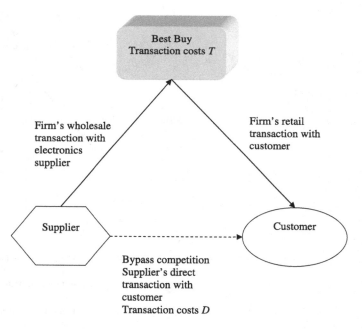

Figure 11.2:   The transaction triangle.

Best Buy, a major retailer of electronic equipment, sells computers made by a number of manufacturers, including Sony and Hewlett-Packard. Some Original Equipment Manufacturers such as Dell Computer sell direct to consumers.

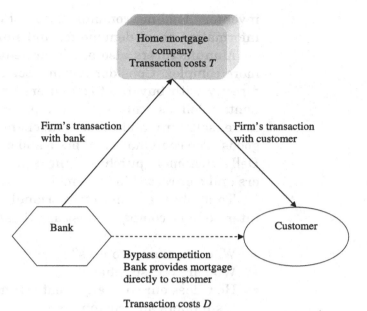

Figure 11.3: The transaction triangle.
Home mortgage companies compete with banks that make direct mortgage loans to household borrowers. Home mortgage companies provide mortgage loans to household borrowers and then resell the mortgage loans to banks and other lenders. To compete with banks as originators of loans, the home mortgage companies must have lower transaction costs than banks.

distributors. Consider the advantages offered by supermarkets. A customer can easily fill a shopping cart with many different products. Customers would need to spend a lot of time assembling the same basket of products if they tried to deal directly with each of the manufacturers individually. The supermarket also provides convenience to manufacturers who do not have to worry about retail marketing and sales since they deal only with the supermarket rather than with individual retail customers. Retail intermediaries include almost any type of retail store, including automobile dealers, appliance stores, electronics stores, hardware stores, book stores, and clothing stores.

To have a transaction advantage over bypass competition, wholesalers and other dealers must offer buyers and sellers lower transaction costs than they would have in direct exchange. Wholesalers provide a wide array of services to businesses, including distribution of goods, management of inventories, communication of price and product information, quality certification, and provision of credit. Wholesalers provide various value-added services including packaging, labeling, bar coding, EDI, product lot tracking inventory controls, logistics, and delivery.[17] E-Chemicals began as an online distributor for chemicals but evolved into a specialized intermediary that provides its buyers and sellers with such value-added services as supply chain management, inventory holding, communication of complex orders, and provision of information about demand for and supply of particular chemicals.[18]

Manufacturers also are intermediaries but their activities can be more complex. Consider a computer company such as Dell that deals directly with buyers of computers and suppliers of computer components. Dell assembles the computer shell, monitor, keyboard, disk drive, software, and other components, much like a market basket of goods. The combination of price and convenience must be better than if Dell's customers purchased the components directly from parts suppliers and contracted for assembly.

To apply the transaction triangle framework, the manager of the intermediary company must ask these questions:

- Who are our customers?
- Who are our suppliers?
- How does our company create transactions with greater value than if suppliers and customers were to bypass our company?
- How does our company create transactions with greater value than competing intermediaries sourcing from our suppliers and serving our customers?

A critical aspect of competing against bypass by suppliers and customers is lowering transaction costs. The manager must understand the transaction costs customers and suppliers experience, the transaction alternatives available to them, and the company's effectiveness in executing and coordinating its market transactions. The transaction triangle framework helps managers evaluate the company's potential transaction advantage.

Intermediaries and Transaction Costs

Intermediaries create more efficient transactions in many different ways, as Table 11.1 suggests. The intermediary provides transaction services to its customers and suppliers. Managing transactions creates value by increasing convenience for the company's customers and suppliers.

Table 11.1: **Intermediary strategies for reducing transaction costs.**

Transaction Costs	Intermediary Strategies
1. Fixed cost of exchange	Realize economies of scale and scope
2. Search costs	Centralize transactions
3. Communication costs	Establish networks
4. Information costs	Gather and provide data
5. Monitoring costs	Build reputation

Fixed Costs of Exchange

Buyers and sellers encounter *fixed costs of exchange* — costs that do not depend on the volume of transactions that are associated with billing, invoicing, providing receipts, and handling payments including cash, checks, and credit. Intermediaries who handle large numbers of transactions benefit from economies of scale and scope by maintaining a place of business such as a store and using record-keeping instruments such as computers, cash registers, bar coding and point-of-sale terminals. The following example illustrates how to apply the transaction triangle framework.

Example 11.1   Transaction Cost Economies

Consider a wholesale market with many buyers and sellers. Each time a buyer and seller meet and complete a transaction, they incur a total cost of $2 per transaction associated with billing, record keeping, arranging for delivery, and so on. An intermediary handles the transactions for the buyers and sellers by investing in cash registers and other computing equipment at a cost of $6,000. The intermediary takes on all the costs associated with billing, record keeping, and arranging for delivery and incurs an operating cost per transaction of $1. Suppose that the intermediary processes 12,000 transactions. Is the intermediary more or less efficient than direct exchange between individual buyers and sellers? The intermediary's fixed cost per transaction is $0.50, which is $6,000 divided by 12,000 transactions. The intermediary's cost per transaction is therefore $1.50, which is the fixed cost per transaction of $0.50 plus the variable cost per transaction of $1. So, the intermediary is more efficient than individual buyers and sellers. Based on the transaction triangle, the intermediary can compete effectively with direct exchange.

Large-scale retailers and wholesalers have substantial economies of scale and scope that allow them to have a lower average cost per transaction than if customers dealt directly with manufacturers. By consolidating purchases and sales, companies can earn returns from these scale and scope economies. Supermarkets and department stores offer a wide variety of merchandise and handle payments through simple standardized checkout counters.

Companies handle a variety of internal transactions as well, allocating capital, human resources, products and services, and technological know-how within the organization. They must consider internal allocation of resources and transfer pricing as alternatives to transactions with suppliers and distributors. In this area, information systems are of growing importance and managers should view them as value creation, not simply as operating costs.

David M. Weiss in *After the Trade is Made* likens brokerage firms' processing of securities transactions to production by a manufacturing firm. The brokerage firm carries out a complicated series of market actions and handles many financial instruments, including stocks, bonds, commercial paper, banker's acceptances, options, and futures. Customers give buy and sell orders to account executives or stockbrokers who pass them on to the firm's order room, which then returns a confirmation of the trade and the execution price. The order room executes the order with the securities exchanges and the over-the-counter securities market. Next, the purchase and sales (P&S) department computes payments and interest (figuration), compares the trade with the opposing brokerage firm's transaction (reconcilement), issues a confirmation to the customer, and books the trade in the firm's records. The reconcilement function is carried out through clearing organizations, such as the National Securities Clearing Corporation, which are also a type of intermediary. Then, the cashiers' department handles the brokerage's transactions with commercial banks, other brokerages, and transfer agents (who reregister a security in the name of the new owner). Five other internal departments handle the margin, stock record, accounting, dividends, and proxy aspects of the transactions.[19]

The brokerage's business is driven by transactions between its customers, the exchanges, and other brokers. Related processes exist for retailers, wholesalers, and manufacturers, but these transaction activities are mingled with the firm's other activities. By recognizing transactions and enhancing their quality and efficiency, companies can win additional customers and improve supplier relationships. It is essential for managers to realize the economic returns from these valuable services.

Search Costs

Intermediaries alleviate *search costs* for buyers and sellers by operating central places of exchange. Since the beginning of civilization, people have met in central marketplaces to trade goods and services. Towns provided a central marketplace in which farmers and buyers could meet conveniently on market day.

If people were to rely on barter, someone who desired apples and had only bananas to trade for them would have to find another person who desired bananas and had apples available to trade. Such a search would be time-consuming. The social contrivance of money saves time for everyone because it solves this problem, which is known as the double coincidence of wants. By using money, people can buy and sell apples, bananas, and other goods and services without looking for the perfect trading partner. But the use of money, and all its modern equivalents such as checks, credit cards, debit cards and even electronic cash, does not solve many of the high time costs of shopping for goods and services. There is still the need to search for merchants and service providers.

It is costly for buyers and sellers to search for each other. A buyer who wishes to purchase a good at a particular time and place has to find a seller who wants to sell that good at the same time and place. The central market place helps buyers and sellers meet. However, the problem is not completely solved because buyers and sellers still have to be at the market place at the same time. Dealers solve both problems of place and time. The dealer solves the location problem by traveling between buyers and sellers or by meeting buyers and sellers in a central location. The dealer solves the time problem by purchasing from various suppliers when they have goods to sell and building inventories. The dealer sells the goods to buyers when they are ready to buy.

Example 11.2   Meeting Your Match

Consider a market for custom-designed equipment.[20] There are two buyers, one with a willingness to pay of $100 and the other with a willingness to pay of $50. There are two companies with established businesses in other markets seeking to expand into custom-designed equipment. One company has a unit cost of $65 and the other company has a unit cost of $40. Although there are very few participants in this market, one can imagine similar markets with many participants in which there are very high costs of finding the right business partner. To represent this possibility, suppose that the two sellers and the two buyers are matched randomly with each other. There are two possible outcomes.

One possibility is that the high-value buyer gets together with the low-cost seller, in which case they create value of $100 − $40 = $60. The low-value buyer and the high-cost seller cannot transact with each other since the buyer's value of $50 is less than the seller's cost of $65.

The other possibility is that the high-value buyer gets together with the high-cost seller, and the low-value buyer gets together with the low-cost seller. In this case, the total value created is only $45! This is because the two transactions yield $100 − $65 = $35 plus $50 − $40 = $10.

An intermediary can enter this market and guarantee that the high-value buyer gets together with the low-cost seller. The intermediary achieves this by offering to buy from the sellers at a price

(Continued)

> **Example 11.2** (*Continued*)
>
> of $60, thus excluding the low-cost seller. The intermediary can offer customers a price of $70, which excludes the low-value buyer. The high-value buyer and the low-cost seller prefer to deal with the intermediary to avoid the risk of a bad match. The intermediary creates total value of $60 because the high-value buyer gets together with the low-cost seller. The intermediary makes a profit by capturing $10 of the value created. By the transaction triangle framework, the intermediary competes effectively with direct exchange.

Retailers provide consumers with a central place of exchange, usually a store or chain of stores. As every retailer will tell you, the keys to success are location, location, location. Centralization in location drove the establishment of department stores and malls. The trend toward superstores reflects an understanding of the value of centralization. Book superstore chains Barnes & Noble and Borders generally carry over 100,000 books. Customers avoid having to search across many smaller bookstores and are assured that the particular book they are looking for will be in stock and that there will be a wide variety of other books to browse through.

The reduction of search costs is best achieved by decentralizing access to a centrally operated network. Thus, a centralized marketplace provided by retailers need not be restricted to a physical location. A good example is the automated bank teller, which can be located in a variety of places including airports and malls. Customers can access their accounts and withdraw cash at any point in the network.

A catalog provides access to a central place of business. Increasingly, customers rely on catalogs for a wide variety of products that are delivered by mail, including clothing, electronic equipment, books, and compact discs. While the venerable Sears catalog has been retired, a new set of large specialized catalog retailers entered, such as L. L. Bean, Lands End, Eddie Bauer, and J. Crew. These retailers offer a wide variety of products with very rapid ordering and delivery, thus achieving immediacy. A customer can comparison-shop through several different catalogs, and complete an order in a matter of minutes, without leaving home. Thus, customers' search costs are drastically reduced.

A phone number alone can provide a central place of business. An increasing number of products can be ordered by mail or by phone, including airline tickets, theater tickets, and fast food. A very high degree of centralization is achieved by businesses that offer a national toll-free telephone number for placing orders. Home-shopping channels on television also provide a national central meeting place for customers, who then order products by phone. Unlike catalogs, immediacy is somewhat restricted in that customers can only observe products when they are scheduled by the shopping channel. However, shopping channels have the advantage of a salesperson that can discuss the product and demonstrate how it operates.

Websites on the Internet are also central places of exchange. Amazon.com's online bookstore provides a single bookstore that serves the entire country, with added sites serving 160 countries including a site in the United Kingdom and in Germany. Amazon states on its website that it offers "Earth's Biggest Selection (TM) of products, including free electronic greeting cards, online auctions, and millions of books, CDs, videos, DVDs, toys and games, and electronics".

Retailers also provide centralized marketplaces for their suppliers. Manufacturers and wholesalers transact directly with the retailer, not the retailer's customers. Retail chains generally increase efficiency by centralizing purchasing at headquarters and centralizing deliveries for many goods at local or regional warehouses. Wholesalers also provide dealer services to retailers and manufacturers by buying and selling products, transferring goods, and centralizing exchange.

Manufacturers also act as dealers by providing a central selling place for customers (whether wholesalers, retailers, or final customers). A failure to understand this can be quite costly. IBM relied on others to market its personal computers. The company finally responded to faltering sales in personal computers by offering a toll-free number for customer orders, something that had already been successful for rivals such as Dell and Compaq. Manufacturers also provide central marketplaces for their suppliers of parts, equipment, and other services.

Communication Costs

Intermediaries reduce *communication costs* for buyers and sellers by establishing networks. Imagine a situation in which many buyers and sellers had to contact each other directly. Not only would they encounter search costs, but the costs of exchanging information often would be significant. Consider again the grocery store. Imagine if customers had to contact all the manufacturers and distributors that supply the store to obtain price and product information. By establishing a network of suppliers and providing the products to customers, the grocery store takes on many of these transaction costs, performing the task more efficiently than individual customers could ever do. The store provides price and product information through its advertising and store displays. The store interacts with each of its suppliers on behalf of its many customers.

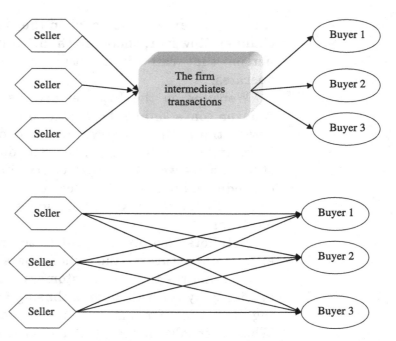

Figure 11.4: Intermediary firms reduce the transaction costs of direct buyer-seller contacts with hub-and-spoke contacts.

Intermediaries provide network economies by replacing the transaction costs of direct buyer-seller contacts with hub-and-spoke contacts (see Figure 11.4). The market with three buyers and three sellers involves only six contacts with the intermediary. The market without an intermediary potentially involves nine contacts between buyers and sellers. Example 11.3 illustrates how intermediaries operate networks that yield economies of communication.

Example 11.3   Network Economies

Consider a market for woodworking tools with four sellers and 10 customers. The sellers provide different types of tools so every customer would like to purchase from every seller. The cost to a seller of contacting any customer to provide information is $3. The cost to a customer of ordering from any seller is $2. So, the total cost of each buyer and seller contact is $5. Direct exchange between buyers and sellers involves 40 possible contacts, so the total transaction costs of direct exchange equal 40 × $5 = $200.

An online catalog company offers sellers an opportunity to list their products. The cost to any seller of providing its catalog to the online catalog company is $8. Customers can visit the online catalog

(Continued)

Example 11.3 *(Continued)*

and simultaneously place orders for the products of multiple sellers. The cost to a buyer of ordering from all sellers at once through the online catalog company is $6. The fixed cost of the online catalog company is $74. Which alternative is more efficient, the online catalog company or direct exchange between buyers and sellers?

The cost of the market with an intermediary is the total cost to all the sellers of providing the catalog to the online catalog company, which is 4 × $8 = $36, plus the total cost to all the buyers of placing an order with the online catalog company, which is 10 × $6 = $60, plus the fixed cost of the online catalog company, which is $74. So, the cost of the market with the online catalog company is $170, which is less than buyers and sellers dealing directly with each other. Applying the transaction triangle framework, the intermediary can compete effectively with direct exchange.

A wholesale distributor lowers the transaction costs for retailers and manufacturers. By serving networks of retailers, the wholesaler creates convenience for manufacturers. The wholesaler interacts with each of its retailers on behalf of the manufacturers it serves. For example, pharmaceutical wholesalers provide medicines to networks of pharmacies, doctors, and hospitals. They obtain pharmaceuticals from networks of manufacturers. These networks save greatly on communication costs in comparison with extensive direct contacts between fragmented retail pharmacies and manufacturers.

Companies that are supply chain managers perform similar network management functions. For example, Solectron Corporation, one of the largest electronics manufacturing services companies, provides outsourcing solutions to leading electronics original equipment manufacturers (OEMs) in such industries as computers, computer peripherals, networking, telecommunications, semiconductor equipment, industrial controls, medical electronics, avionics, and consumer electronics. Solectron views itself as a "a supply chain facilitator" — an enterprise that promotes speed, efficiency and cost-containment throughout the entire supply chain. Solectron maintains a vast global network of customers (OEMs) and suppliers (component manufacturers). It reduces communication costs by applying electronic data interchange, centralizing data collection and analysis, and jointly monitoring manufacturing and fulfillment processes.[21] The centrally managed network allows outsourcing transactions to be managed at far lower costs than if the OEMs and component suppliers were to transact directly with each other.

Exhibit 11.1 The Airline Computer Reservation System (CRS)

The airline computer networks are excellent examples of reducing communication costs.[22] The computer reservation systems (CRS) coordinate staggering numbers of transactions. They are capable of handling millions of flights and all the details of passenger reservations on those flights. The CRS reduces communication costs for customers and travel agents. An airline can adjust its offerings on the CRS relatively easily and update its internal reservation information directly, so its communication costs are lowered as well.

The airlines initially established their systems as a way of keeping track of many thousands of airline fares and all of the available seats and passenger reservations. The CRSs were intended as a means of internal accounting for the airlines to aid in booking passengers. The four main systems are Sabre (American Airlines and US Airways), Apollo/Galileo (United Airlines), Worldspan (Delta, Northwest, and TWA), and Amadeus (Continental and European airlines). The CRSs are important not only because they serve the large travel market but also because they were the forerunners of many types of Internet commerce.

Airlines provided the systems to travel agents to simplify their ticketing and communication. The computer reservation systems also listed the flights of other airlines, which were charged a fee for the service, as a matter of convenience for the travel agents. The convenience was so great that almost all travel agents signed on to a CRS and practically all tickets were written on a CRS. The airline providing the computer system, known as the "host", initially listed its flights first on the screen.

Airlines soon discovered that this listing practice resulted in substantial diversion of passengers to the host airline and significant incremental revenues. Some airlines charged that travel agents tended to favor the host airline in booking flights, known as the "halo effect", partly as a result of special commissions. Non-host airlines also said that the CRSs were slow to list their fares and allowed the host airline to adjust its fares in response. Sabre and Apollo have dominated the CRS market, while other systems had smaller shares, and American Airlines and United Airlines (which operate those systems) were accused of taking unfair advantage of their CRSs. On November 14, 1984, the Civil Aeronautics Board passed a number of regulations governing access and pricing of CRSs, including rules against "onscreen

(Continued)

Exhibit 11.1 (*Continued*)

bias", and the Transportation Department put forward additional rules in 1991.

As the popularity of CRSs with travel agents grew, airlines observed the important consequences for sales that listing on the CRS had. What had begun as an accounting device began to be understood as a powerful competitive tool. Airlines realized that they could adjust prices and seat availability rapidly in response to sales, a process known as yield management, which allows pricing and availability to reflect current market conditions. Moreover, the CRSs are a source of important market information.

The CRSs grew to become broad travel networks, providing booking for hotels, rental cars, and other services in addition to air travel. In 1990, American's Sabre system formed an alliance with the Amadeus Global Travel Distribution, a network set up by Air France, Iberia, Scandinavian Airlines, and Lufthansa, to provide international services. United Airlines' Covia, which owns the Apollo system, merged in 1992 with Galileo, which was formed by nine European carriers (Aer Lingus, Air Portugal, Alitalia, Austrian Airlines, British Airways, KLM Royal Dutch Airlines, Sabena, Swissair, and Olympic). A third international system called Worldspan was formed by merging Pars, owned by Transworld Airlines and Northwest Airlines, and the Datas II system owned by Delta Airlines. The CRS business proved to be profitable even as the airline business itself suffered losses. For example, American Airlines' Sabre earned a quarter of a billion dollars in 1993 even as the company's airline operations lost money.

The CRSs adapted to online reservation systems, allowing passengers direct access to information through the Internet. American's Sabre operates the Web-based service Travelocity. American Express offers the Internet Travel Network (itn.net and Getthere.com) using United's Apollo CRS. Worldspan operates Worldspan.com and partners with Microsoft's Expedia travel service. Amadeus established AmadeusLink (amadeus.net) to provide international travel information although bookings are made through online travel agencies. A variety of independent travel agencies offer online bookings and many of the airlines have their own websites that allow reservations and ticketing, such as American Airlines (aa.com), United Airlines (united.com), and Southwest Airlines (southwest.com).

The Orbitz website, (orbitz.com), jointly owned by United, Northwest, Continental, Delta and American, with 29 other airline

(Continued)

Exhibit 11.1 *(Continued)*

affiliates, represented a significant attempt to present an online reservation system with comprehensive information. Orbitz offered access to over 450 airlines worldwide, as well as to hotels, car rental agencies, and cruise lines. Jeffrey G. Katz, the chairman and CEO of Orbitz, testified before the Senate Committee on Commerce, Science, and Transportation in July 2000 that Sabre/Travelocity and Microsoft/Expedia together controlled over 70 percent of Internet airline reservations, although only about 4 percent of total reservations were made over the Internet that year. According to Katz, Orbitz's "objective is to provide absolutely unbiased display of every airline's flights and fares, whether they are investors in Orbitz or not, and whether they are associates of Orbitz or not. We want every airline to be fully and equally displayed in Orbitz, and we want Orbitz to treat every airline the same". Katz continued, "We do this not simply out of charity or good will, but because we have to provide that level of quality information if we are to have any chance of winning consumers away from these two dominant websites." The Orbitz system featured advances in computer hardware and software to supplant the mainframe-based legacy CRS systems. Katz observed that "between City A to City B and back there are typically somewhere between half a billion and a billion possible combinations of airlines, schedules, and fares" and traditional CRSs only search 5,000 to 10,000 possible combinations. Orbitz proposed to search all possible combinations and present them to customers based on various criteria such as lowest-cost first or time of travel.

The CRSs supplemented the brokerage services of travel agents by providing a central place of exchange for all airlines, hotels, and other travel services. By simplifying the interaction between passengers and providers of travel services, CRSs greatly reduced transactions costs and created immediacy for airlines and passengers. The CRSs came full circle: They began as the earliest application of electronic commerce long before the World Wide Web, and moved on to advanced data processing techniques. In the process, they expanded from basic reservations to comprehensive travel services. The profitability of the CRSs demonstrates the high economic returns to managing transactions and enhancing communication between buyers and sellers.

Information Costs

Intermediaries mitigate the impact of *information costs* for buyers and sellers by gathering and providing many types of market data.

Companies spend substantial resources informing their customers about the prices and features of their products. Companies also communicate information to their suppliers about the quantity and types of inputs that they need. Wal-Mart built a satellite-based data network to track sales in its stores, and it communicates the relevant sales information to its thousands of suppliers so they can make the necessary manufacturing and delivery decisions.

Internet companies provide information while connecting buyers and sellers. Yahoo and other websites provide significant amounts of information to consumers, including news and weather, access to online reference materials, Internet search engines, and extensive directories of sellers. Internet companies connect advertisers to potential customers, just as do other media companies, such as radio and television broadcasters and newspaper and magazine publishers. Overture Services ranks advertisers on the basis of the amount they bid to be displayed by search engines and splits the revenues with search engines that display the advertisers.[23]

Monitoring Costs

Intermediaries reduce problems stemming from *monitoring costs* by building their own reputation for trustworthiness. Rather than dealing directly with each other, buyers and sellers place their trust in the intermediary, thus avoiding insurance and other costs stemming from risky trading partners. Intermediaries substitute their ability to make commitments for those of individual buyers and sellers.

For example, some retailers test and certify the quality of products obtained from manufacturers. The customer benefits by relying on those retailers with a good reputation, and manufacturers benefit from the retailer's quality certification. Customers purchase the house brands of retailers even if they do not know the manufacturers of those generic products. Customers rely on the reputation of pharmacies when they purchase generic pharmaceutical products. Customers purchasing used cars from dealers rely on the quality certification and warranties of the dealer, which give the used car dealer an advantage over individuals selling their own used cars. Online auctioneer eBay gathers information about the performance of buyers and sellers in carrying out their agreements and posts the ratings, thus reducing the costs of monitoring to auction participants.

11.3 *Market Makers*

Companies that are *market makers* achieve an important type of transaction advantage by creating and managing their markets.[24]

A *market* is a centralized exchange where buyers and sellers meet each other directly or trade through dealers. Market makers provide these institutions of exchange by coordinating interaction between many buyers and many sellers. Market makers operate the market by adjusting prices and balancing purchases and sales. In contrast to intermediaries, who match up a particular buyer with a particular seller, market makers add up the demands of many consumers and the supplies of many producers and create a market by combining demand and supply. Market makers create markets by standardizing products, from basic commodities to manufactured parts.

Market making creates new opportunities for creating value. Companies that bring buyers and sellers together raise the efficiency of markets by lowering transaction costs. Market makers generate benefits for customers and suppliers by improving transactions. Market makers perform several major functions. First, they coordinate exchange between buyers and sellers, reducing costs of search and bargaining. Second, they adjust prices to influence buyer demand and to give incentives to sellers, while providing information to the market. Third, they adjust purchases, sales, inventories, and production to balance supply and demand, avoiding inefficient discrepancies. Finally, they allocate sales and purchases of products across market segments in response to differences in buyer valuations of the products and productive efficiency of suppliers. Market-making by a firm is represented in Figure 11.5.

Coordinating Purchases and Sales

In financial markets, where the term originates, market makers are dealers that stand ready to buy and sell a particular financial asset if

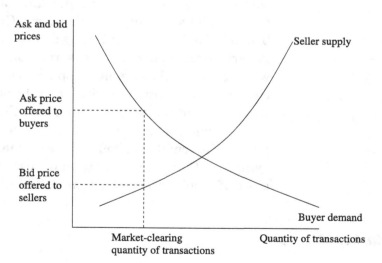

Figure 11.5:   The market-making firm chooses the market-clearing ask price offered to buyers and the market-clearing bid price offered to sellers.

a buyer or seller is not available.[25] Thus, market makers provide *liquidity* by keeping a stock of a particular asset for possible sale as well as cash on hand to buy that financial asset. Investors know that they can enter the market to buy or sell at any time, since they will always find someone to transact with. More investors are willing to invest in a bond or security knowing that it will be possible to cash out. In turn, the more buyers and sellers participate in a market for a financial asset, the more liquid that market becomes.

In a similar way, intermediaries reduce risk for buyers and sellers in product markets by standing ready to purchase or sell. By selling when customers are ready to buy and buying when suppliers are ready to sell, firms provide *immediacy* to product markets that is similar to liquidity in financial markets.[26] Retailers perform market-making functions in product markets. Consumers are able to make shopping plans with confidence if they know that gasoline will be available at the service station or specific products will be on the shelf at the grocery store. Customers rely on retailers having a wide variety of goods in stock, including food, clothing, furniture, hardware, appliances, computers, or other general merchandise. Customers dealing with the retailer have convenient access to the products of many different suppliers without having to deal directly with the suppliers. Suppliers to the retailer have convenient access to the retailer's many different customers without having to deal directly with the customers.

Wholesalers also perform market-making functions. Manufacturers are able to plan production with greater confidence if they know that their supplier of chemicals, energy, or parts is standing ready to serve them. Holding inventory is costly, not only due to the costs of storage, but also due to the capital costs of holding the unsold goods. Manufacturers can lower inventories if they expect just-in-time purchase and delivery. Wholesalers often aggregate the products of many manufacturers and aggregate the demands of many buyers to facilitate such just-in-time functions.

Market makers include large-scale companies that create worldwide markets. Oil companies Exxon-Mobil, BP, and Royal Dutch Shell manage major portions of the worldwide oil markets for crude oil, natural gas, petrochemicals, and refined gasoline. Companies such as Cargill and Archer Daniels Midland create worldwide markets for all types of grain and agricultural products. Dealers that act as market makers include major retailers and general merchandisers such as Wal-Mart, Target, Home Depot, and Best Buy. Major wholesalers such as Genuine Parts, W. W. Grainger, TruServ, Wesco International, Ace Hardware, and Ingram Micro aggregate demand and supplies for manufactured goods. Financial firms such as banks, insurance companies, and diversified financial companies create markets for financial assets and products.

Manufacturers also act as market makers, although their transaction activities are intertwined with their productive activities. Nokia manages a large portion of the worldwide market for mobile phones, creating generic platforms and adjusting components so that mobile phones have features that meet local needs. Auto manufacturers, including GM, Ford, Toyota, DaimlerChrysler and Volkswagen, create worldwide markets in platforms. Auto-parts makers such as Delphi and Siemens create international markets for standardized parts and components. Purchasing productive inputs and selling outputs are critical activities for most companies. By understanding the relationship between these activities, managers can enhance the company's market-making activities. It is useful for managers to perceive the company's buying and selling activities as a *service* that coordinates the transactions of its customers and suppliers.

Adjusting Prices

Pricing is the crucial market-making activity of firms. While it is sometimes convenient to say that "the market sets the price", price setting is not carried out by an invisible auctioneer. Rather, pricing often is an integral part of competitive strategy. Moreover, a firm's pricing is an essential aspect of the intermediation between its customers and its suppliers. Because pricing is at the heart of competition, the prices managers choose reflect strategic interaction between the company and its competitors.

A basic example illustrates how competitive pricing is connected with market-making. Consider a flower stand that buys and sells only roses. The flower stand sells the flowers by the dozen at a price of $15 and buys the flowers at $10 a dozen. The flower stand's revenue equals the price times the quantity sold and its cost equals the price times the quantity bought. If the flower stand succeeds in selling all that it has purchased, its profits are equal to the *price spread* of $5, which is $15 minus $10, multiplied by the quantity it sells.

The flower stand faces a number of competitive challenges. A competitor with the same supplier may offer the flowers at $14 a bouquet and lure away all of its customers. The stand's manager must be alert to all prices of competitors and be prepared to vary its output price accordingly. A competitor may find another supplier offering roses at $9 a dozen and then offer its roses at $10, effectively driving the stand out of business. The stand's manager must be aware of alternative sources of supply and be prepared to secure lower prices from its supplier or to seek an alternative supplier. In practice, the stand constantly faces fluctuations in both the sales and purchased prices of roses and must be constantly prepared to respond to variations in the spread.

Now suppose that the flower stand purchases and sells a wide variety of flowers, including not only roses but tulips, daffodils, impatiens, carnations, chrysanthemums, geraniums, daisies, and marigolds. The stand now relies on a variety of suppliers and is subject to wide variations in prices and seasonal availability. Moreover, the flower stand offers its customers mixed bouquets combining the flowers in an almost infinite variety of ways. Since flowers are perishable, the manager worries about the possibility of unsold inventory. It becomes necessary to monitor carefully the prices of competitors' flowers and bouquets and the relative prices of alternative sources of supply. The flower stand constantly assembles and reshuffles a portfolio of flowers taking into account the likelihood of sales and the expected prices. The manager's decisions now must respond to the changing price spreads for each type of flower. Since the flowers are mixed in some bouquets, all of the stand's purchases depend on each of the spreads. Retailers offer their customers a wide variety of products, often obtained from many suppliers. Manufacturers produce products assembled, like bouquets, from a variety of inputs. Whether reseller or manufacturer, firms live on the price spreads between the goods they sell and the inputs they buy.

Managers should pay careful attention to the relative levels of purchase costs and sales prices. The difference between the firms' asking prices for outputs and bid prices on inputs determines the earnings of the firm. The price spread gives a good indication of the firm's value added. The spread between the bid and ask prices in financial markets reflects the transaction costs of an asset and provides returns to the market-making activities of firms. In retail and wholesale product markets, the spread between the purchase and resale price also reflects transaction costs and provides a return to market-making activities.

Managers should coordinate their company's purchasing and sales efforts. They must react quickly to changes in price spreads whether from the input or output side or both simultaneously. This concept is familiar to financial firms and other participants in the markets for securities, bonds, derivatives, futures, and currencies. The speed with which product prices can change has increased significantly. Managers should try to control the destiny of the firm by coordinating the choices of prices they offer to the firm's customers and suppliers. Managers adjust pay levels for new and existing workers. Managers make price offers to the company's suppliers, engage in bargaining with suppliers, or take bids from one or more suppliers.

Companies often face high transaction costs in setting prices, so that getting it right becomes all the more important. A typical supermarket chain, such as Jewel or Dominick's, sets prices on 50,000 to 70,000 stock-keeping units. Not only that, the prices vary across the stores in the chain reflecting different costs and different local competitors. Also, many of the prices change daily or weekly as the supermarket responds to seasonal availability of fruits and vegetables, changes in customer tastes, manufacturer promotions, and prices set by rival chains. Keeping track of this data consumes company resources. Many chains pay outside tracking services to report on a small number of rival prices that serve as benchmarks.

Adjusting prices also involves transaction costs since it is costly to print new menus, change catalogs, and revise advertising. Managers should understand that computing and adjusting prices are activities that create value. Effective pricing requires investment in training and information processing equipment, such as scanners, computers, and digital networks that allow close monitoring of current sales and reduce the costs of updating prices.

Companies should communicate price information clearly to buyers and sellers so that they can make the necessary adjustments in their plans. A simple statement of the product price in an advertisement is valuable to customers since it allows them to compare advertised prices. Clear price information is one of the secrets of success of catalog companies such as Land's End. The consumer can compare the costs of different items within the same catalog and calculate the cost of the total purchase, something that may be quite difficult to do in the hustle and bustle of a store. Consumers' need to know price information for comparison-shopping and budgeting creates new opportunities for companies that can provide this information at the lowest cost, whether through superstores, catalogs, computer networks, or interactive video systems for home shopping.

Prices are generally the most informative part of marketing claims, before product characteristics and brand image. Discounters in particular stress price comparisons with competitors: "We're the low price leader", "We won't be undersold", "Always the lowest price, guaranteed." By pricing low, discounters signal the efficiency of their operations and their high volume of sales. They let customers know that they are committed to maintaining low prices in the future. Customers understand that the low-price leader is the winner of the market and anticipate that many others will shop at the same store. A continued high volume of sales then allows the store to maintain its low prices in the future, thereby confirming the customer's decision to shop at the low-price leader.

Not all companies understand the value of clearly communicating their prices. Auto dealers traditionally have tried to disguise the price of vehicles by not being explicit about the level of discounts offered from the list price. Few auto dealers advertise their price or quote prices over the phone. Indeed many auto dealers try to raise the cost of learning the price by extended delays, often keeping customers waiting for hours while sales personnel make endless trips to the "back room," ostensibly to check their calculations with their manager. The purpose of such delays is in part a traditional bargaining technique familiar to anyone who has visited a Middle Eastern bazaar. While salespeople drag out the process, they learn more about the customer's income, preferences, and willingness to pay for the car, so they can adjust the discount accordingly to achieve first-degree price discrimination. In fact, the customer's impatience is itself useful as the customer weighs the cost of his time versus the cost of making an early price concession. Moreover, by raising the cost of bargaining through delay, the customer becomes stuck at a particular dealer and may not have the time to visit a rival dealer. The exhausted customer may simply agree to a price without a comparison because he or she is disheartened at the prospect of repeating the bargaining process.

While sometimes effective at raising dealer profits, this rather inefficient technique increases the transaction costs for the automobile market as a whole, thereby reducing overall demand for cars. The costs of search and bargaining are so high under the old system that in many cases it may affect the consumer's choice of which type of car to purchase. Certainly it can reduce a customer's loyalty to a dealer despite other efforts at customer service. This situation has created opportunities for entry for dealers who post prices clearly, such as dealers selling GM's Saturn.

Clearing the Market

Market-making companies *clear the market* by balancing supply and demand, which reduces wasteful inefficiencies that would occur if the total amount supplied were greater than total demand. This also helps avoid inefficiencies associated with unmet needs that would occur if the total amount demanded were greater than total supply. A market clears when the total amount consumers demand equals the total amount sellers supply. When prices clear a market, buyers can purchase all they want at the ask prices in the market and sellers can sell all that they want to at the bid prices in the market.

Retailers are concerned with matching sales to inventories. If stock-outs occur, with sales demand exceeding inventory, the retailer misses valuable sales and may lose important customers since availability is vital for maintaining customer loyalty. On the other hand, excess inventory is not desirable either. The cost of holding inventory is not simply the cost of warehousing and shelf space, it is the interest cost of the investment in inventory. Retailers wish to achieve a high rate of product turnover. Turnover simply means the rate at which the retailer can sell products and use the proceeds to restock. The faster inventory turns over, at any given price, the greater the rate of return on the retailer's investment. To increase turnover, retailers adjust prices, as we discuss in the next section.

In addition to price adjustment, retailers take many actions that help the market to clear. Retailers continually adjust the quantity of inventories purchased based on demand projections and daily monitoring of the pattern of sales to increase turnover. They frequently reorder items that sell quickly and order slower-selling items less often. Advances in computerized inventory management and product bar coding have enabled retailers to monitor sales as they occur, allowing shorter reordering times and closer adjustment of inventories to demand patterns. Retailer thus can keep less inventory at the store or the warehouse while reducing the likelihood of stockouts. The amount retailers sell and the amount they purchase track each other more closely over time, bringing the quantities bought and sold in the market closer to being in balance.

Retailers share demand information with suppliers, allowing suppliers to adjust production and shipments to satisfy demands without excessive inventories as well. Wal-Mart uses direct telecommunication and computer links with its suppliers to provide them with detailed information about the sales of their products in individual stores. Some manufacturers restock store shelves themselves or through jobbers, and thus obtain frequent information about the pace and mix of their sales. For example, bakeries restock the bread aisle in supermarkets and adjust deliveries to the pattern of sales.

Another method retailers use is to adjust their sales efforts to move inventories, advertising and prominently displaying items when excess inventories build up. An interesting example of market clearing can be seen on televised home shopping networks, which inform viewers of the remaining numbers of an item that are available, and count down as sales occur. Promotion of the item ceases at the moment supplies are exhausted.

Another way to synchronize purchases and sales is by pooling inventories. Chain stores use central warehouses to pool inventories across stores. This serves to smooth out the demand pattern by pooling the risks from demand fluctuations across stores. The chain store can provide product availability while keeping a much smaller total inventory than it would have to if inventories were kept at each store.[27] Retailers that take orders by mail or by phone and serve customers from a central national location benefit by taking orders from a large pool of customers, which reduces the risk of demand fluctuations in comparison with a decentralized system of local stores and regional warehouses. Wholesalers and other intermediaries earn profits by serving multiple retailers from centralized warehouses, thus effectively pooling inventories.

Pooling is familiar to managers of electric utilities that utilize large-scale electric power pools such as the New England Power Pool. To avoid power curtailment or interruption, utilities must build sufficient productive capacity to meet peak demand, which is quite costly because the incremental capacity lies idle during off-peak periods. Utilities give customers various price incentives to smooth out demand fluctuations, but significant random variations in demand persist. Utilities have found that by buying from the pool when their demand is high and selling to the pool when they have excess capacity, they can reduce the need for costly generating capacity. So, by pooling their capacity, utilities are able to bring the supply and demand for electrical generating capacity closer together.

Manufacturers play an important market-making role by adjusting production to changing demand information. Dell and other computer makers assemble computer systems to order, thereby adjusting instantly to demand patterns. Increasing automation of inventories allows manufacturers to reduce their inventories of parts and materials. The use of just-in-time inventory management has often been promoted as a cost-cutting activity. By reducing the time inventories sit idle, manufacturers reduce the time cost of investment in inventories. However, just-in-time inventory management should also be seen as an important *market-making* activity. Manufacturers' reduction of parts inventories means that their demand for parts is more closely synchronized with the supply of parts, thereby providing clearing services to the market for parts.

Even more dramatic is the development of the flexible factory that eliminates delays both in production and distribution. By producing multiple products on the same assembly line, the flexible factory allows a manufacturer to tailor more closely its production to demand patterns without building substantial inventories or operating multiple production systems with excess capacity. The flexible factory allows a multi-product firm to achieve the benefits of economies of scope. Like just-in-time inventories, just-in-time production and the flexible factory also perform market-clearing roles by improving the adjustment of production to market demand information.

George Stalk, Jr., and Thomas M. Hout of the Boston Consulting Group describe Toyota's response to the problem of delays in distribution and sales. Toyota found that only 10 percent of the time a customer had to wait for a car was due to production whereas 90 percent of the delay was due to distribution and sales. Toyota merged its manufacturing and sales companies in 1981 and established an information network that by the end of the 1980s linked several hundred wholesalers and over 4,000 independent distributors. Toyota changed its strategy from "sell the customer what we have" to "sell the customer what he wants".[28]

Retailers such as the Gap improved market clearing in apparel by reducing the ordering cycle for clothes. This change allowed the Gap to adjust quickly not only to the sales patterns at their stores but also to respond rapidly to fashion changes in colors, fabrics, and styles. The Gap was able to attract business from those department stores that had longer ordering cycles and were thus less in tune with changing customer demands. Companies that supplied clothing to the Gap were able to adjust their manufacturing plans, hiring, and materials purchases in response to more up-to-date information about demand. These new developments represent increased efficiency in clearing markets for clothing.

By adjusting inventories and production to customer demand patterns, firms provide critical services to their customers and help bring the market into balance. Improvements in computer and communications technology allow firms to closely track customer demand patterns. Just-in-time inventories and flexible production techniques then allow rapid adjustment to customer requirements while reducing costly inventories and production capacity. These developments are critical to the market-making functions of firms. The market is closer to being in balance because demand and supply are continually brought closer together. Matching demand and supply more closely is not just a question of lowering inventory costs. Managers should use market demand and supply information in adjusting the quantity and variety of the company's products.

Allocating Goods and Services

Companies provide another valuable market-making service by allocating their products across different markets. Companies create value by allocating their products to markets with the highest-value users. Companies respond to customer needs and supplier availability through sophisticated distribution and supply chain management. Retail chains allocate product deliveries to regional warehouses and to individual stores in response to sales. Wholesalers allocate goods to their customers based on orders and availability. Airlines practice yield management by constantly changing the proportion of discounted and full-fare seats on flights. Manufacturers allocate production to retailers, wholesalers, and other manufacturers in response to orders and production requirements. International businesses must make complex allocation decisions that coordinate their deliveries to different countries based on demand projections, transport costs, trade regulations, and competitive considerations.

Companies rely on such devices as delivery lags to ration customers during periods of tight supplies in many industries such as textiles, mill products, paper and allied products, steel, electrical and non-electrical machinery, and fabricated metals.[29] In addition to decisions about how much of their products to supply, companies allocate products by varying their prices across market segments and geographic locations. In addition, marketing and sales efforts across markets and the choice of distribution channel the allocation of products.

In some industries the allocation of scarce capacity is built in to product offerings. For example, Federal Express offers many categories of mail delivery, including next business morning, next business afternoon, and second business day, which correspond to different priority levels. Customers pay the most for the highest priority, next business morning, which gives their package first access to the firm's transportation and delivery capacity. If capacity is available, lower-priority mail can arrive early. When capacity is scarce, lower priority mail is shifted to later in the day when more capacity becomes available. Similar forms of priority pricing have been used to allocate electricity generation and transmission capacity. Industrial customers can sign up for interruptible or curtailable service at a discount. In peak periods, such as hot summer days when everyone is turning on their air conditioners, the marginal cost of generation is high and capacity shortfalls occur. These customers must then reduce their power usage so that higher-priority users can be served.

Companies also allocate resources internally. Managers allocate resources through their investment, production, and distribution decisions. Managers of multi-divisional firms allocate scarce capital across the firm's divisions. The corporate office evaluates the expected returns and strategic value of the projects and monitors their progress, thus performing capital market functions. Managers allocate resources and products through divisional budgets and transfer pricing. They allocate human resources through internal labor markets consisting of job categories, promotional procedures, compensation rules, and performance evaluations. Personnel are continually reallocated through the formation of project teams, and the reassignment of employees and managers. Technology is allocated within the firm through decisions about R&D expenditures, investment in the design of new products and production processes, and assignment of technical personnel. Managers should view such allocation decisions as market-making activities with the potential to create value.

11.4 *Entrepreneurs and Transaction Innovation*

Any specific transaction advantage is necessarily temporary; maintaining it requires a continual process of innovation. In a dynamic economy, a succession of competitive firms introduces new business methods and puts together new combinations of buyers and sellers.

Companies constantly face bypass by customers and suppliers or replacement by competing intermediaries. Business methods are easily copied and applied. To maintain its intermediary position, a company must create better transactions than its customers and suppliers can devise in direct exchange with each other. Moreover, the company must outperform competing intermediaries. Transaction innovations lower transaction costs or generate new forms of transactions.

Entrepreneurs are entrants that create innovative transactions, often by making new combinations of customers and suppliers. Maintaining a transaction advantage requires continually identifying new transaction opportunities. Managers must be prepared to forge new market relationships with customers and suppliers. Adam Smith long ago observed that innovations are made by "men of speculation, whose trade is not to do anything but to observe everything; and who, upon that account, are often capable of combining together the powers of the most distant and dissimilar objects".[30]

Entrepreneurs

The entrepreneur discovers new forms of business organization, new types of contracts, new sources of supply, unmet customer needs, and new applications for research and development. Entrepreneurs act as intermediaries between suppliers of technology and those who derive benefits from employing the new technology, whether they are consumers or firms.

Entrepreneurs create new combinations of products, processes, and transactions. Established companies or entrants create entrepreneurial innovations by bringing together their suppliers and customers in novel ways. Companies developing and applying entrepreneurial innovations need not be the original creators of product, process, or transaction innovations. Rather, they bring one or more of those innovations to the marketplace. Entrepreneurial innovation creates competitors with original combinations of products and services, manufacturing technology and transactions.

The noted economist Joseph A. Schumpeter described an entrepreneur as someone who introduces technological change to the marketplace. He gave the now-classic example of an individual who first introduced the power loom to some segment of the textile industry during the industrial revolution in Britain. Prior to that time the textile industry had relied on manual labor. The entrepreneur need not be a manufacturer either of the capital equipment or of the final product. The profit the entrepreneur earns is not a return to ownership of capital equipment, nor is it a return to ownership of the firm through the provision of finance. The entrepreneur's profit is not a return to the research that produced the innovation. Finally, the entrepreneur need not be the risk bearer, unless he chooses to self-finance.[31]

The entrepreneur earns a return from making the deals needed to introduce the innovation and by founding new businesses. By introducing the power loom to the textile industry, the entrepreneur earned the initial returns due to the increased productivity in textile manufacturing. Generally, the entrepreneur earns the initial returns to the new combination. After a time, competitors imitate or surpass the innovation, eroding the returns to the new combination.

Entrepreneurs discover arbitrage opportunities. Arbitrage refers to profit-making activities that connect suppliers and customers. The firm resells or transforms products and services obtained from suppliers to meet the needs of its customers. Arbitrage profits depend on rapid action since arbitrage opportunities can erode in the blink of an eye in the most competitive markets. Managers must have up-to-date knowledge of shifting supplier capabilities and changing customer requirements. Managers must develop the ability to recognize these opportunities and to execute the necessary transactions rapidly.

What makes entrepreneur arbitrage possible is the existence of substantial market frictions. Without these frictions, arbitrage opportunities would be negligible. Transaction costs are pervasive, including the costs of gathering and comparing price and product information and the costs of searching for customers and suppliers. The ability to create better transactions is offset by the potentially high costs of negotiating, writing, and monitoring contracts. Moreover, managers have different perceptions and possess asymmetric information about technology, customer characteristics, and supplier capabilities. Natural constraints on the cognitive capacity of market participants limit their ability to discern complex market opportunities, even with the aid of advanced data processing techniques.

Ultimately, the entrepreneur's discovery of new types of combinations is an art. This is why so much effort is devoted to understanding the psychology, motivation, and life experiences of entrepreneurs. It remains a puzzle how certain creative individuals perceive opportunities that so many others failed to understand. The complexity of markets and high costs of transactions suggest that there will always be a need for entrepreneurs to create new combinations. Entrepreneurs are often misunderstood and face criticism and even ridicule because people tend to resist new ideas. Case studies often describe entrepreneurs as persistent, optimistic, and driven people, perhaps because such personality traits are necessary to develop a new type of business.

There are many entrepreneurial success stories; the many who tried but did not succeed are less heralded, of course. Debbi Fields launched her first cookie store in Palo Alto, California and later franchised Mrs. Fields Cookies, which grew to over 700 locations, over 65 of which were international. Debbi Fields observed, "The important thing is not being afraid to take a chance. Remember, the greatest failure is to not try. Once you find something you love to do, be the best at doing it".[32]

Entrepreneurs create all types of innovative transactions in all kinds of industries. Martha Stewart started with her successful book *Entertaining* and her popular magazine *Martha Stewart Living*, but it was an innovative product licensing deal with K-mart that launched her company, Martha Stewart Living Omnimedia. Wayne Huizenga, the founder of Waste Management and later Blockbuster Video, said that he applied a basic principle to all of his businesses: to always rent products to customers since they could be rented over and over, earning a stream of returns for the firm. Michael Dell launched his computer company from his college dorm room by recognizing that by ordering standard parts and farming out assembly and distribution, he could compete against large vertically integrated computer makers.

An entrepreneur brings products to market in an unconventional way, combining buyers and sellers differently. Sometimes that can take the form of new retail stores. Sam Osman had little education, having dropped out of the eighth grade after being held back for two years. He began work as a street peddler with a pushcart on Manhattan's Lower East Side and opened his first store in the 1930s. The store eventually became the Job Lot Trading Company, an emporium with a constantly changing array of thousands of items cast off by other stores, including "items from unsuccessful sales promotions, salvage merchandise from ships and goods bought from bankrupt companies and United States Customs Service auctions". Sam Osman once observed, "Our business is other people's mistakes." The Job Lot Trading Company offered the original combination of a wide variety of greatly discounted cast-off merchandise in a general retail setting frequented by customers from all income levels. The store earned revenues per square foot many times the average for retail stores.[33]

By applying new transaction technologies, entrepreneurs redesign existing markets to reduce transaction costs. A notable example is Pierre Omidyar who founded eBay in 1995 after his fiancé expressed an interest in getting together with other collectors of Pez candy dispensers. Omidyar redesigned the market for collectibles and created a universal auction mechanism for individuals to buy and sell almost any type of product, whether new or used (see Exhibit 11.2). Jeff Citron, along with Josh Levine, created Island, an electronic communication network for executing stock trades that bypassed securities market makers and fundamentally changed securities trading (see Exhibit 11.4).

Exhibit 11.2   eBay

Meg Whitman, CEO of eBay, set up customer focus groups to evaluate the experience of buying and selling through the company. The company proposed charging 10 cents for sellers to delay the start date of their auction but checked with users first. Such a small charge met with considerable resistance from users and was quickly dropped. Whitman, who came to eBay from the FTD flower

(Continued)

Exhibit 11.2 (*Continued*)

network, emphasized the returns to listening to customers: "It's far better for a car dealer to hear about eBay from another car dealer than from our sales force to call on them." She continues: "Buyers attract sellers, who in turn attract buyers, and so on. That's been the success formula in every category on eBay so far".[34]

eBay bills itself as the "world's largest auction". Sharing many features of the traditional swap meet, eBay is far more than an electronic flea market. In its first several years of operation, sellers put over 30 million items up for sale and received well over 100 million bids. With nearly one million items for sale at any one time, the site receives over 170 million hits per week. Each of the 1 million registered users can create a personal Web page for free. eBay combined Internet technology with traditional high-end auctions in a novel way.

What distinguishes eBay from the flea market is not its sheer scale, although that is certainly impressive. It is the fact that eBay is an online community. It brings buyers and sellers together not only to transact, but also to interact with each other and share common interests. Visitors to chat rooms are urged to converse with other collectors about their favorite area but "No business please!" Category-specific chat rooms are devoted to such areas as antiques, Beanie Babies, coins, computers, dolls, Elvis, glass, jewelry, photo equipment, pottery, stamps, toys and trading cards. Meg Whitman's focus has been on the needs of the community of users.[35]

eBay focuses on convenience. The system is designed to allow buyers and sellers to search easily across categories and to participate in auctions with as little friction as possible. Within any category, such as computer hardware, auctions in any subcategory can easily be identified along with featured auctions and hot items, which have received over 30 bids. Online tutorials lead newbies through a simple four-step process: register, find stuff, bid, and sell. Users can customize the service, keeping track of bids made, offers received, and account status. Buyers and sellers can connect to companies providing escrow services (Tradesafe and i-Escrow). Because eBay does not certify the sellers or the quality of their products, it operates on the principle of *caveat emptor*, let the buyer beware, but it offers online tutorials and extensive tips on safe transactions. After an auction is completed, the seller and the high bidder exchange e-mail addresses and deal directly with each other.

(*Continued*)

Exhibit 11.2 (*Continued*)

Because eBay brings buyers and sellers together in new ways, it has created a huge number of markets for resold goods of all types, dramatically cutting the time needed for millions of buyers and sellers to find each other and transact. eBay's market value went public at $18 per share and its share price rose to $147 within two months. By that point, eBay had already attained a market value of over $25 billion.

Another way that eBay created a new combination of buyers and sellers was by acquiring Butterfield and Butterfield Auctioneers, the largest auction house in San Francisco and the third largest in the United States. Butterfield auctions high-cost items such as art, antique furniture, and rare collectibles, and competes with the larger international auction houses Sotheby's and Christie's. eBay combined buyers and sellers in a new way by bringing Internet technology to the high-end auction business for the first time.

eBay brought its 2.4 million buyers and 6.5 million users to Butterfield, whereas the top 10 traditional auction houses including Butterfield previously had a total customer base of 500,000. The companies created a site called Internationalgallery.com so that bidders can view auctions and participate from remote locations. Moreover, they set up online-only auctions with considerably lower fees than standard live auctions. Sellers pay a fee of 5 to 10 percent of the purchase price as compared to traditional auctions where sellers pay 10 to 25 percent and the buyer pays 10 to 15 percent.[36] Online auctions of high-priced objects constitute a new service. The costs of providing such a service are lower because the Web lowers the costs of marketing the auction and exhibiting the objects to be auctioned. Transaction costs are lower since buyers can monitor the auction and submit bids from home on their PCs. However, eBay did not create these product, process, or communications innovations. eBay acted as an entrepreneur by combining all three innovations in a new way.

External Analysis

Transaction innovation begins with an evaluation of transaction technology. Novel forms of communication allow the creation of new types of economic transactions. The commercial impact of advances in communication technologies is a fundamental driver of transaction innovation. Consider the effects of the postal service, telegraph, telephone, radio, television, Internet, and wireless devices. The Internet stimulated the explosion of diverse electronic commerce applications, from retail websites to business-to-business intermediaries. Wireless devices such as wireless phones and Internet-connected laptop and palm-top computers led to the development of mobile commerce. Managers must monitor advances in communications so as to develop new types of transactions based on those advances.

Similarly, fundamental advances in computers and software underlie new forms of transactions. Advances in data processing capabilities led to the development of point-of-sale devices such as bar code scanners, which in turn lead to a host of transaction innovations. Advances in software that allow more advanced data processing change the nature of transactions. For example, Amazon.com applies advanced statistical analysis to information about customer purchases to suggest products in which its customers might be interested. Advances in extensible markup language (XML) standardized requests for proposals that allow companies to streamline their procurement process (see Exhibit 11.3). Managers should consider changes in IT to determine the potential for developing new forms of transactions.

The technology of transactions includes much more than computers and communications networks. Ideas from economics, finance, and management translate into new types of transactions. Financial markets have benefited from academic discoveries about portfolio management, option contracts, and market design. Managers can incorporate these ideas in the design of creative contracts and services.

Entrepreneurial innovations combine customers and suppliers in new ways, through new products and services, or by adding new customers and suppliers. To create entrepreneurial innovations requires managers to continually monitor markets for potential arbitrage opportunities. The tools of external analysis of customers, suppliers, competitors, and partners outlined previously are essential in discovering new combinations.

Managers should ask some of the following questions. Are there innovative products and services that have not been brought to market that would serve customer needs? Are there customers whose needs are not being addressed by competitors? Are there advances in transaction technology that permit the introduction of new goods and services? Carrying out such an analysis requires developing managerial powers of observation. For established companies, discerning transaction opportunities depends on encouraging of creativity within the organization so managers can become "intrapreneurs".

An important source of entrepreneurial returns is to bring existing technology to new applications.[37] In addition, entrepreneurs bring new technologies to market by acquiring the technology from inventors and reselling it to companies that will apply it to produce goods and services. Identifying technology for acquisition requires monitoring innovative start-up companies, university and government laboratories and other technology sources.

Internal Analysis

Pursuing transaction innovation requires continual appraisal of the company's business processes. In addition, the company must evaluate the ability of its organization to monitor market conditions. The manager should determine the company's ability to monitor external technological change and the company's ability to create innovative transactions.

To innovate as an intermediary, the company must continually refine and improve its business processes. Managers need to determine the quality of the company's business processes and the company's ability to develop innovative transaction technologies.

Managers should take stock of the company's transaction technologies. How sophisticated are the company's customer transactions, including sales and customer service? How effective are the company's supplier interfaces? How efficient are the company's back-office systems? How well integrated are the company's demand side management systems and supply chain management systems with the company's enterprise management systems? The manager should evaluate how well these transaction systems perform in comparison with market alternatives.

The manager should determine how well the company's transaction systems perform in comparison with market alternatives. In addition, the manager should consider the organization's ability to carry out improvements in transaction technology or whether it should rely on customers, suppliers, or partners for such enhancements.

To continue to innovate as an entrepreneur, the company must continually refine and improve its market monitoring capability. Managers need to determine the company's ability to identify new technologies and customer applications. It is commonly recognized that organizations can become less entrepreneurial over time. Successful companies are less willing to take risks or to embrace different approaches to the business. As companies grow, management becomes more bureaucratic, responding more to internal rules and procedures rather than taking advantage of market opportunities. To overcome organizational inertia, some companies establish separate units to identify new business opportunities.

Exhibit 11.3 Transaction Innovation and Computer Languages

A key technology expected to facilitate business-to-business e-commerce is the application of extensible markup language (XML). The language is a refinement of the standard generalized markup language (SGML), which is a language used to define languages. The XML language allows documents to be treated as data so computers can exchange data more effectively, thus aiding the automation of data exchange between companies. Standardized XML data-description "tags" are being developed for different market applications in order to capture the types of data that are most important in each market (sizes, prices, material grades, colors, delivery methods, etc.).

If the standards are generally adopted, they will help manufacturers, suppliers, and distributors to exchange commercial information without creating customized formats for each partner.[38] Hypertext markup language (HTML) speeded the development of Internet websites by establishing a standard format for documents that allows users to use a standard browser to view styled text, graphics, and hyperlinks to other Web pages. The XML standard promises similar advantages for exchange of information between businesses, including data on sales, inventory, production, payments, and transportation.

Using XML, a person can receive and analyze data obtained on the Internet without the need to go back to the host server on which the data is stored, thus vastly speeding up communication, making more efficient use of scarce server capacity, and reducing Internet traffic. For example, a user can obtain information from a travel agency about flights on a given day and then compare airfares and itineraries without relying on data processing by the travel agency's server.[39] Unlike HTML, XML also handles multiple human languages and computer languages within the same document. Because of this flexibility, XML is compatible with industry-specific tags inserted into documents that allow the development of standardized commercial documents. Accordingly, the use of XML promises to allow each industry to establish simple standards for the exchange of data and the execution of commercial transactions.

Innovation Strategy

Transaction innovation is an essential component of the company's strategy process. To achieve and maintain a transaction advantage, the company should continually monitor transaction technologies. The company needs to make sure that it has a transaction cost advantage over competitors and that it will not be bypassed by its customers and suppliers. To achieve and maintain a transaction advantage as an entrepreneur, the company must possess a superior capability in identifying and realizing innovative transactions.

Intermediary innovation requires the development of higher-value transactions. Innovations in transaction technology, such as enhanced communications and data processing techniques, allow the design of new types of transactions. The transactions must be such that the company creates greater value than competitors. This means that lower transaction costs are used to provide greater customer value, supplier value, and shareholder value. Improvements in transactions have little purpose in themselves; they must be used to deliver value to the market.

It is not enough simply to apply the new technology without also having an updated business model. The significant shakeout of Internet startups known as the dot-coms demonstrates that businesses are not founded simply on applications of information technology to business. The demise of such companies as Pets.com demonstrated the need for more careful strategic analysis by the new businesses and their venture capital backers.

Office-supplies retailer Staples created a separate business unit, Staples.com, that quickly grew to annual revenue of over \$1 billion. The company reversed course and folded the online unit back into the company to take advantage of a multi-channel transaction strategy. A typical customer who visits only the store spends about \$650, but that amount doubles for a customer who uses the Staples catalog as well as visiting the store. The customer who shops online in addition to using the store and the catalog spends about \$2,500. Staples' online business offers the advantage of 200,000 items at a time when stores are reducing the range of products that they carry.[40] The website offers not only office supplies, but also furniture, electronics, and access to business services. Staples created a sales force to encourage customers to order online and by catalog, while also stressing direct sales to corporate and business customers.[41] The company's multi-channel approach allowed it to coordinate the way it handled different types of transactions with its customers. Staples thus made innovative use of Internet technology by integrating online transactions with its catalog, in-store, and direct sales.

Improvements in transaction technology have an inevitable by-product: the process of competition speeds up as well. For example, the ease of searching the Internet means that competitors are only a mouse-click away. Quick comparison-shopping implies that companies cannot hope to attract and retain customers on the assumption that information about competing alternatives is limited. Old strategic advice urging companies to erect entry barriers no longer applies — the moat around the castle is easily crossed. Companies have to continually move faster than competitors to attract customers. Services must be more readily accessible that those of competitors. Companies must provide immediacy; goods and services must be available when customers need them.

Innovations that reduce transaction costs can also reduce companies' response time, further speeding up competition. Catalogs of products, services, and prices are easily altered online. This slashes "menu costs", which are the costs that restaurants face every time they want to offer new dishes or change their prices. Price changes can become more frequent and product cycles can be shortened. Companies can respond quickly to changing traffic patterns on their website. Traditional retailers and catalog merchants have slower response times because of the costs of changing inventories and posting new prices. Because online merchants can easily change offerings and prices, the speed of competitive interaction changes accordingly.

As competition speeds up, competitive strategies go through more rapid transformations as well. For example, within the space of several years, Microsoft shifted its Internet strategy many times: from the proprietary content service Microsoft Network (MSN), to separate Internet services such as Carpoint car buying service and Expedia travel services, to the multimedia website or portal MSN.com that doubles as the home page for Microsoft's Internet Explorer, to portals for mobile computing.

Transaction costs are not easily reduced; even the so-called new economy has turned out to be far from friction-free. The proliferation of websites quickly recreated the problem of time-consuming search in the virtual world. Search engines were imperfect and frequently generated a blizzard of scattered or irrelevant information. Specialized "infomediaries" began to sort through the confusion and provide tailored information and business connections.

Entrepreneurial innovation is market-directed innovation. Entrepreneurs emphasize the demand side of the market in their external analysis; they bring to market technological innovations that address customer needs. Entrepreneurs are not motivated by the demands of technological change itself, nor are they limited by reliance on invention within the organization. Companies such as Hewlett-Packard, IBM, and Corning switched from being technology-driven to being market-driven. They apply technological advances to new products after evaluating customer needs. Corning chairman Roger G. Ackerman observed, "We used to be like a casino — invent something and then roll the dice ... But now we realize that, if you don't understand market dynamics, you'll crash."[42]

Entrepreneurs frequently use the market as a source of technological change. By using technology markets as a source of innovation, established companies mitigate the difficulties inherent in achieving continual innovation.[43] Many companies invest in start-up ventures as a method of monitoring technological development. For example, Eastman Chemical invested in new ventures to learn about their technology, business models, and industry analysis.[44] Often, companies obtain new technologies by acquiring the companies of entrepreneurs who have blazed the trail. Cisco Systems brought many new technologies to market by buying both start-ups and established companies. Cisco's strategy of "innovation by acquisition" included the purchase of two-year old Cerent for $6.9 billion to obtain equipment that allowed transmission between copper wire telephone networks and fiber optic networks.[45]

Entrepreneurs use the market to assemble new combinations of technologies. Corning had limited success capitalizing on its invention of fiber optic cable. The company decided to offer the market a package of technologies. Corning acquired Oak Industries for $1.8 billion to obtain its pump laser technology, purchased the optical components unit of Pirelli for $4 billion, and spent $400 million to acquire a number of smaller companies for their optical switch technologies. The company formed partnerships with Honeywell's Optical Polymer Group, Optivac, NZ Applied Technologies, NetOptix, and British Telecom's Photonics Technology Research Center. Moreover, the company set up a research center in Rochester, New York, to attract scientists from Eastman Kodak and Xerox and established a plant in Nashua, New Hampshire, near Boston's Route 128 high-tech corridor.[46]

Relying on markets as a source of innovation potentially speeds up technological change, in contrast to a system in which companies base their products on internal R&D. If companies seek the best available technology and compete against each other to obtain it, they are using technology in its highest-value applications. This raises the returns to innovation and stimulates inventive activity, thus increasing the rate of technological change.

Entrepreneurial innovation uses market transactions to assemble the appropriate technologies to meet customer needs — requiring continual adjustment of the technological mix to meet evolving customer requirements.

Exhibit 11.4 Island

Island electronic communication network (ECN) urges investors to "discover a better marketplace". Founded in 1997, Island grew within several years to over $3 trillion in annual volume.

(*Continued*)

Exhibit 11.4 (Continued)

According to *Red Herring*, the company applied basic technology in an innovative manner to securities markets:

"Hidden behind stacks of servers and cables you'll find a $2,500 Dell desktop computer the size of a large briefcase. Although it's the same off-the-shelf technology found in countless homes and offices, this computer has the Wall Street establishment shaking in its collective wingtips. Linked to 292 brokerage firms around the country, this solitary desktop does the job of a thousand traders, matching orders from buyers and sellers of 200 million shares of NASDAQ-listed stocks each day. One out of eight NASDAQ trades is now handled by this machine."[47]

Island offered rapid trade execution and bypassed the specialists that are typical of the traditional open-outcry auction format of the New York Stock Exchange. Moreover, it made information more transparent to individual investors since all unfilled buy and sell orders were displayed online, rather than the traditional practice of only reporting completed trades. Individual investors' trades were matched even through they fell within the bid-ask spread offered by NASDAQ securities dealers. Instinet Group, a Reuters subsidiary that handles trades for institutional investors, acquired Island to form a company with an order flow that was one-fifth that of NASDAQ's.

11.5 *Overview*

Transaction advantage goes beyond production cost advantages and product improvements. Managers must coordinate the company's demand-side distribution, supply-side procurement, and enterprise systems, while increasing convenience for customers and suppliers. A company has a transaction advantage if innovative transactions allow the company to consistently outperform competitors and earn greater economic profits.

Companies that act as intermediaries to match buyers and sellers obtain a transaction advantage by lowering transaction costs, not only in comparison with competitors, but also in comparison with bypass competition from customers and suppliers. For example, wholesalers must lower transaction costs in comparison with direct transactions between manufacturers and retail firms. Achieving and maintaining a competitive advantage as an intermediary requires continual improvement in transaction efficiency.

Companies that act as market makers obtain a transaction advantage by creating and operating markets more efficiently than competitors and more efficiently than decentralized interaction between buyers and sellers. Market makers aggregate and balance demand and supply, adjust and communicate prices, coordinate exchange, and allocate products.

Transaction advantage also recognizes the role of the firm as an entrepreneur, that is, as a creator of new combinations of customers and suppliers. Managers must continually identify new market opportunities and have the flexibility to adjust to shifts in customer and supplier characteristics and technological change. Achieving and maintaining a competitive advantage as an entrepreneur requires continually creating new types of transactions as competitors erode arbitrage profits.

Questions for Discussion

1. Identify an inconvenience that you have encountered in shopping or in purchasing a product or service from a company (for example airline, automobile dealer, restaurant, supermarket, clothing store, telephone, or cable TV company). How could the company that you dealt with have increased your convenience? How did the inconvenience affect your purchasing decision?

2. There is a heated debate over whether computers have increased workplace productivity. Statistical analysis of productivity lends support to both sides. Based on your experience in a recent job situation, evaluate the productivity benefits of using computers. Take into account the costs of the equipment and employee training.

3. Select a specific company that you are familiar with. Identify the company's six main transaction costs and those of its customers and suppliers (fixed cost of exchange, search costs, communication costs, risk costs, information costs, monitoring costs).

4. Consider an industry that you are familiar with and describe a situation in which lower transaction costs might give a company a competitive advantage.

5. Apply the transaction triangle framework to diagram the activities of a commercial bank.

6. Consider Starbucks Coffee. Apply the transaction triangle framework to determine how the company functions as an intermediary by coordinating transactions with its main customers and transactions with its main suppliers.

7. How do venture capital investors act as entrepreneurs?

8. Identify a newly established company described in a business publication. Did the company bring together a new combination of buyers and sellers?

9. Companies that have tried to adopt electronic commerce approaches to procurement have encountered some resistance from purchasing managers. Why might some purchasing managers be less prone to adopt electronic ordering systems than other purchasing managers?

10. How do business methods that lower transaction costs tend to increase the speed at which competing companies react to changes in each other's strategies? Do lower transaction costs intensify competition?

Endnotes

1. Recall from Chapter 7 that the value created by the firm is the sum of customer value, supplier value, and the value of the firm. As we saw, this implies that the value created by the firm equals customer willingness to pay minus the costs of using the firm's own assets and supplier costs.

2. This chapter draws from David Lucking-Reiley and Daniel F. Spulber, "Business-to-Business Electronic Commerce," *Journal of Economic Perspectives*, Winter 2001, and Daniel F. Spulber, "Clockwise," *Business 2.0*, March 2000, pp. 212–216.

3. This discussion draws upon Spulber, "Clockwise."

4. Introduction by John T. Dunlop and Jan W. Rivkin in *Revolution at the Checkout Counter: The Explosion of the Bar Code*, ed. Stephen A. Brown (Cambridge, MA: Harvard University Press, 1997), p. 5.

5. Ibid., p. 9.

6. Jones, Kevin, "E-commerce Liposuction Key to Stealth Strategy," *Forbes ASAP*, June 4, 1999. The quote is attributed to Alf Sherk, the founder of e-Chemicals.

7. Margot Slade, "Sales? The Internet Will Handle That. Let's Talk Solutions," *New York Times*, June 7, 2000, Special Section on E-Commerce, p. E1.

8. EDI is an e-commerce technology older than the World Wide Web that involves point-to-point communications done over proprietary networks, rather than over the Internet. Relative to EDI, Internet commerce offers considerable advantages in terms of cost and convenience. Internet commerce typically makes use of open standards and off-the-shelf technology on a global network, whereas EDI relies on customized hardware and software. However, despite the relative advantages of Internet commerce, the installed base of EDI connections will likely coexist with the Internet for some years into the future (Leif Eriksen, "Online Vertical Markets: Not a One-Size-Fits-All World," *The Report on Manufacturing*, Boston, AMR Research, March 2000, p. 14). According to the Commerce Department, EDI is currently used at more than 250,000 companies in the United States, processing an estimated $3 trillion in transactions in 2000 (Charles Phillips and Mary Meeker, *The B2B Internet Report: Collaborative Commerce*, New York: Morgan Stanley Dean Witter Equity Research, April 2000, p. 25).

9. Phillips and Meeker, *The B2B Internet Report*, p. 31.

10. Scott Alaniz and Robin Roberts, 1999, *E-Procurement: A Guide To Buy-Side Applications* (Little Rock, Arkansas: Stephens Inc., December 27, 1999), p. 13.

11. "Internet Economics: A Thinker's Guide," *The Economist*, April 1, 2000, pp. 64–66.

12. Robert M. Solow, "We'd Better Watch Out," *New York Times Book Review*, July 12, 1987, p. 36.

13. See Robert J. Gordon, "Does the 'New Economy' Measure up to the Great Inventions of the Past?" *Journal of Economic Perspectives* 14 (Fall 2000).

14. Dale W. Jorgenson and Kevin J. Stiroh, "Raising the Speed Limit: U.S. Economic Growth in the Information Age," Working paper, May 1, 2000, Federal Reserve Bank of New York.

15. Stephen D. Oliner and Daniel E. Sichel, "The Resurgence of Growth in the Late 1990s: Is Information Technology the Story," Mimeo, Federal Reserve Board, May 2000. See also Paul A. David, "The Dynamo and the Computer: An Historical Perspective on the Modern Productivity Paradox," *American Economic Review* (Papers and Proceedings) 80, no. 2 (1990), pp. 355–361.

16. Jack Triplett and Barry Bosworth, "Productivity in the Service Sector," Mimeo, The Brookings Institution, January 5, 2000.

17. *U.S. Industrial Outlook 1994* (Washington, D.C.: U.S. Department of Commerce, 1994).

18. http://www.eyeforchem.com/print.asp?news=3044.

19. The account of the securities firm in this paragraph draws on David M. Weiss, *After the Trade Is Made: Processing Securities Transactions*, 2nd ed. (New York: New York Institute of Finance, 1993).

20. This example draws on work by Thomas Gehrig, "Intermediation in Search Markets," *Journal of Economics & Management Strategy* 2 (1993), pp. 97–120; Daniel F. Spulber, "Market Microstructure and Incentives to Invest," *Journal of Political Economy* 110 (April 2002), pp. 352–381; and Daniel F. Spulber, *Market Microstructure: Intermediaries and the Theory of the Firm* (New York: Cambridge University Press, 1999).

21. http://www.solectron.com.synapta.net/about/index.html.

22. The CRSs charge fees to airlines for listing on the systems and for bookings that are completed on the system. Travel agents pay for CRS services as well.

23. Saul Hansell, "Google's Toughest Search Is for a Business Model," *New York Times*, April 8, 2002, p. C1.

24. The material in this chapter is adapted from Chapter 3 of Daniel F. Spulber, *The Market Makers: How Leading Companies Create and Win Markets* (New York: McGraw Hill/ Business Week Books, 1998).

25. Dealers play a prominent role in financial markets. Individual investors buy and sell securities through dealers in the over-the-counter market. Operated by the trade group for major securities dealers, the National Association of Securities Dealers Automated Quotation (NASDAQ) service provides a record of individual stock transactions current prices and

posts median prices. These prices reflect the bid and ask prices and completed transactions of many individual dealers and their customers. The NASDAQ market competes for listings and for investors with auction markets for securities such as the New York Stock Exchange (NYSE) or the American Stock Exchange. Also, *within* the NYSE and other auction markets, there are dealers known as *specialists* who buy and sell particular stocks by matching buy and sell orders and trading on their own account, subject to rules made by the exchange. These types of dealers are referred to as market makers.

26. Armen Alchien observed that specialists provide information at a lower cost than search, in "Information Costs, Pricing, and Resource Unemployment," *Western Economic Journal*, 1969, pp. 109–127. The immediacy in financial markets is discussed by Harold Demsetz, "The Cost of Transacting," *Quarterly Journal of Economics* 82 (1978), pp. 33–53.

27. See Daniel F. Spulber, "Risk Sharing and Inventories," *Journal of Economic Behavior and Organization* 6 (1985), pp. 55–68.

28. This is quoted from George Stalk, Jr., and T. M. Hout, *Competing Against Time* (New York: Free Press, 1990), p. 69. For the account of the Toyota distribution system, see pp. 68–69.

29. See Dennis W. Carlton, "Equilibrium Fluctuation When Price and Delivery Lag Clear the Market," *Bell Journal of Economics* 14 (Autumn 1983), pp. 562–572. Dennis Carlton observes that, with random demand, there are costs to using the price system since a price that is too high will reduce sales and result in excess inventories, whereas a price that is too low will require rationing customers and forgoing sales. He suggests that many firms use non-price rationing methods, and that "firms and organized markets are competitors in production 'allocations'". See Dennis W. Carlton, "The Theory of Allocation and Its Implications for Marketing and Industrial Structure: Why Rationing Is Efficient," *Journal of Law and Economics* 34, no. 2, part 1 (October 1991), p. 257.

30. In 1776, Adam Smith noted that innovations, in particular improvements in machinery, have been made not only by those who use the machines or make the machines, but also by these "men of speculation". Adam Smith, *An Inquiry Into the Nature and Causes of the Wealth of Nations* (Washington, D.C.: Regnery Publishing, 1998), p. 8.

31. Joseph A. Schumpeter, *The Theory of Economic Development* (New York: Oxford University Press, 1961) (first published 1934).

32. http://www.mrsfields.com/history/.

33. Douglas Martin, "Sam Osman, 88, Founder of Job Lot Trading," *New York Times*, February 18, 2000, p. A29.

34. The quotes and information about Meg Whitman are from Saul Hansell, "Meg Whitman and eBay, Net Survivors," *New York Times*, May 5, 2002, Section 3, p. 17.

35. Ibid.

36. Saul Hansell, "Internet Auctioneer eBay to Add Land-Based Rival," *New York Times*, April 27, 1999, p. C2.

37. On the role of companies as knowledge brokers, see Andrew B. Hargadon, "Firms as Knowledge Brokers: Lessons in Pursuing Continuous Innovation," *California Management Review* 40 (Spring 1998), pp. 209–227.

38. Ian Mitchell, "A Universal Proposition," *Chicago Tribune*, August 9, 1999, Business Technology, p. 1.

39. Jon Bosak and Tim Bray, "XML and Next Generation Computing," *Scientific American*, May 1999.

40. Information about Staples.com is from Glenn Rifkin, "New Economy: The Staples Merger of its Website and Catalog Business Offers a Lesson in How to Reevaluate Online Strategies," *New York Times*, June 25, 2001, p. C4.

41. Joseph Pereira, "Staples Inc. Pulls Back on Its Store-Expansion Plans," *The Wall Street Journal*, March 13, 2002, p. B4.

42. Claudia H. Deutsch, "The Horse and the Cart, In Order," *New York Times*, January 7, 2001, Money and Business, p. 1.

43. On the difficulty of continuous innovation, see I. Nonaka and H. Takeuchi, *The Knowledge Creating Company* (New York: Oxford University Press, 1995). On the innovation process see F. Kodama, *Emerging Patterns of Innovation* (Cambridge, MA: Harvard Business School Press, 1991); and G. Basalla, *The Evolution of Technology* (New York: Cambridge University Press, 1988).

44. Rick Mullin, "Innovative Growth," *Chemical Week* 164, no. 11 (March 13, 2002), pp. 17–22.

45. David Bunnell, *Making the Cisco Connection: The Story Behind the Real Internet Superpower* (New York: Wiley, 2000).

46. Deutsch, "The Horse and the Cart, In Order."

47. John Birger, "Isle of Might," *Red Herring*, July 2000, p. 416.

CHAPTER 12

ENTRY STRATEGY

Contents

Competition is a dynamic process, with market structures shaped by the entry and exit of firms. Companies entering the market devise strategies to compete with established companies. Companies already established in the industry in turn devise strategies in anticipation of entry. Established firms and entrants alike match their organizational abilities to market opportunities as they try to achieve competitive advantage. This chapter considers the competitive strategy of entrants and established firms.

The process of entry is costly: companies incur substantial costs of learning about the market, researching products and production processes, establishing facilities, contacting suppliers, and raising finance capital. An *entry barrier* is any competitive advantage that established firms have over potential entrants. An established firm, also known as an *incumbent*, benefits from an entry barrier because the barrier presumably prevents competition from entrants. Yet, as with competitive advantage generally, entry barriers tend to be *temporary*. An incumbent firm should not count on building an invulnerable fortress against entry because no specific entry barrier is sustainable for very long. Companies trying to enter markets can seize opportunities created by technological change and by shifts in customer demand. This chapter examines entry barriers and strategies that entrants can employ to overcome them.

There are three categories of entry barriers, each corresponding to one of the three types of competitive advantages. There may be cost-based entry barriers if an incumbent has a cost advantage over potential entrants, due perhaps to various first-mover advantages or cost efficiencies.[1] There may be product-differentiation-based entry barriers if an incumbent has a differentiation advantage over potential entrants resulting from better products or a strong brand. Finally, there may be transaction-based entry barriers if an incumbent has a transaction advantage over the entrant due to lower transaction costs or established relationships with customers, suppliers, and partners. These entry barriers are likely to be temporary because they depend on costs, products, and transactions, all of which are subject to changing market forces.[2]

Companies can apply creative entry strategies to surmount most entry barriers and enter markets. When managers of a potential entrant perceive that incumbent firms have cost, differentiation, or transaction advantages, they must devise appropriate strategies to overcome these advantages. Technological change allows entrants to lower production costs, introduce new products, or arrange novel transactions. Growing market demand or changes in consumer tastes also generate opportunities for entry.

Chapter 12: Take-Away Points

Managers of incumbent firms should formulate strategies that anticipate innovative entry; managers of entrants should formulate strategies to address potential advantages and strategies of incumbents:

- Barriers to entry are based on competitive advantage and tend to be temporary since they are eroded by innovation.
- Companies should concentrate their strength against the weaknesses of their competitors.
- Entrants who face incumbents that have overwhelming strength can pursue indirect entry strategies by adding value to underserved markets.
- A potential entrant deals with incumbent cost advantages by process innovation, increased efficiency, outsourcing, and contracts with customers.
- A potential entrant deals with incumbent differentiation advantages by product innovation and adapting to customer preferences.
- A potential entrant deals with incumbent transaction advantages by innovation in transaction technology, development of new forms of transactions, and creation of new combinations of buyers and sellers.
- Companies pursuing a go-between entry strategy try to become intermediaries between existing buyers and sellers.
- Companies following a bring-together entry strategy consolidate fragmented supply chains or distribution channels and lower transaction costs.

12.1 *Entry and Market Structure*

Managers make decisions about whether or not the company should enter or exit a market based on their external analysis of the market's customers, supplies, competitors, and partners. Their entry decision also depends on internal analysis of the company's ability to compete effectively for the market.

Dynamic Aspects of Competition

Markets evolve as companies enter and exit from an industry. Companies' strategic choices determine *market structure*, which is the number and size of firms that operate in the market. The market structure at any moment reflects individual companies' past choices about what markets to enter and exit. The strength of competition in the market is not determined by the number of competitors, but rather by their strategies and ability to carry them out.

Established companies, referred to as *incumbents*, not only compete with other firms that are in the industry but also take account of the possibility of entry of new firms. Companies contemplating entry into an industry, referred to as *potential entrants*, must determine whether the market opportunities justify the costs of entry. Companies operating in the industry continually examine whether their ongoing profitability justifies continuing operations or exiting from the industry.

The market structure in any given industry is likely to change over time. The number of competitors changes based on the decisions of established firms about continuing to operate or to exit the market and the decisions of firms to enter the market. The size of companies changes as they compete for market share and as total industry sales rise and fall. Pradeep Chintagunta and Dipak Jain examined strategic interaction in the soft drink, beer, pharmaceutical, and detergent industries. They show that there is a strategic interplay between the marketing efforts of competing companies, including their advertising and sales force levels. Managers take into account how market demand and their company's sales will respond to the marketing efforts of their firm and its competitors. Decisions about brand characteristics and order of entry into a market also will depend on strategic interaction and the anticipated effects on demand and sales.[3]

In some industries, companies battle for market leadership, a process referred to as *Schumpeterian competition*. Everyone is familiar with the children's game called "King of the Hill." Someone takes the top of the hill and everyone tries to push them off. Whoever stays on top is the king of the hill. As the game progresses, the king is pushed aside by other players who must then defend their position. In many industries, a single company may dominate its market with a substantial market share. This need not indicate an absence of competition because other companies are climbing the hill to try to unseat the king through innovation, superior marketing, and better service. Start-ups can appear suddenly and rapidly replace the market leader. Sleeping giants, aroused by creative managers, can supplant today's prominent company.

Competing for the top spot in an industry often accompanies technological change. The incumbent firm can be tied to an out-of-date technology due to organizational inertia that makes it slow to recognize or to adopt a new technology. An entrant that adopts a more effective technology may then supplant the incumbent. The incumbent may suffer decreased sales or exit the industry entirely, retiring obsolete facilities and products. Thus, entry and exit of firms often accomplish the process of creative destruction.

Market leaders sometimes become complacent and forget that they must win the market again every day. High prices attract lower-cost competitors. Low product or service quality stimulates competitors to apply total quality management techniques. Sticking to outmoded production methods or obsolete products motivates rivals to leap-frog the current leader's technology. The most intense competition sometimes occurs with a small number of companies vying for market leadership. Creative destruction can occur as companies update their own facilities and technology, as market shares are passed back and forth between competitors.

Dynamic competition can lead to a *shakeout*, a situation in which many established firms rapidly exit the industry. Shakeouts can occur for many reasons. A drop in demand or a rise in costs may drive out highest-cost firms, leading to survival of the fittest. An innovation race will cause the companies to sink substantial expenditures into R&D. To the extent that the winning firm can protectively patent its inventions, the losers will not share in the economic returns when the race is over. Similarly, companies racing to outspend each other for capital equipment to manufacture similar products will end up in a costly shakeout if market revenues cannot support operating all of the new facilities.

Some shakeouts follow rapid entry of many firms into an industry, often to take advantage of opportunities created by product or process innovations. Why do firms voluntarily enter into such shakeouts? Companies may have imperfect information about their rivals' costs or capabilities. Each therefore believes that it has a chance of success, so its profit expectations are positive. Competition reveals which company has the cost, product, or transaction advantages.

Another type of entry followed by a shakeout occurs due to uncertainty about the size of the market. Companies continue to enter until an additional firm drives expected profits below zero. As entry takes place, those in the industry discover the size of the market. A low realization of demand would cause some firms to exit while a high realization would result in additional entry if existing firms face capacity constraints. A similar effect occurs if companies are using an untried technology. They are then uncertain about *their own costs*. As they begin to operate and the technology turns out to be expensive to run or the learning curve is not steep, the less productive companies exit the business.

Managers evaluate their company's strengths and weaknesses in comparison with the characteristics of their competitors. Recall that the manager's internal analysis yields information about the abilities of the organization and the manager's external analysis yields information about competitors. The combination of these sources of information is necessary to evaluate the company's *relative* strengths and weaknesses. After identifying critical differences with their competitors, companies devise competitive strategies based on them. This information is particularly important for entry strategy.

The entrant should try to concentrate its strengths against the weaknesses of incumbent firms.[4] Lower-cost entrants use their cost advantage against incumbents. Entrants with better products use their differentiation advantage. Entrants with lower transaction costs or innovative combinations of buyers and sellers use their transaction advantage. Entrepreneurial entrants try to apply their speed and flexibility against slower or more bureaucratic incumbents. Focused entrants concentrate their management efforts and employee skills on a market segment whereas large diversified incumbents cannot effectively respond due to their diverse interests. Large established companies entering new markets benefit from their experience, R&D capabilities, and economies of scale in competition against smaller incumbents. International businesses employ their global brands, global sourcing capability, or superior technology to compete against domestic incumbents who may be limited by their concentration on the domestic market. Companies such as Unilever and Nestlé have entered many different national markets with their global brands.

Incumbent Strengths and Indirect Entry Strategies

In some industries, entrants may find that incumbent firms have overwhelming strength by virtue of the incumbents' experience, financial assets, brand recognition, technological expertise, and other attributes. Then, potential entrants should postpone head-to-head competition if possible. An entrant faced with stronger incumbents should pursue *indirect entry strategies* by entering markets that are not served or not served well by incumbent firms. In this way, the entrant avoids costly battles with the incumbent and builds strength for a future direct challenge to established firms.

If the managers of the potential entrant believe that the incumbent firms have competitive advantages that cannot be profitably overcome, they can pursue an indirect entry strategy by avoiding or delaying direct confrontation with incumbents.[5] The entrant builds bridgeheads in market niches that competitors are less likely to observe or understand. The entrant maximizes value added by serving neglected markets, meeting unmet needs, increasing product variety, and providing innovative goods and services.[6]

The firm's objective is not competition itself; the goal is winning markets. Head-to-head competition is an expensive proposition involving lower revenues from price reductions and higher costs for promotion, advertising and enhanced services. Indirect strategy increases profits by reducing the costs of direct confrontation. The entrant can concentrate its abilities, management effort, marketing, and sales on a targeted market segment while neutralizing the competitive advantages of an established incumbent. By serving markets with less competition, entrants build strength in preparation for direct competition with established incumbents.

Entrants pursuing an indirect strategy choose markets that are not served, or not served well, by established competitors. Such market segments may yield low revenues but require correspondingly lower costs, so profits are higher than in hotly contested markets. In some cases, a market that is difficult to serve can be profitable if the costs of service have deterred other firms from providing service.

Direct conflict has a high cost. It is measured in terms of economic returns that companies could have earned by investing their capital and resources in other markets. The key to avoiding direct conflict is to discover opportunities to serve customers not perceived by others. In this way, the firm will earn arbitrage rents without costly price wars.

A competitive battle with an entrenched incumbent could be highly costly in terms of marketing and product design expenditures. The entrant might better invest money in developing new products and services for other markets. A price war can lead to economic losses and industry shakeouts. Managers should only enter into such a direct confrontation if they are sure their company has the necessary cost efficiencies, superior product performance, or transaction efficiency.

By serving undefended markets or market segments, a start-up company builds strength. The entrant develops expertise in purchasing, distribution, marketing, production, product design, and R&D. The company identifies talented managers and employees, forms market networks of suppliers and distributors, and develops relationships in financial markets. The entrant eventually broadens its markets and takes on established rivals.

Economic frictions resulting from transaction costs make it possible to pursue indirect strategies. The economy is full of imperfections and asymmetric information. Changes in consumer tastes and technology continually create new opportunities. It is easy to imagine that no other manager has yet observed a particular market opening. Finding such unmet needs takes creativity and talent, which are not available in unlimited supply in business, as in any other endeavor. Moreover, identifying opportunities is costly, requiring market research and technical know-how. If companies differ in their research costs, those with greater research efficiencies will find different opportunities. There is also a large element of chance, so discerning an unmet need may be fortuitous. For all these reasons, there are always many unexplored opportunities. If the firm can bring buyers and suppliers together in a new way, selling products and services to customers that have not been served before or purchasing inputs for new types of production or distribution, the firm captures economic rent.

Identifying unmet needs is a crucial management skill. The undefended territory can be simply bringing new products and services to a geographic region not served by others. Alternatively, it can be a customer group, distinguished by age, gender, or other aspects, whose needs for a particular service are being ignored. The point of entry may be a low-cost generic service or a high-quality service, but it need not be a niche market. A sufficiently innovative product or service for a mass market can be supplied without initially encountering rivals.

Exhibit 12.1   Wal-Mart

The idea of targeting markets that are not served by major competitors is best illustrated by the history of Wal-Mart. The late Samuel Moore Walton, known to everyone as Sam, began with a single variety store in a small town. From humble beginnings, he built a $100 billion a year company with about 2,000 stores. He drove an old Ford pickup truck and lived simply, but became the richest man in America by the mid-1980s. Every management guru has offered an explanation for the success of Wal-Mart Stores, Inc. There certainly are many reasons for the company's dominance, not least being the creativity and focus of Sam Walton.

However, a key to making Wal-Mart the number one retailer that has not been given the attention it deserves is the expansion of the company in Midwestern and Southern small towns. This was neglected territory, served by small statewide retail chains and local merchants but overlooked by large retailers such as Sears Roebuck and J. C. Penney.

(Continued)

Exhibit 12.1 *(Continued)*

Of course, much strategy is serendipity. The location of Wal-Mart reflected Sam Walton's origins. He grew up in small-town Missouri and graduated from the University of Missouri in 1940.[7] Upon completing school, he declined an offer from Sears and went to work for J. C. Penney in Des Moines, where he began to learn about retailing. Sam Walton later met and married his wife, Helen, in her home town of Claremore, Oklahoma, and served for two years in the Army during the war. Helen told Sam that since they had moved 16 times during his stint in the Army, they would have to settle down in a city with no more than 10,000 people. Heeding this request, he began his retail career in earnest operating a franchise for the Ben Franklin variety store in Newport, Arkansas, a town of 7,000 in the Mississippi River Delta. The second franchise store that he ran was in Bentonville, Arkansas, which was closer to his wife's hometown. As Walton put it,

> "I wanted to get closer to good quail hunting, and with Oklahoma, Kansas, Arkansas, and Missouri all coming together right there it gave me easy access to four quail seasons in four states."

He began to open a series of Ben Franklin franchises in the area. By the early 1960s, Walton had decided that the future of retailing lay in discounting, and in 1962, he established the first Wal-Mart in Rogers, Arkansas, down the road from Bentonville. Rogers was a town of only 6,000 at that time.

That is not to say that no one else had the same idea. Sam Walton observes that 1962 was "the year which turned out to be the big one for discounting". S. S. Kresge started Kmart, F. W. Woolworth established Woolco, and Dayton-Hudson began Target. As Walton marveled,

> "It turned out the first big lesson we learned was that there was much, much more business out there in small-town America than anybody, including me, had ever dreamed of."

During this time, Wal-Mart was "too small and insignificant for any of the big boys to notice."[8]

Competition certainly was present in the small towns from regional chains and local merchants. Initially, Walton's stores thrived based on his creativity, innovativeness, and relentless discounting of prices. At some juncture, however, his competitive

(Continued)

Exhibit 12.1 (*Continued*)

tactics moved from intuitive understanding to grand strategy. After only a few Wal-Marts had been established, Walton's overall conception was clear. He understood where he wanted to go and had few doubts about the potential success of Wal-Mart. The company accelerated its expansion in regional small towns, adding new stores at an increasing rate. By the mid-1980s, over half of the chain's stores were located in towns of 5,000 to 25,000, with a third of the stores in towns or counties not served by competitors.[9]

The other discounters, such as Kmart and Target, focused attention on larger towns. Sears did not embrace discounting until decades later. Wal-Mart escaped scrutiny, even though it reached a half-billion dollars in sales by the mid-1970s and two and a half-billion dollars in sales by 1981. Continually building strength, the company was well positioned by then to enter larger and larger towns and challenge the larger chains directly.

Early on, Wal-Mart began construction of a computerized central warehousing system. It implemented cross-docking, a system that allows a warehouse to simultaneously receive and send out pre-assembled orders for individual stores. With some resistance from its founder, the company nonetheless embraced advanced computer and communications technology as soon as they became available. These technological advances made expansion feasible and conferred competitive advantages on the company, but these operational or logistical strengths should not be confused with strategy.

The strategy was to avoid head-to-head confrontation, to serve areas not served by others and to have the "lowest prices always". The company continued to grow, reaching 330 discount stores by 1980, at a time when Kmart had almost 2,000 stores. Within 10 years, Wal-Mart had far surpassed Kmart in sales. Other discounters were less successful. Sam Walton noted that 76 of the 100 top discounters that were in business in 1976 disappeared by the early 1990s. The disappearing act continued unabated into the 1990s, particularly among regional chains. In the Northeast, for example, Ames with 307 stores, Hills (164 stores), Caldor (166 stores), and Bradlees (136 stores) filed for Chapter 11 bankruptcy, and Jamesway (90 stores) liquidated entirely. Yet Wal-Mart had grown from $1 billion in sales in 1980 to $100 billion in sales 16 years later.

(*Continued*)

Exhibit 12.1 (Continued)

Wal-Mart grew out of the discount category to become America's number one retailer. Were other retailers asleep at the wheel? Perhaps they did not think it was worth paying attention to Wal-Mart. After all, by the time Sears opened the Sears Tower in Chicago in 1973, Wal-Mart had only been a public corporation for several years. Sears had long ago cut off its small town roots when it shifted in the 1920s from a mail-order to a chain store. Moreover, retailers had expansion plans of their own. In the 1980s, Sears was diversifying into real-estate sales and financial services. Nethertheless, at some point large retailers ignored changes in the retail marketplace. They did not remodel stores or embrace discount prices. The large retailers did not pay sufficient attention to the advantages of up-to-date communications and inventory management technology. By the early 1990s, as Sears fell behind Wal-Mart and Kmart, it reversed its financial supermarket strategy, divesting most of Coldwell Banker real estate and Dean Witter Financial Services Group, and part of Allstate insurance, to focus on the retail business, its main competitor being more clearly in focus.

Kmart, the largest retailer at the start of the 1990s, fell to a third the size of Wal-Mart by the middle of the decade amid rumors of Chapter 11 bankruptcy.[10] According to Kmart's then-CEO, Floyd Hall, about half of Wal-Mart's customers drive past a Kmart to get to Wal-Mart. His predecessor, Joseph Antonini, had diversified dollars and management attention away from minding the store toward other ventures: OfficeMax, the Sports Authority and Builders Square. Like Sears, such diversification made Kmart take its eye off Wal-Mart.[11]

By the year 2000, Wal-Mart was the largest retailer in the world with sales of $165 billion. Its international division had sales approaching $30 billion. Wal-Mart operated in the United States and many other countries, with a total of 2,373 discount stores, 1,104 supercenters and 512 Sam's Clubs. Would Wal-Mart's philosophy of continuous improvement be sufficient to sustain its industry leadership?

Perfect Competition

Perfect competition is the economist's ideal case, although it is an abstraction that is seldom observed. A perfectly competitive market has a large number of firms, each with little or no effect on price.

Market prices are determined by forces beyond the control of individual buyers and sellers, the so-called invisible hand. The market-clearing price causes the total amount consumers demand to match up exactly with the total amount firms supply. Individual companies take this price as a given because they would have no customers if they were to price even slightly higher than the market price. Furthermore, companies can sell all that they want at the going price; their sales decisions do not affect the price, so they have no reason to price lower than the going price. Therefore, in the standard textbook model, companies make their purchasing and production plans based on the market price so they are said to be *price takers*.

The notion that an auctioneer adjusts the market price upward to reflect scarcity and downward to eliminate surpluses might describe how organized exchanges for securities or commodity futures might function. However, in those markets, the auction mechanism is tangible and is provided and operated by the exchange. In the market for securities, the market institutions include auction markets such as the New York Stock Exchange or the American Stock Exchange, an over-the-counter market operated by securities dealers with a national, computerized quotation system (NASDAQ), and a network of brokerage firms. The firms that provide market-making services do so for a profit. Specialists on the exchange provide market-making services. Dealers in the over-the-counter market earn a spread on sales by arbitraging between buyers and sellers. Stock brokerage firms earn commissions from processing transactions.

In most markets, the institutions of exchange are created by companies. Although markets for goods and services generally do not have organized exchanges, the underlying mechanism for price adjustment often reflects the activities of brokers, dealers, wholesalers, and other companies that adjust prices in response to the pattern of their own sales and purchases. Companies provide market-making services as an important part of their regular activities. To view a large corporation such as Procter & Gamble simply as a producer of soaps and other products misses the complex forms of market-making it carries out through marketing and sales activities. A company such as Procter & Gamble has a vast distribution network, operates a complicated system of product pricing and advertising, and engages in substantial collection of information about customer demand.

Few markets resemble the perfect competition ideal. Most markets contain a few large companies rather than many infinitesimal firms. Prices are set by competing companies. Far from being price takers, most companies carry out complicated pricing strategies. Products are differentiated by unique features and brand names.

Exhibit 12.2   Agricultural Markets

Does any market fit the economists' definition of perfect competition? Consider the case of agriculture. Markets for agricultural products include millions of farms, each very small relative to the size of the market. Grains such as wheat or corn not only have a national market, they also have a global market, so every producer is quite small relative to the size of the total market. The best that farmers can do is to monitor price fluctuations and make the best decision possible as to how much to plant and when to sell. Also, farmers take input prices as given, purchasing farm machinery, fuel, fertilizer, and seeds at market prices. They work from dawn until the cows come home to put food on the nation's table.

Yet, even this idyllic picture of agriculture is misleading. Governments are major players in agricultural markets, doling out subsidies, supporting crop prices, and erecting tariffs and other barriers to international trade. Moreover, many large businesses make up the farm sector, helping to adjust prices and clear markets.

Although there may be many small farms, there are many large ones as well. The average farm in the United States has about 470 acres, but almost half of the farmland is in farms having 2,000 acres or more. In terms of cropland harvested, over 43 percent is in farms of 1,000 acres or more (encompassing only about 170,000 farms). While this is still a large number of farms, not all of them operate in the same market. There are, of course, many specialized agricultural markets including dairy products, meat products, grains, feed, cotton, wool, tobacco, fruits, and vegetables. Since transportation is costly for some products, there are also geographic market divisions.

The consolidation of farming is apparent in that the total number of farms in the United States fell to about 2 million by the mid-1990s.[12] Megafarms owned by agri-businesses are industrializing their production techniques and further increasing their scale. Dean Kleckner, executive director of the Farm Bureau, which represents 4.5 million farmers, noted the fundamental shift to a marketing strategy: "Our goal is no longer to sell what we can produce but to produce what we can sell."[13]

Even so, the prices of farm products do not appear by magic. The market for farm products cannot be described as having many small farmers on the supply side and many households buying food on the demand side. Prices are set through the intermediation of major agri-business companies as well as organized exchanges.

(Continued)

Exhibit 12.2 (*Continued*)

Many prices are discovered on organized futures exchanges such as the Chicago Board of Trade or the Chicago Mercantile Exchange. Trading is highly centralized. Jeffrey Williams observes that in world markets "the same commodity is rarely actively traded on more than one exchange". Thus, single exchanges serve as market makers for specific commodities.[14]

The major agri-business companies have substantial market power. They effectively create and operate market bridges between farmers and consumers of farm products. Companies like Cargill ($48 billion), ConAgra ($24 billion), and Archer Daniels Midland ($14 billion) are not small price takers.[15] Instead they are market makers, creating agricultural markets and setting prices for many products. These businesses understand their economic functions very well. With 85,000 employees in 60 countries, Cargill termed itself as "an international marketer, processor and distributor of agricultural, food, financial and industrial products and services". ConAgra has called itself a "mutual fund of the food chain". Archer Daniels Midland advertised itself as "supermarket to the world."[16]

Companies like these consolidate the demands of many buyers and the supplies of many sellers. Their sales and purchasing activities help determine prices in the food business. For example, IBP with $14 billion in sales is known as the world's largest butcher, slaughtering about one-third of U.S. cattle and one-fifth of U.S. pork.[17] With over 100,000 employees, Kraft is the largest U.S. food company. The Swiss company Nestlé (about $50 billion in sales) is the world's largest food company, selling a host of brands including *Perrier*, *Taster's Choice Coffee*, and *Baby Ruth* candy bars. Large dairy companies and cooperatives provide over 70 percent of dairy products in the United States.[18]

There are major market makers in farm inputs as well. Seed companies such as the market leader Pioneer Hi-Bred International make long-term contracts with farmers. Companies like John Deere are major players in farm equipment. Franchise contracts for farmers, in which the parent company supplies seed, chemicals, marketing, and food-processing capabilities, have increased.[19] Farmers in turn subcontract work such as pesticide spraying, planting, and harvesting. As franchise operations, farmers are part of production networks set up by the prominent agri-business companies.

(*Continued*)

> **Exhibit 12.2   (*Continued*)**
>
> The example of agricultural markets illustrates how different actual markets are from the perfectly competitive ideal. While there are indeed many farms, the notion that prices are determined in small marketplaces through the uncoordinated interaction of many buyers and sellers is perhaps appropriate only as a historical description of rural life in the 19th century. In modern agricultural markets, major companies create and operate markets and manage significant transaction volumes. Prices are adjusted by large firms that continually monitor the spread between the prices paid to farmers and the prices of food and other agricultural-based products.

12.2 *Cost Advantage and Entry Strategy*

Potential entrants face a barrier to entry into an industry if they must incur costs that incumbent firms can somehow avoid.[20] Several types of cost differences can confer strategic advantages on incumbents. However, innovation tends to erode incumbent cost advantages. This means that creative entrants can devise competitive strategies to overcome cost-based entry barriers.

To challenge entrenched incumbents, managers of potential entrants must anticipate the actions of established competitors. Managers should try to determine what the future strategy of the incumbents will be. The incumbent will not stand still in the face of entry. Incumbent company managers will anticipate entry and attempt preemptive strategies, such as price cuts or new product introductions. Managers of entrants must make their entry plans in anticipation of these strategic moves. Moreover, managers of entrants must plan ahead for challenges that they in turn will face from future entrants.

The incumbent's performance provides useful information about the costs and earnings of a company operating in the industry. However, managers of companies contemplating market entry should not base their decision to enter on the incumbent's profitability. Instead, the entering firm's managers should consider the expected profits that the entrant will make in competition with the incumbent.

Sunk Cost

One of the key strategic aspects of entry is the need to make irreversible investments in setting up new operations, including production facilities, marketing, market research, and R&D. Companies that expand their operations also make irreversible investments. These irreversible investments are known as *sunk costs*. Sunk costs can be perceived as a barrier to entry if entrants need to make irreversible investments in capacity while incumbents have already incurred these costs.[21]

Sunk costs appear to confer a strategic advantage on incumbents because the decision of an incumbent differs from that of an entrant. Having already incurred the costs of entry, an incumbent writes off those costs; that is, the incumbent does not base future decisions on the costs of entry because the costs are sunk and thus unaffected by later decisions. The incumbent is only concerned about the benefits and costs resulting from future decisions. The incumbent will continue to operate as long as revenues are sufficient to recover operating expenses.[22]

An entrant, in contrast, must decide whether or not to incur the costs of entry. The entrant must anticipate that its earnings after entry will exceed not only operating costs but also the irreversible costs of establishing its facilities. The entrant will only enter if expected revenues are sufficient to recover both sunk costs and operating expenses.

An entrant might be deterred by sunk costs under some circumstances. An entrant might anticipate that competing with the incumbent will not generate sufficient revenues to cover entry costs, particularly if the entrant is concerned that a price war will break out after entry. Shakeouts have been identified in a wide range of industries. For example, shakeouts occurred with the entry and exit of the Internet dot-com start-ups, which discovered that even electronic commerce entails the need to sink costs. Such price wars are part of the everyday struggle of companies seeking to win markets.

For example, suppose that the sunk costs of entry are $100 and both the entrant and incumbent have operating expenses of $2. The potential entrant will not commit to the industry unless the entrant anticipates revenue of at least $102. No matter what prices are before entry, the entrant is concerned that prices will fall below the level necessary to generate revenue of $102 after having entered the market. The incumbent need only earn revenue of little more than $2 to stay in business. In fact, after entering, the entrant also will continue to operate with revenue above $2. If a price war occurs, prices will fall all the way to $2. If prices fall below $102, the entrant would fail to recover the sunk cost of entry.

Example 12.1   Sunk Costs

Entering into an industry entails the need to make an irreversible investment of $500. After entry, firms operate for one period. All prices and costs are in current-value terms. How would the sunk cost of $500 act as a barrier to entry?

(Continued)

> ## Example 12.1 (*Continued*)
>
> After entry, companies in the industry have $60 in per-unit operating costs. If the market price were $75, and the incumbent could produce and sell one unit, would the incumbent continue to operate? An established company, having already incurred the entry cost of $500, has written off that cost since it is unaffected by any strategic decisions. The incumbent needs a market price of at least $60 to stay in business. Accordingly, if the market price were $75, the incumbent firm would cover its per-unit operating costs and continue to stay in business.
>
> If the entrant expected the market price to be $75, would the entrant choose to enter? Before entering the market, a company must anticipate earning $560 in revenue to cover its total costs. If the expected market price were $75, the potential entrant would not choose to enter the market.

In practice, however, the need to sink costs is not necessarily an effective barrier to entry. All competitive markets involve some degree of irreversible investment — whether in buying specialized capital equipment, establishing a brand, or carrying out R&D. Entrants normally commit capital resources in markets where they expect to earn at least a competitive return on their investments. Entrants incur sunk costs after incumbents simply as a matter of timing. It is evident that since sinking costs to enter an industry is a routine part of doing business, and since companies are continually entering industries, sunk costs need not be a barrier to entry.

Potential entrants should expect price wars only if entry will result in excess capacity in the industry. If the irreversible investment is embodied in productive capacity, that capacity often stays in service even if the firm that originally constructed the capacity exits the industry. Incumbents and entrants may battle for market share if each believes it can serve customers at the least cost. Duplication of investment and the entry of excess capacity can take place when there is uncertainty regarding costs, technology, or market demand. Companies may find that they have over-invested after they observe increases in industry costs or reductions in market demand. When entry creates excess capacity, vigorous competition and industry shakeouts that reduce the profits of incumbents often result.

However, entrants should not necessarily be deterred by an incumbent's threat of a price war. Such threats are not always credible. After the entrant has invested in facilities, it is on the same footing as other incumbent firms, and competitors will only cut prices to marginal cost if it is in their interest to do so. If the incumbent and entrant have limited production capacity, it is unlikely that they will pursue a price war for market share that would call for output in excess of that capacity.

Many factors lessen the likelihood of a price war after entry. If the incumbent and entrant offer differentiated products, price competition tends to be reduced. Both the incumbent and entrant will have the opportunity to earn profits in post-entry competition. Because a lower price than a competitor causes only some customers to switch their purchases, the incumbent and the entrant will not have an incentive to engage in an all-out price war. Since the incumbent and the entrant earn positive profits in competition after entry, it is more likely that the entrant can earn a sufficient margin above operating expenses to recover the sunk costs of entry. Other factors that lessen price wars are customer-switching costs, customer brand loyalty, different convenience features, and imperfect information. If these factors are present, the entrant can expect a reduction in the severity of post-entry competition, allowing for the recovery of sunk costs. Therefore, with product differentiation and other factors, sunk costs are less likely to be a barrier to entry.

If the entrant has a differentiation advantage over the incumbent, it has a better chance of recovering its sunk costs. An entrant could offer products that deliver sufficiently greater value to the customer than the products of established companies. In return, the entrant will earn margins that allow for the recovery of sunk costs incurred in entering the market.

An entrant with a transaction advantage over the incumbent need not be deterred by the need to sink costs. Through innovative intermediation between buyers and sellers, the entrant can earn operating profits after entry. By reducing transaction costs, the entrant will earn returns that allow for the recovery of sunk costs. Accordingly, entrants can make investments in information technology, communications systems, customer support, supplier connections, and back-office processes, which are recovered through transaction advantages over incumbents.

Generally, *with technological change*, the need to sink cost is not an insurmountable barrier to the entry of new competitors. If an entrant employs new technologies to reduce its operating costs, it can enjoy a cost advantage over an incumbent operating outdated technology. Even if the incumbent and entrant compete on price, an entrant with an operating cost advantage over the incumbent will earn positive margins that allow for the recovery of sunk costs.

Moreover, sunk costs need not be an entry barrier because *the entrant's sunk cost is a matter of strategic choice*. The entrant makes various decisions about how much to spend on facilities, marketing, R&D, and so on. In particular, the entrant can offer different products from those of the incumbent, thus changing the entrant's production costs. The entrant can serve different groups of customers than the incumbent, thus changing the entrant's need for distribution facilities and marketing expenditures. The entrant can adopt different production or distribution technology than the incumbent, often drastically changing the mix of investment and operating costs.

Consider telecommunications for example, where the incumbent local operating companies such as Verizon or SBC Communications have made substantial sunk investments in building legacy wireline networks. An entrant into local telecommunications need not replicate these wireline networks but can offer a different product such as mobile communications, which uses a different technology of wireless transmission. Wireless transmission technology involves substantially lower sunk costs than wireline systems in local exchange telecommunications but provides services that are increasingly able to compete with traditional systems. See Exhibit 12.3 for additional discussion.

Even with similar products and technology, entrants can reduce the risk associated with making investment commitments in a variety of ways. The entrant can lessen the risk of post-entry competition by forming *contracts with customers* before it makes irreversible investments. The entrant can compete with the incumbent for customers before deciding to enter the market and then only incur entry costs if the customer contracts will generate sufficient revenues. The company can find out if its product will be successful before making substantial investments in facilities. For example, aircraft manufacturers such as Boeing and Airbus sign up prospective customers on a contingent basis before starting a production run on a new plane.

The success of the contracting strategy also depends on the level of transaction costs. If the transaction costs of contacting with customers are relatively low in comparison with sunk costs of entry, then testing the waters through contracts is worthwhile. The entrant can use contracts to establish prices and customer orders before entering the market, thus reducing the risk of irreversible investments and avoiding price wars after entry. Thus, efficiencies in contracting can mitigate the impact of entry costs. Accordingly, entrants can use contracts as an entry strategy when sunk costs are substantial.

Exhibit 12.3   Entry into Local Exchange Telecommunications

Telecommunications provides a good example of how the sunk cost argument is applied in antitrust policy. Closer consideration of the industry shows the limits of sunk cost as a deterrent to entry. AT&T controlled both long-distance and local telecommunications until it was broken up in 1984 under a consent decree between the company and the Justice Department. The company was broken up into regional bell operating companies (RBOCs) and the AT&T

(Continued)

Exhibit 12.3 *(Continued)*

long-distance unit. The long-distance unit retained manufacturing (Western Electric) and research (Bell Labs), which were eventually spun off as a company called Lucent.

As part of the breakup, the RBOCs were barred from entering long-distance telecommunications service. Judge Greene, who supervised the Consent Decree with AT&T, stated: "Although monopoly power may be inferred from a firm's predominant market share, size alone is not synonymous with market power, particularly where entry barriers are not substantial."[23] However, in justifying restrictions placed on the RBOCs, Judge Greene stated that other companies are deterred from entry into the local exchange because they would have to incur "an enormous and prohibitive capital investment" to duplicate the "ubiquitous local exchange networks."[24] Motivated by similar concerns, the 1996 Telecommunications Act was intended to speed the entry of competitors into the local exchange by requiring the RBOCs and other local exchange companies to open their networks by reselling wholesale services and leasing facilities known as unbundled network elements.

These regulations were intended to make the incumbent companies' facilities (for which costs were already sunk) available to entrants, who would then not have to sink costs as heavily into facilities. Was regulated access necessary?

Entrants into telecommunications did not have to duplicate the RBOC's entire transmission and switching systems to enter the market profitably. The entrant only needed to enter portions of the market where the expected revenues exceed the costs of providing new service. Thus, an entrant only had to sink the costs required to serve its specific customers. The size of the incumbent's sunk costs was irrelevant to the entrant, who only cared about the entrant's own investment. By serving only part of the local exchange market, the entrant's irreversible investment was substantially reduced.

An entrant did not need to duplicate the technology of the incumbent RBOC's transmission and switching systems. The technology of telecommunications transmission changed substantially in a manner that alters the types of investment required to establish a network. The transmission wires portion of the traditional local exchange company (LEC) transmission technology represented the primary irreversible investment the LECs incurred. In contrast, many new technologies (such as cellular, mobile radio technology, and satellite transmission) substantially reduced the

(Continued)

Exhibit 12.3 *(Continued)*

wired portion of the transmission system. These new wireless technologies, by avoiding the wire-based transmission technology, substantially lowered transaction-specific irreversible investment. By serving areas using wireless transmission, particularly for the "last mile" to the customer's location, the new technologies were not tied to specific customers through investment in transmission.

In fact, the transmission technology was not customer-specific since a transmission tower and receiver can serve *any* customer in a given geographic area. By avoiding investment in wires to the home, wireless technology avoided customer-specific sunk costs. Moreover, radio transmission facilities themselves do not necessarily represent sunk costs, since they could be physically moved to serve other markets. These features of some of the new transmission technologies eliminated the need for significant transaction-specific or customer-specific investments.

Technological change altered the economic design of telecommunications systems. Telecommunications systems provide transmission services using switches and transmission. Technological progress in computers had major implications for the relative application of switches and transmission in the production of telecommunications services. The increased power and reduced cost of computer chips correspondingly increased the power and reduced the cost of telecommunications switches. To the extent that switches could be substituted for transmission lines, a fall in the price of switches relative to lines implied that the cost-minimizing input mix would involve a greater reliance on switches as opposed to transmission lines. Increased productivity of switches allowed a reduction in the number of switches and lines required to produce a given level of transmission capacity. For example, the private branch exchange (PBX) switching technology allowed a reduction in the number of lines required to provide a given level of capacity to a customer's premises.

A reduction in the number of lines required to provide a given level of capacity implied that this potential source of sunk costs was reduced for a new entrant into the local telecommunications market. Moreover, the switches operated by a telecommunications company did not have to be irreversible transaction-specific investments. The switches were not tied to a particular customer location, as were traditional phone lines. Moreover, the switches could be shifted to other applications and thus did not represent sunk costs at all, as transmission lines to customer premises do.

(Continued)

Exhibit 12.3 *(Continued)*

And, of course, the switches owned and operated by *customers* certainly did not represent sunk costs for a telecommunications supplier. Thus, as technological change significantly reduced sunk costs, the argument that sunk costs created a barrier to entry into the local exchange ceased to apply.

Entrants could achieve lower operating costs due to progress in telecommunications technology (including computerized switching, fiber optics, and wireless transmission). Further, the performance characteristics of existing and new technologies differed substantially. New access technologies offered various benefits that were not available using the copper-wire access technology in the local loop. These benefits included the mobility of radio services, the increased bandwidth of fiber optic services, and the television transmission capabilities of coaxial cable. If the incremental revenues that could be obtained from the provision of value-added services (such as mobility or data transmission) were sufficient to cover the costs of establishing and operating a new system, then it is irrelevant to the prospective entrant that the RBOC enjoyed a supposed advantage from its existing copper wire and other investments in the local loop.

Of course, the RBOCs could invest in the new technologies as readily as any entrant could, in the absence of any regulatory hurdles. However, the incumbent and the entrants would then be on an equal footing. The need to sink costs in a new technology would fall evenly on the incumbent and entrant and thus could not constitute a barrier to entry. Competitors found many ways of entering into local telecommunications, including long-distance companies, dozens of competitive access providers, cable companies, cellular companies, and other wireless transmission suppliers.[25] After making investment commitments, entrants became incumbents, and competition in the local telecommunications markets was launched.

Scale Economies and Absolute Cost Advantage

Economies of scale are sometimes alleged to confer a cost advantage on incumbents that creates a barrier to entry. When production technology has economies of scale, an established firm with a higher output than an entrant will have lower unit costs than the entrant. Were this situation to persist, the incumbent firm would be a cost leader and the entrant might be priced out of the market. However, there are various entry strategies to overcome the effects of scale.

When economies of scale are critical, challengers need to invest in the required productive capacity and to undertake the necessary marketing to build a sufficient level of sales. For example, in producing automobiles, a challenge to General Motors, Ford, or Toyota requires a start-up scale in the millions of cars per year. Yet, the opportunity to realize economies of scale is available to entrants just as it is to incumbents. New companies are just as capable of discovering technological efficiencies as are established companies. Entrants are equally able to apply management skills to achieving production economies and to build large-scale facilities as are established companies. Other sources of economies of scale such as automation and the use of information technology can be applied by entrant and incumbent alike. Entrants and incumbents can also both derive advantages for employee training and assigning employees to perform specialized tasks in which they develop expertise. Little that is inherent in increasing-returns technology acts as a barrier to entry.

If the entrant and the incumbent operate the same technology with equal efficiency, greater output results in greater scale economies. However, scale economies do not confer any inherent advantage in attracting customers. By pricing competitively, the entrant has the opportunity to attract customers away from the incumbent, thus reducing differences in sales and allowing the firms to have similar unit costs of production. Even if the scale economies were on the marketing and distribution side rather than in production, entrants could benefit from those same economies to mount a substantial sales effort.

The entrant may have different technology than the incumbent. Even if the incumbent benefits from significant scale economies, an entrant that operates with greater efficiency could price to build market share and successfully enter the market. For example, if the incumbent produces 100 units at a unit cost of $5, an entrant that operated more efficiently might produce 100 units at a unit cost of $3. Such an entrant could profitably price below the incumbent's cost and attract the 100 units of sales away from the incumbent. Moreover, a more efficient entrant might reach a lower unit cost at a smaller output, thus achieving lower cost than the incumbent with lower sales.

As a small-scale new entrant into the steel industry, Nucor's enterprising adoption of a new steel production technology, known as thin-slab casting, surprised incumbents such as larger-scale, traditional steel producers USX and Bethlehem. SMS, a German supplier of steel-making equipment, developed the technology and widely publicized it within the steel industry. Other steel producers were slow to adopt the new technique despite its cost savings. Nucor established mini-mills that achieved efficiencies from small-scale production of steel from scrap.[26]

The impact of economies of scale is also offset by product differentiation. If the entrant offers a superior product, it obtains a differentiation advantage that allows it to charge a price premium over the incumbent. A sufficient price premium would allow the entrant to overcome any advantage the incumbent derives from higher sales and lower unit cost.

The impact of scale economies can also be offset if entrants gain a transaction advantage. The Barnes & Noble superstores seemed to offer significant scale that would be difficult for entrants to challenge. Online booksellers, including Amazon.com and a host of smaller bookstores, were able to enter the retail book business by offering convenient transactions through the Internet.

If economies of scale are significant relative to the size of the market, there may only be room for one or two firms in the market. This means that it is more efficient for the market to be served by one or two large firms with economies of scale than by a fragmented industry of small firms with high unit costs. This does not imply, however, that significant economies of scale function as an entry barrier because entrants can displace incumbent firms. If an entrant has greater cost efficiency, superior products, or lower transaction costs, it can successfully enter at the expense of incumbents.

Example 12.2   Natural Monopoly

Consider the case in which the efficient scale of the incumbent firm is sufficient to serve the entire market demand when price is equal to unit cost, a situation known as *natural monopoly*. The market demand at a price of $5 equals 100 units and the market demand at a price of $4 equals 120 units. Suppose that with economies of scale, the incumbent can produce 100 units at a unit cost of $5, and 120 units at a unit cost of $4.50. Then, the incumbent may be said to have a natural monopoly. The incumbent could sell 100 units at a price of $5 and exactly break even. However, suppose that a more efficient entrant comes along who can produce 120 units at a unit cost of $3. The entrant could then displace the incumbent by selling 120 units at a price of $4 and still make a profit. Therefore, natural monopoly does not prevent the entry of more efficient competitors.

When an incumbent has an *absolute cost advantage* over entrants, that is lower unit costs, entry is deterred unless entrants can overcome that advantage with product differentiation or innovative transactions. Yet, cost leadership by incumbents is not a permanent entry barrier. As with all competitive advantage, such deterrents are temporary. Technological change in manufacturing technology, product design, and transaction methods allows entrants to address cost advantages and challenge incumbents. If the incumbent's cost advantage is difficult to overcome rapidly, potential entrants can find ways to avoid head-to-head confrontations through indirect strategy, as will be shown later in the chapter.

Michael E. Porter introduces the concept of *strategic groups* as groups of firms within an industry that follow similar strategies.[27] For example, some firms in a given industry may pursue low-cost strategies while other firms pursue product differentiation strategies, thus creating two strategic groups. An industry might be divided into a group of large firms and a competitive fringe of smaller firms. Some firms may be specialized with a narrow product scope while others may offer a full line of products and services. Some firms may be vertically integrated while others may outsource much of their manufacturing and other operations. Companies within the same industry may differ along many dimensions at once.

A key premise of management strategy is that within the same industry the actions of firms can differ considerably. The complexity of competition implies that there are many alternative strategies to choose from. Company differences also reflect the different abilities and resources within their organizations. Such differences are due to diversity in historical experiences as well as the divergent perceptions of managers. With global competition, company differences often reflect each company's country of origin, which affects business practices, management strategy, and corporate structure.

Porter further defines *mobility barriers* as barriers within industries that prevent companies from entering strategic groups. Within an industry, some firms may have competitive advantages over other firms, leading to strategic groups. These advantages may result from cost advantages, differentiation advantages, or transaction advantages. Yet, just as all competitive advantages are temporary, so are mobility barriers. Innovation and creative strategies help entrants overcome entry barriers so companies within an industry can surmount mobility barriers.

With growth in customer demand, the advantages of scale can be increased or reduced. If incumbents have considerable excess capacity, an expansion in demand allows average costs to be lowered even further. However, if incumbents are operating near full capacity, expanding production can raise average costs, creating opportunities for entrants. In such a situation, entrants do more than take up the slack; they can take a substantial market share by price competition with incumbents. By carving out a market niche, entrants can avoid some of the costs of full-service incumbents, allowing entrants to overcome the incumbents' apparent economies of scale and absolute cost advantages. For example, Southwest Airlines entered the market with a different route structure that allowed it to compete with the major airlines (see Exhibit 12.4).

Exhibit 12.4   Southwest Airlines

Southwest Airlines is the Wal-Mart of the air. Its slogan, "*The* low price airline", indicates its leadership of the discount carriers. Chairman and CEO Herb Kelleher started the airline from scratch as a regional carrier, and took it to the top of the airline market with substantial earnings while the major airlines, including American, United, and Delta, were posting losses in the early 1990s. Investment analysts compare the potential of the airline's stock to that of Wal-Mart.[28]

Southwest's flights began in the early 1970s from Dallas, competing with local carriers but not directly challenging the majors. Southwest did not appear on competitors' radar screens for many reasons. It flew directly between city pairs, avoiding the hub-and-spoke system that the majors had embraced when the industry was deregulated. Unlike the majors, Southwest followed a regional strategy, flying short-haul routes and sticking to secondary airports, including Love Field in Dallas, Midway in Chicago, and the Detroit City Airport.[29] Indeed, its main competitors were passenger cars and other ground transportation.

While the majors fly a diverse assortment of planes to fill out different portions of their route structure, Southwest stuck to 737s. Southwest offered no-frills, low-price service as compared to the majors, who sell a wide range of fares from super-saver to first class, carrying over an emphasis on service from the days of regulation when airlines coordinated fares rather then competing on price. Finally, Southwest does not sell through airline computer reservation systems, thus avoiding listing fees at the expense of reduced travel agent sales.

Low costs eventually allowed Southwest to challenge the established major airlines. However, in the early years, Southwest created a stealth airline, visible to passengers but less well understood by the major airlines who saw little need to resist the niche carrier. By the time Southwest was perceived as a worthy competitor, the airline had already built its strength, in terms of brand recognition, management expertise, a skilled workforce, and investor confidence.

The initial response of the majors was to imitate Southwest. United introduced a lower-cost clone called Shuttle by United. Yet, this approach still missed the point. The Southwest strategy was to build a transcontinental airline that would challenge the majors. Fighting back with regional clones only postponed the need for the majors to reconfigure their operations.

12.3 *Differentiation Advantage and Entry Strategy*

Managers of entering firms may perceive that an incumbent firm has a differentiation advantage that acts as an entry barrier. A differentiation advantage allows incumbent firms to command a price premium while deterring entry because customers recognize the incumbent's brand and product quality. Established firms have well-known brand names whereas start-up entrants are necessarily unknown. Yet many entry strategies are available to address a perceived differentiation advantage. Entrants have successfully launched new brands that compete with those of incumbents and often displace them.

Product Differentiation and Brand Advantage

Incumbent firms maintain highly recognized brand names by consistent delivery of high-quality products. They build brand equity through marketing and customer service. A company contemplating entry into the beverage industry must contend with the *Coke* and *Pepsi* brands. A company trying to enter into insurance must deal with such brands as *Allstate*, *State Farm*, and *Prudential*. Nestlé and Unilever offer internationally recognized food brands.

A potential entrant may face a situation in which technological advances and imaginative designs contribute to the attractiveness of the incumbent firm's products. Entrants into the microprocessor industry must take account of the high quality of Intel's microprocessors. Entrants into electronics must deal with Sony's product designs.

Yet despite these clear advantages, product differentiation in itself need not be an entry barrier. Entrants are limited only by their creativity in devising new products, marketing their services, and building customer trust. Start-ups can distinguish their products from the familiar characteristics of the incumbent's brand by offering better quality, higher durability, greater safety, or other enhanced features. The strength of an incumbent's brand name does not guarantee competitive advantage.

Indeed, product differentiation is the key to competitive entry. Entrants can launch a new brand through investment in marketing. By creating a new brand, they are able to distinguish their offerings from the incumbent's. If customers perceive that the new brand has more attractive features, then they will switch from the incumbent's product to that of the entrant. Price discounts and promotions can attract early interest. Over time, entrants can build up confidence in new brands. Even though IBM had a brand synonymous with computers, and personal computers are still referred to as IBM-compatible, companies such as Compaq and Dell were able to launch successful brands. The American big three automakers, GM, Ford, and Chrysler, also seemed invulnerable until the Japanese automakers successfully developed such brands as *Toyota*, *Nissan*, *Honda*, *Lexus*, and *Infinity*.

There are few limits to creativity in product differentiation strategies. The strengths of an incumbent firm's brands sometimes help entrants identify weaknesses. Because leading brands have clearly defined features, new brands can easily distinguish their products relative to those brands. If incumbents offer "one-size-fits-all," entrants can offer customized services. Such a strategy can go undetected until the incumbent discerns that the entrants are pursuing a mass customization strategy. If incumbents offer confusing contracts, entrants can cut through the complexity and offer clarity. In long distance telecommunications, pricing plans offered by AT&T, Sprint, and MCI/Worldcom became increasingly complicated, allowing entrants to attract customers by offering simpler pricing plans. If incumbents sell strictly through stores, entrants can offer their product through new channels such as catalogs and Internet shopping.

Companies with established brand equity in other markets can expand into new markets under their existing brand. For example, although there were many established brands of coffee in the supermarket such as *Folgers* and *Hills Brothers*, *Starbucks* built its brand name through its coffee shops and then was able to enter the supermarket as an established brand in coffee. Alternatively, an entrant can acquire an established brand and invest in further improvements in product performance and sales.

Companies with an established brand name may use that name as an umbrella to enter into related markets. For example, Nestlé, the world's largest food company, sells a variety of products around the world under locally known brands, with the Nestlé brand name as an umbrella brand. In 1985, Nestlé acquired *Carnation Milk*, which had been marketed since 1907 as the "Milk from Contented Cows". Nestlé markets the company's products as Nestlé's Carnation Evaporated Milk, Dry Milk, and Condensed Milk.[30]

If the entrant believes that the incumbent has a significant product differentiation advantage, the entrant may elect to pursue a price-leadership strategy. By entering a market at the low end, companies may find they face few initial challenges from higher priced high-end competitors. The entrant can avoid directly challenging established competitors by offering the low-priced product in a different market segment than the incumbent's product.

Companies that produce generics are a good example of low-end entry. Supermarkets have higher margins on some lower cost generic products, from prepared foods to detergents. Costs are kept lower through reduced marketing and sales expenditures as well as lower production costs. In many categories, the major producers of brand-name products did not see the generics coming. The major brands did not believe that generics posed a significant challenge to the sales strength and reputation of their brands. Eventually, companies such as Procter & Gamble had to adapt to the incursions of generic household products by consolidating brands, reducing price variability associated with promotions, and lowering prices consistently.

An entrant can distinguish its product offerings from that of an incumbent in many different ways. For example, companies can enter the market by concentrating on one aspect of the business and executing better than incumbents. Domino's Pizza became the leading pizza delivery company because it only delivers, deciding early on that combining delivery with restaurants would lower the service quality of delivery. This strategy allowed Domino's to compete effectively against incumbents that offered a combined restaurant-delivery approach.

Entry and Position

Managers of potential entrants may be concerned if incumbents offer a great variety of brands. Established companies may offer complete product lines with many different product features. Also, established companies may offer diverse product lines suited for different market segments. The existing brands may cover the range of prices and features serving customers with different willingness-to-pay levels.

Brand proliferation, when established companies crowd the brand space, need not create a barrier to entry. The established firm obtains a differentiation advantage only if an entrant finds it difficult to improve on its set of products. A brand proliferation strategy can be vulnerable to entry strategies.

If a company offers many brands, the customer base for each brand will tend to be narrow. The company's own brands will compete with each other, a process known as fratricide. In addition, crowding the brand space is costly. The established company will have high costs of designing, manufacturing, and marketing many different brands. The established firm may obtain economies of scope from producing multiple brands, but it may forgo economies of scale from producing higher outputs of a smaller variety of products. Moreover, distributing many brands has high costs and resellers may have limited shelf space. Finally, managers may not be able to devote sufficient attention to critical market segments if they are handling too many brands.

Therefore, a company that offers too many brands can be vulnerable to an entrant that offers fewer brands. The entrant can lower costs by taking advantage of standardization. The entrant can also offer greater simplicity and lower transaction costs to retailers and wholesalers. In addition, an established company with too many brands will be vulnerable to companies that target specific segments and serve them better. An incumbent that tries to be all things to all people can face competition from specialized entrants.

Established companies often add and retire brands. Procter & Gamble (P&G) divested *Oxydol* because the brand "no longer provided a strategic fit" (see also Exhibit 12.5).[31] Entrants also should avoid a brand proliferation strategy. When food giant ConAgra launched its $1 billion flagship brand *Healthy Choice*, the brand appeared almost simultaneously throughout the grocery store, taking on canned soup, dairy products, frozen foods, and other products. Although sales jumped, the company faced targeted retaliation by companies dominant in each segment, creating lower returns. Entrants should target specific market segments for the launch, and then progressively extend the brand to other market segments.

Exhibit 12.5   Procter & Gamble

Variety may be the spice of life, but there is such a thing as too much product variety. Procter & Gamble (P&G) found itself spread too thin, with many brands serving ill-defined market segments, often overlapping and competing for the same customers. Providing so much variety is costly for manufacturers in terms of manufacturing, marketing, distribution, and transaction costs. The proliferation of stock-keeping units not only increased the costs of multi-product manufacturers, they also increased transaction costs for retailers and within-store search costs for consumers.

P&G faced competition from its own distribution outlets when supermarkets and other stores offered house brands (generics). Companies such as P&G, which sold brand name products, failed to perceive growing competition because the generic products were coming from the low end of the market. The brand-name companies seemed to believe that national advertising and a quality differential were sufficient not only to distinguish their products from the generics but also to maintain their market share. However, price-conscious consumers increasingly turned to low-priced alternatives, posing a challenge to P&G, Kraft, and other leading suppliers of brand-name products. For those consumers, generics offered significantly lower prices, enough to outweigh perceived quality differences between generic and branded products.

Retailers also benefited from the generics because they were able to obtain sufficiently lower wholesale prices so their margin on the generics was higher than on the name brands. A higher turnover on generics also helped the retailer realize a higher return on their limited shelf space. Moreover, supermarkets and other retailers were able to put their own name on generic products, thereby enhancing the market recognition of the retailer's

(Continued)

Exhibit 12.5   (*Continued*)

own brands. A stroll through any supermarket was sufficient to confirm the growth of house brands.

The initial entry of retailers into generics was an indirect strategy because they did not face a strong response from the brand-name companies. Then, as the brand-name manufacturers began to realize the increasing impact of generics, they began to formulate defensive strategies. They lowered prices to increase consumer value added and made additional marketing expenditures to enhance the perceived value of their brands.

To concentrate marketing and managerial effort where it would have the greatest payoff, P&G decided to shed brands. By the mid-1990s, P&G had cut the total number of its products by one-third, including reducing the number of its shampoos and other hair-care products by one-half.[32] It accomplished this by eliminating minor variations of the same product, standardizing packaging, reducing new product introductions, and divesting brands.[33] P&G sold off *Bain de Soleil* sun-care products, its stake in the pain-killer *Alleve*, *Lestoil* household cleaner, *Lava* soap, and other brands.[34] The company also divested its oldest brand of detergent, *Oxydol*, as well as its bleach detergent *Biz*.

By realizing cost savings from reduced variety, P&G could operate profitably at lower prices, increasing its competitiveness with the generics offered by supermarkets. It could also be more effective by concentrating its marketing efforts and management attention on fewer brands. By consolidating brands, P&G was in a better position to defend the market share of its remaining brands.

In addition to consolidating brands, P&G met the generics by simplifying its pricing. By offering discounts and promotions that varied over time, P&G and other manufacturers had encouraged retailer chains to purchase products based on wholesale discounts rather than demand patterns. Thus, retailers arbitraged against manufacturers by building warehouses to store the purchases made during promotions, reducing purchases and relying on inventories when a manufacturer's promotions were not available. By switching to "everyday low prices", retailers could smooth out their purchases and adjust them to demand patterns while at the same time reducing their inventories of manufactured goods, thus avoiding the costs of holding unnecessary inventories.

P&G's defensive strategies targeted not only consumers, through lower prices and increased marketing, but also their retail intermediaries such as supermarkets and discount chains such as

(*Continued*)

Exhibit 12.5 *(Continued)*

Wal-Mart. P&G looked for ways to reduce retailer transaction costs for retailers and to enhance retailer margins. By concentrating on fewer brands, it boosted sales of existing brands, thereby raising turnover for retailers and justifying its shelf space.

P&G's experience shows that even if crowding the brand space were possible, it would be a weak defensive strategy. Companies that try to crowd the product space by offering many brands will not necessarily discourage entry. Crowding the brand space is difficult, if not impossible, because there are so many potential product variations. Entrants have an easier time concentrating their efforts on particular market segments when an incumbent is spread too thin. P&G defended its market-leading brands by reducing its product variety.

Network Effects

The phenomenon of critical mass appears to influence many types of economic and social behavior. Neighborhoods improve or deteriorate due to self-confirming expectations, social conventions become self-enforcing, and prophecies become self-fulfilling.[35] Do critical mass effects imply that consumer choices collectively create market inefficiencies? Are market outcomes subject to the phenomenon of *tipping*, in which inferior products are purchased by all due to problems of coordination? Critical mass effects create benefits to consumers that have access to a network. The more consumers that join a communications network, for example, the greater the benefits from access because a consumer can potentially interact with many other people.

Network effects have been said to constitute another type of entry barrier arising from the demand side of the market. One type of network effect arises from complementary goods. Companies selling computers benefit from the availability of compatible software and companies selling software benefit from sales of computers that use the software. The more users that adopt a particular type of computer, the better off they will be because more software will be made available for that type of computer.[36] Do leading firms gain an advantage from network effects?

Leading firms often offer products that set technology standards, creating substantial value for customers because of the benefits of product standardization. Also, companies offering complementary products benefit from conforming to a widely accepted standard. However, when such standards are proprietary, competing companies often must either obtain licenses from the established firm or offer alternative products that conform to a different standard. The question of whether such technology standards reduce competition is the subject of some controversy.

In the computer industry, technological standards are referred to as *platforms*. Platforms have shown market concentration and persistence, particularly the IBM compatible platform with Microsoft's Windows/DOS personal computer operating system and Intel microprocessors. The popularity of Windows sets a standard in operating systems. Such a standard benefits consumers because personal computer makers and software application designers can reduce their production costs by standardization. Moreover, computers can communicate more easily and consumers need only learn to use one system. However, competitors have complained about barriers to entry resulting from the widespread use of the Windows operating systems. Although the Windows operating system retained a leading position in operating systems, many different types of companies supplying many varieties of both hardware and software have entered the computer industry.[37] Standardization of technology and product features not only creates economies of scale and scope, it also stimulates the growth of product variety and the supply of complements. Standardization is likely to have increased entry in computer hardware and software.

Because standardization creates value, it is sometimes viewed as a source of competitive advantage. Managers may perceive substantial difficulties in entering a market where the incumbent owns the technological standards. Microsoft maintained its position in operating systems and continually enhanced the functions and features of its Windows program. Yet, such advantages are often temporary because as technology changes, so do product standards. The greater the rate of technological change, the faster opportunities arise for entrants. For example, the analog standard in cellular phones gave way to digital transmission, allowing the entrant Nokia to surpass the established market leader Motorola. Moreover, the development of digital personal communications services, with new paging and telephone equipment and newly released spectrum, created opportunities for a wide range of companies to supply digital communications services and all sorts of wireless devices.

History is replete with examples of dominant technological standards being surpassed and replaced, yet another illustration of the competitive process of creative destruction. As the advantages of the alternative become apparent to consumers and exceed the costs of switching, a new product can rapidly overtake a market leader. The VHS standard for video cassette recorders appeared dominant, but it was rapidly replaced by digital video disks (DVDs). Technological change of this type creates opportunities for entrants to supply the new product (DVD players) and complementary products (DVD disks and peripherals).

Technology standards need not be set by the first mover. Entrants can learn from the limitations of the product offered by established companies and change the market standard. Thus, standard-setting sometimes benefits entrants, casting doubt on whether standard-setting and network effects are entry barriers. Sometimes later entrants innovate more effectively, such as the makers of liquid crystal display (LCD) screens that compete effectively against the older cathode ray tube (CRT) technology. Sometimes incumbents get things right by improvements to existing technology. For example, flat-panel screen makers faced entry from Candescent, a firm offering a next-generation technology called thin-CRT and backed by Sony, Compaq, and U.S. government grants. However, the makers of LCD screens improved their product by making larger, brighter displays, and the entrant Candescent was not successful. Some new technologies are disruptive but many are not.[38]

The market for handheld electronic organizers had a series of incumbents, including Psion in 1984, Sharp in 1987, Atari in 1989, and Apple with the notorious Newton in 1993. Despite these established companies, Palm Computing's 1996 Palm Pilot sold over a million within a period of 18 months, selling "faster than the VCR, the color television, the cell phone and the personal computer."[39] Palm's operating system set a market standard adopted by applications developers and phone designers. Despite being the market leader, Palm itself faced challenges from Pocket PCs offered by Hewlett-Packard, Compaq, Casio, and others based on Microsoft's Windows CE operating system and the advent of handheld devices with wireless communication. Nokia attempted to promote its operating system for advanced mobile phones by promoting an open standard; that is, by making its source code available to other companies.

If proprietary standards are seen to benefit incumbents, entrants can pursue a strategy of promoting open standards. Such a strategy puts pressure on incumbents to open their proprietary standards and to seek alliances with prospective entrants and suppliers of complementary products. IBM pursued such a strategy by promoting open computing and investing in open computing standards for hardware and software, such as spending $1 billion in a single year to support the Linux operating system. Microsoft's CEO Steve Balmer understood the strategic threat, noting that "Linux is our enemy No. 1."[40]

Network effects often benefit entrants rather than incumbents. Although incumbents can realize demand growth attributed to network effects, such growth also helps entrants. For example, companies in the personal computer industry realized substantial growth due to standardization. However, the advent of the Internet created growth opportunities for a vast number of new companies providing everything from infrastructure hardware and software to e-commerce. The network effects that stimulated the growth of the Internet created expansion in demand for the services of both entrants and incumbents. As with other types of competitive advantages, network effects need not create barriers to entry because they can be overcome by creative strategies.

Exhibit 12.6 Java

Sun Microsystems' Java software surprised Microsoft and a host of other software companies as well as computer makers. According to Sun Microsystems' chairman and CEO Scott McNealy, "the beauty of the fact that not everybody buys into what we're doing is that it gives us a head start."[41]

How did the strategy come about? Bill Joy, co-founder of Sun and its vice-president for R&D, set forth a two-part strategy that consisted of preventing any company or product from dominating the Internet and changing the way documents and programs work on computers that are linked to the Internet.[42] He clarified Sun's innovation strategy in contrast to the standard-setting activities of leading firms:

> "To me, the beauty and significance of the Net is that it is, by its very design, a decentralizing force. Not only does it defy being controlled by any one entity, it doesn't discriminate. There are no 'wrong' types of computers or software for the Net, as long as they follow some very basic communications rules… It's a liberating atmosphere that encourages more innovation than central planning or strict standards ever could and it's something we don't want the Internet to lose".[43]

The strategy was not accidental. Joy and Eric Schmidt, who later served as Sun's chief technology officer, had long ago developed "Berkeley Unix", a form of AT&T's operating system software they adapted to connect with the Internet.

According to Schmidt: "We always knew that microcomputers made the most sense not in isolation but when they were connected in networks."[44] Thus, each of the 1.5 million computers sold by Sun Microsystems had the hardware and software capability for Internet usage.[45] The problem was how to implement the innovation strategy successfully, particularly for a company that was primarily a hardware manufacturer with software resources overshadowed by Microsoft's legions of programmers.

The innovation of the Java software language must be measured against this backdrop. The Java language is constructed to be

(Continued)

Exhibit 12.6 (*Continued*)

read by practically any computer operating system, whether IBM, Apple, Unix or something else, thus transcending operating systems. While the operating system is running the individual machine, the language connects individuals to Web documents. It thus becomes a universal *interpreter*, that is, an intermediary between the computer's operating system and the Internet. In this position, Java becomes the channel for all transactions between individuals and the Web. Sending and receiving information must go through the interpreter. This supersedes the functions of operating systems that are meant to underlie applications, the software that carries out tasks. In fact, the language ultimately provides an alternative operating system for the PC.

Moreover, the language changes the nature of applications themselves. Java supplies small software programs called "applets" that allow a computer user to observe an animated Web page, or even to carry out jobs specific to the Internet site, such as ordering products, calculating the cost of an order, or receiving some type of information-based service.

Java also changes the world of computer operating systems regardless of the Internet because programs run on a Java "virtual machine". This means that programs written for Java can run on top of any machine's operating system (OS). Therefore, applications software developers can avoid having to adapt a program to every operating system. They can lower costs by writing the program only once for Java. Therefore, while other operating system software depends on being run on a compatible machine, Java transcends the equipment.

One way to understand Java is that it is the computer equivalent of an intermediary. Operating systems are software intermediaries because they stand between a computer and the software applications running on it. Operating systems interpret the computer code in software applications so that they can run on the computer. Thus, every computer must have an operating system. Microsoft benefited from a contract with IBM to provide a Disk Operating System (DOS) for its PCs. Providing the operating system is a software go-between strategy, that allows computers and applications to interact through the operating system.

Java takes the go-between strategy further, by intermediating between applications and *most* operating systems. Through its virtual machine, Java interprets applications for the resident operating system. This means that computer users can simply add

(*Continued*)

Exhibit 12.6 *(Continued)*

Java onto their existing machines to be ready for Java applications, without purchasing additional equipment.

The Java strategy is indirect because it does not appear to compete with other operating system software, but instead works together with it. According to *Byte*, "Java is a stealth platform that propagates entirely in software and co-exists peacefully with the native OS."[46] Apple, IBM, Microsoft, Novell, Silicon Graphics, and others adopted the Java interface, known as a "run-time environment."[47] Thus, Java virtual machines are included in later versions of Windows, Mac's OS, Unix, OS/2, Netware, and other systems, as well as in IBM's corporate computers.[48] Also, Web browsers such as Sun's HotJava, Netscape Navigator, and Microsoft's Internet Explorer include a Java run-time environment.[49] Thus, competitors helped propagate Java onto existing machines, much as the Greek's gift of the Trojan horse was welcomed into the gates of Troy.

Sun Microsystems emphasized that Java is an open system. Companies can write applications for Java as a platform and use the Java language in writing software. However, if Java were to become more popular than the underlying operating systems, Sun Microsystems, the owner of the Java brand, would stand to gain as a supplier of tools for software developers. This gain would potentially occur at the expense of competing brands of operating systems and microprocessors, which could become even more interchangeable. As software developers embraced Java features, Microsoft responded with a different approach that was part of its so-called .Net strategy. Microsoft would provide Web-based services using the XML language. With Microsoft's approach, software developers could write their programs in many different languages but users would depend on Windows operating systems for servers.[50]

12.4 *Transaction Advantage and Entry Strategy*

Managers of entering firms may believe that they face an entry barrier if the incumbent has a transaction advantage. Recall that by acting as an intermediary, the company lowers transaction costs for its customers and suppliers. By acting as an entrepreneur, the company creates new combinations of buyers and sellers. Transaction advantages are likely to erode quickly, limiting their potential effects as entry barriers. Entrants devise strategies to address the incumbent's transaction advantage. To surpass incumbent advantages, an entrant must lower transaction costs relative to incumbents or create greater value for suppliers and customers. Entrants create their own innovative transaction methods or they identify new combinations of buyers and sellers.

Transaction Cost and Entry Strategies

At the most basic level the transaction technology itself may offer economies of scale and scope. Retail stores have fixed costs of transactions; that is, costs do not depend on the volume of transactions, such as information processing equipment like computers, cash registers, bar coding, and point-of-sale terminals. These cost economies need not translate into barriers to entry. As with production cost advantages, the entrant can apply innovations in transaction technology to produce transactions at a lower cost. For example, an entrant could apply new types of enterprise software, point-of-sale equipment, or communications devices as means of lowering transaction costs.

Transaction technologies such as back-office information technology or point-of-sale systems can involve a significant level of sunk costs. Entrants may perceive an entry barrier if incumbent firms have made irreversible investments in such transaction technology. However, sunk costs in transaction technology can be overcome by continued innovations. Moreover, entrants can pursue different distribution channels that lower transaction costs.

Kohl's discount chain designs its stores in the shape of a racetrack with a wide aisle that goes all the way around the store and a large center aisle that provides a shortcut. According to CEO R. Lawrence Montgomery, "Our whole philosophy is: How can we have shoppers buy more *and* spend less time in our stores?" The company, which opened in 1962 and developed its store design concept in the 1990s, successfully entered the market in competition with incumbents such as Sears and Wal-Mart.[51] The company reduced the transaction costs of its customers through the convenience of its store design, thus gaining a transaction advantage that it used to enter the market.

A critical transaction advantage stems from identifying innovations and bringing them to market faster than competitors. However, incumbent firms that achieve success from such a strategy often build their business by producing products based on a particular generation of technology. The successful incumbent has an incentive to stick with a particular generation of technology to provide service to its installed base of customers. The incumbent may choose to incrementally improve its products because continually changing its basic technology would involve substantial investment and costs of adjustment. As a result, entrants can gain a transaction advantage by embracing later generations of technology. For example, Cisco Systems appeared to have a distinct transaction advantage in incorporating new technologies into its products, but entrants were successful in penetrating the market for Internet routers (see Exhibit 12.7).

Entrants may perceive that the incumbent firm has a transaction advantage resulting from supplier and customer relationships that are difficult to duplicate. Moreover, the established firm may have experience in coordinating its supplier and customer transactions. To overcome such advantages, entrants must offer different types of transactions that improve upon existing types of exchange. For example, Amazon.com was able to enter the retail book business by selling through the Internet even though established bookstores had longstanding relationships both with customers and with publishers.

Exhibit 12.7 Entry into the Market for Internet Routers

Cisco Systems appeared to be invulnerable to entry because its founders had pioneered the network router, a device that directs traffic on the Internet. Cisco Systems obtained a dominant market share in routers and was a market leader in network switches.[52] The company had a *transaction advantage* over potential entrants. First, it was adept at making new combinations with its strategy of innovation by acquisition that involved purchasing companies with innovative technologies. Second, Cisco Systems reduced transaction costs through its design of a virtual company with over half of customer orders going directly to Cisco's contractors and savings of over 30 percent from outsourcing manufacturing as compared to in-house production.[53]

However, entrants were able to overcome this apparent transaction advantage. For example, Juniper Networks, which was established in 1996 and went public in 1999, entered the market for routers by producing a particular type called a core router, used in the backbone networks of Internet service providers. By 2000, Juniper Networks had 28 percent of the market and Cisco System's share had fallen to 71 percent for core routers. Moreover, Juniper Networks identified innovations that had not yet been incorporated into Cisco Systems' routers and shipped the next generation of core routers (a 10 Gbps core router) about a year earlier than Cisco Systems.[54]

Cisco Systems' acquisition of innovation is exemplified by its major purchases in 1999 of optical networking technology companies, Monterey Networks for almost half a billion dollars, and Cerent Corporation for almost $7 billion. Within two years however, entrants such as Avanex had penetrated the optical

(Continued)

Exhibit 12.7 *(Continued)*

switching market; Cisco Systems had shut down Monterey, and its Cerent investment was threatened. At the same time, chip makers such as Broadcom and Applied Micro Circuits entered the market by incorporating functionalities that were in Cisco's routers directly into more advanced chips.[55]

Intermediary Strategies

One type of entry strategy, the *go-between* strategy, requires interposing the company between buyers and sellers who are currently transacting. For the strategy to be effective, the company must improve on the existing transaction costs that buyers and sellers face. The go-between strategy competes with the direct transactions between buyers and sellers, as Figure 12.1 illustrates.

To carry out the go-between strategy, managers of a potential entrant begin by identifying existing combinations of buyers and sellers and targeting particular market transactions. For example, the entrant can become a wholesaler by helping manufacturers sell directly to retailers. An entrant can become a supply chain manager by helping manufacturers manage their purchases from suppliers. Using the transaction triangle framework, the entrant tries to replace direct transactions with intermediated transactions, as in Figure 12.2. This approach requires that the total transaction costs of intermediated exchange are less than the transaction costs of direct exchange.

Then, the entrant contemplating a go-between strategy surveys its own resources and attempts to create innovative ways to carry out the transactions at lower costs. An important part of the go-between strategy is to get close to the customer. The entrant improves and bypasses

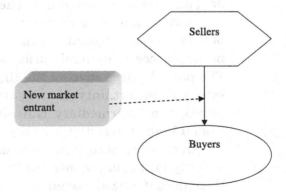

Figure 12.1: The firm employing the go-between strategy intermediates between existing buyers and sellers.

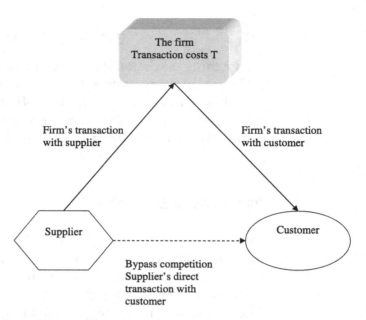

Figure 12.2:   The go-between strategy requires improving the efficiency of transactions in comparison with direct exchange between buyers and sellers.

competing marketing channels and converts established sellers into their suppliers. By controlling access to the customer, the go-between is in a position to receive economic rents from suppliers seeking to sell their products. For example, Charles Schwab entered the market for brokerage between mutual fund companies and individual investors (see Exhibit 12.8).

The go-between strategy involves many different methods of improving transactions. By consolidating the demand of many buyers and many sellers, the entrant reduces marketing and distribution costs for sellers and shopping costs for final customers. If the entrant can handle a greater volume of transactions, it can achieve economies of scale in the back office that are not available to buyers and sellers that deal directly with each other. The entrant pursuing a go-between strategy may be able to offer improved market information. If sellers and buyers rely on negotiation, the entrant might offer the convenience of posted prices or dynamic price adjustment methods such as auctions. The posted prices save on the time costs of bargaining and remove the inherent uncertainty in negotiation. Therefore, by clearly posting prices, the intermediary can attract customers and suppliers away from a high transaction cost market that involves bargaining.

An entrant using a go-between strategy can create markets when it is costly for buyers to monitor the quality of products offered by sellers. With decentralized exchange, there can be variations in product quality. Consumers are uncertain about the efficacy of pharmaceuticals, the durability of appliances, or the quality of automobiles. Suppliers generally have better information about the products they offer than their customers. The entrant can develop innovative methods for testing products and certifying product quality.

By dealing with many buyers and many sellers, the entrant following a go-between strategy can offer immediacy and other market-making services to improve upon existing transactions. By consolidating transactions, the entrant gains an advantage over direct exchange between buyers and sellers. For example, the Internet auctioneer eBay entered the market by improving upon direct transactions between buyers and sellers in markets for collectibles and resold goods. The leading international reseller of computers and other technology products, Ingram Micro, entered the computer market by going between manufacturers and resellers of computer hardware and software. Ingram Micro purchases and resells about 280,000 products from 1,700 manufacturers. Through its operations and affiliates in 35 countries, the company serves over 175,000 resellers in more than 100 countries.

Exhibit 12.8   Charles Schwab

Discount broker Charles Schwab & Co. entered the market for securities brokerage by offering a transaction advantage over existing full-service brokers. To understand how the company was able to enter successfully, it is useful to review some of the basics of investment and the intermediaries with whom Schwab competes.

There are many types of financial intermediaries in the securities markets: specialists and dealers, brokers, mutual funds, and families of mutual funds. The ultimate intermediary in securities is the stock market itself, where buyers and sellers of stock trade through specialists on the New York Stock Exchange or through dealers in the over-the-counter market. Individual investors must rely on brokers to trade in the stock market.

Everyone knows the first rule of investing: Don't keep all your eggs in one basket. Stocks are risky, so the best way to invest is to diversify by holding a portfolio of financial assets including stocks. This is simple in principle, but very difficult to carry out in practice. A small investor would incur almost prohibitive transaction costs in trying to assemble a portfolio of assets that is large enough to diversify sufficiently.

The investor must incur the research costs to track the performance of companies in the portfolio and other companies that might be added. The investor must also keep track of purchases and sales for investment planning and tax purposes. Individuals

(*Continued*)

Exhibit 12.8 *(Continued)*

can purchase shares directly from companies or a broker can serve as a go-between, providing the individual investor with investment advice and management of investment records.

However, the broker does not fully solve the investor's problems. The individual investor still must monitor the contents of the portfolio, and evaluate and act on the broker's advice. In addition, the investor incurs brokerage fees every time a stock is purchased or sold, making it costly to adjust the portfolio to account for new information. Despite the broker's services, it is still prohibitive to assemble a fully diversified portfolio of shares.

Enter the mutual fund. The mutual fund serves as a go-between in purchasing a portfolio of stocks. The investor need only buy shares in the mutual fund to have a portfolio with the same mix of stocks as the mutual fund. The mutual fund is able to earn returns as a go-between as a consequence of its economies of scale in gathering information about the companies whose shares it owns or may purchase. The mutual fund also has economies of scale in making trades because it buys and sells stocks in large blocks by aggregating the investments in the fund. Finally, the mutual fund has economies of scale and scope in record keeping, allowing it to provide investors with accurate and complete records of the fund's performance and other useful information. The mutual fund also supplies fund managers that pursue various investment philosophies in acquiring and divesting stocks.

Although a mutual fund significantly lowers the investor's transaction costs, the problem of diversification again arises if the investor wishes to purchase more than one mutual fund. Funds differ significantly: Some are centered in one economic sector, such as high-tech companies or transportation companies. Others are based on market indexes, holding portfolios that mirror the stocks used to calculate the Dow Jones Industrial Average or the Standard and Poor's 500. Still others seek different combinations of risk and return, striving for income or growth. The individual investor must monitor the performance of a host of mutual funds, seek out each fund individually, and assemble a portfolio of funds.

This creates a role for another type of go-between, the *mutual fund family*. Companies such as Fidelity, Magellan, or Vanguard maintain a number of different mutual funds so an investor can easily assemble a portfolio of mutual funds and move money among them with relative ease. Yet even this innovation is not

(Continued)

Exhibit 12.8 (*Continued*)

sufficient, particularly if the investor wants to purchase funds from different families.

Charles Schwab became the ultimate go-between with its creation of the *mutual fund supermarket*: a family of families. Through Schwab's Onesource, investors have access to a family of no-load mutual fund families. The funds themselves pay the fees for the transactions, instead of the investors paying Schwab directly. Through its Mutual Fund Marketplace, Schwab offers hundreds of additional no-load and low-load funds, but charges investors' fees. Thus, Schwab becomes a fund *retailer*, and the funds it offers take the position of wholesale suppliers.

With a single account, the investor can choose among literally hundreds of mutual funds. The investor can easily move money across the funds purchased without withdrawing money from the account, providing flexibility and convenience and offering advantages for retirement accounts where the money remains in the account for a long period of time. The investors receive a single statement summarizing all account activity.

Schwab has created its own fund family and even funds of funds, which compete with the funds offered by other companies. It has dozens of its own funds including securities funds, index funds, bond funds, and money market funds. For example, the company established the Schwab 1000 Fund, which is a no-load fund that attempts to match the performance of the Schwab 1000 index of stocks. Schwab also offers "Onesource Portfolios," which are funds that invest in other mutual funds. The fund manager monitors the performance of hundreds of mutual funds and continually updates the fund mix in response to changing market conditions. One fund, called the Growth Allocation fund, primarily focuses on securities funds, with some bond and money market funds, while another, called the Balanced Allocation fund, holds three-fifths securities funds.

Schwab, while not offering advice, still offers plenty of information to investors. The most important information is simply in its lists of available mutual funds, which considerably reduce investor search costs. Moreover, investors can obtain printed information on individual companies or funds and some basic investment information from account managers. Schwab also works closely with thousands of independent investment advisers, who provide their customers with financial planning while using Schwab's back-office services to manage their customers' accounts. Thus, in the market

(Continued)

Exhibit 12.8 (*Continued*)

segment of investors that seek advice, Schwab is an intermediary between the advisers and the financial markets. At the same time, Schwab is an intermediary between the advisers and their customers through a referral service called AdviserSource. Schwab receives fees from the independent advisers for referrals.

Schwab has garnered millions of customers to become the largest discount brokerage. With its mutual fund supermarket, the company is attracting still more investment assets. It is continuing to expand by offering a broader range of investment services. Schwab reaches investors through its over 200 branch offices. It has established an Internet online investment service for handling investment communication and customer accounts. It also has a 24-hour telephone service, called TeleBroker. For active traders with a minimum level of assets and frequency of commission trades, the company offers additional brokerage services.

Schwab competes not only with other discount brokerages but also with full-service brokerages as well, including Salomon Smith Barney and Merrill Lynch. In its alliance with independent financial planners, Schwab competes for customers seeking information with financial planning firms such as American Express. Schwab also competes with the suppliers of fund families such as Vanguard and Fidelity, which offer access to other mutual funds in a manner similar to Schwab.

The chairman of Fidelity, Edward C. Johnson 3d, told the *New York Times*: "The best place to be is in the distribution business because you have access to everybody else's business."[56] Accordingly, Fidelity withdrew many of its funds, including its Magellan, Contrafund, and Growth and Income funds, from the Schwab supermarket and those of other discount brokers, offering them only through direct sale.[57] This was a fundamental change in strategy in which the company places its retail distribution services ahead of its wholesale fund sales business, a direct challenge to Schwab's approach.

Charles Schwab's mutual fund supermarket got close to the customer by offering variety, convenience, and significantly lower transaction costs. Schwab entered the mutual fund business with its OneSource service by going between investors and mutual fund families. It handles a high volume of transactions by offering diverse financial assets, yielding economies of scope in distributing mutual funds. By offering many families of mutual funds under one roof, Schwab gives its customers the convenience of one-stop shopping and product variety that specialized mutual fund families do not offer.

Another type of entry strategy that attempts to overcome incumbent firms' transaction advantages involves consolidation of fragmented distribution networks or supply chains. Entrants pursuing a *bring-together* strategy attempt to reduce industry transaction costs by acquiring small or specialized distributors or by rationalizing supply chains. Such consolidation creates economies of scale and scope that allow the entrant to lower transaction costs relative to established firms. Moreover, by consolidating distribution, the entrant may enhance efficiency by greater coordination of purchases and sales.

When distributors resell the goods of many diverse sellers, the bring-together strategy offers a supermarket approach to distribution; the seller offers a wide variety of products and services. Customers gain from one-stop shopping where they can select from a wide variety of products, choosing the best by comparison-shopping at one central location. Customers who purchase multiple products also gain from the lower costs of putting together a market basket of goods and services. Thus, the entrant can improve on the offerings of distributors that handle only a few products or services.

The bring-together strategy allows the entrant to achieve economies of scope by handling multiple products. The company saves on the overhead and other costs that multiple distributors would incur. This increased efficiency allows the company to lower prices relative to independent distributors. Consolidating distribution can require a substantial initial investment to provide a broad range of products and services by contracting with a wider range of suppliers or customers than competing firms. Entrants can merge with or acquire existing distributors to bring their markets together. Alternatively, they can carry out the strategy through targeting a market segment and diversifying the firm's offerings over time. Dole Food Company entered the fresh-cut flower industry by consolidating part of the wholesale distribution channel (see Exhibit 12.9).

The returns to the bring-together strategy are evident in the growth of superstores and category killers. Book superstores such as Barnes & Noble and Borders brought together in one place the books offered by a host of specialty bookstores. By adding recorded music, Barnes & Noble attracted customers who buy both books and music, and thereby competed with independent music stores. Electronic superstores such as Best Buy combined a variety of products, outperforming specialized stores selling computers, cameras, or stereos. These companies took advantage of scale and scope in distribution to outperform specialized suppliers that were already established in the market.

> ### Exhibit 12.9   Dole Fresh Flowers
>
> Dole Food Company brought its knowledge of international distribution of perishable foods to a similar industry: fresh-cut flowers. In 1998, Dole established a flower division by Sunburst Farms, Inc., the largest importer and marketer of fresh-cut flowers to the United States. To expand the division, Dole also acquired Finesse Farms, a premier rose importer; Four Farmers, Inc., a bouquet company; and CCI Farms, a major fresh-cut flower producer. In addition, Dole also acquired 60 percent of Saba Trading AB in Sweden, a leading importer and distributor of fruit, vegetables, and flowers.[58] These purchases consolidated an important portion of international flower distribution.
>
> The Netherlands is the world's leading flower exporter. Colombia is the second largest exporter, growing two-thirds of the fresh-cut flowers that are sold in the United States, which represents over 80 percent of Colombia's total output. Other major flower exporting countries are Ecuador, Israel, India, Kenya, and Zimbabwe.[59] Dole purchased four of Colombia's largest flower growers, representing 25 percent of the country's total flower production. Dole's subsidiary Americaflor, operating in Colombia and Ecuador, is the world's largest grower of fresh flowers. Dole entered the market for fresh-cut flowers through its acquisition of growing capacity and its consolidation of a segment of international flower distribution.

12.5 *Overview*

Entrants perceive the presence of entry barriers if they believe that incumbent firms have competitive advantages. However, cost advantages, differentiation advantages, and transaction advantages tend to be temporary. They are subject to the effects of technological change in production processes, product features, and transaction methods. Moreover, changes in customer preferences create opportunities for entrants to provide new types of products and services.

Managers of companies that are contemplating entry should identify specific competitive advantages of incumbents and devise counter-strategies based on adopting product and process innovations, adapting to changing consumer preferences, or creating innovative transactions. Although the process of entry is costly, there are many strategies that an entrant can employ in attempting to overcome barriers to entry. Cost advantages are addressed by embracing technological innovations before incumbents, by introducing products that add sufficient value to surpass cost advantages, or by transaction innovations.

Product differentiation advantages can also be addressed by applying technological innovations or by introducing product varieties that satisfy unmet needs. Transaction advantages can be dealt with through innovative transaction methods, by creating new combinations of buyers and sellers, and by reorganizing supply chains and distribution channels.

Questions for Discussion

1. How did the emergence of jet airlines change the demand for alternative forms of transportation in comparison with automobiles and rail travel?

2. How is the Internet a disruptive technology? Did it displace any technologies?

3. Give some examples of industries that have experienced creative destruction; that is, consider industries where the products and production facilities of established companies were rendered obsolete by technological change.

4. Telephone companies invested a lot of money to set up pay phones, including the cost of the phone lines and the phone booths. With the widespread use of mobile phones, fewer people use pay phones. Discuss how substantial sunk costs invested in the system of pay phones failed to prevent the entry of mobile phones. Consider how mobile phones are different products than pay phones. Consider the difference in customer convenience between pay phones and mobile phones. With the arrival of Internet-capable mobile phones, how do the enhanced features of mobile phones affect the need for pay phones?

5. Cable News Network (CNN) entered the market by offering television news 24 hours a day. It was able to enter the market in competition with established news organizations at the big broadcast networks, ABC, CBS, and NBC. How did the company enter despite the well-established brands of the broadcast networks? How did CNN overcome a perceived differentiation advantage?

6. Many people purchase airline tickets and make hotel and car rental reservations online, relying less on travel agents. How did travel services such as Expedia, Travelocity, Priceline, or Orbitz compete with established travel agencies? What was the role of transaction costs in customer decisions to purchase tickets online? How did these companies successfully enter the market for travel reservations?

7. Airbus and Boeing are the major competitors in the market for jet airplanes. Brazil's Embraer and Canada's Bombardier entered the market for smaller regional jets. The regional jet airplanes were priced lower than larger aircraft. They were also differentiated

products because they had fewer seats and, being more fuel-efficient, were less costly to operate. The companies producing smaller airplanes compete with the companies producing larger airplanes since an airline can purchase two or three smaller planes to replace a larger plane, or it can change its route structure to favor more direct flights with less reliance on hub-and-spoke routes. What does this example illustrate about entry barriers and entry strategy?

8. Many new restaurants open every year. Although many do not succeed, some are very successful. The new restaurants offer different types of food, different locations, or different types of service in comparison to existing restaurants. What does this example illustrate about entry barriers and entry strategy?

9. Select an industry you are familiar with. Consider the leading companies in that industry. Imagine that you are an entrepreneur considering entry into that industry with a start-up company. How would you go about competing with the incumbent firms? What types of entry barriers would your company have to overcome? What challenges would your company face? What entry strategy might work best given the characteristics of the incumbent firms?

Endnotes

1. Proponents of the structure-conduct-performance school identify three basic *entry barriers*: economies of scale, absolute cost advantages, and product differentiation. See particularly J. S. Bain, *Barriers to New Competition* (Cambridge, MA: Harvard University Press, 1956). According to this view, if industries have a concentrated market *structure*, the companies in that industry must have monopoly *conduct*. In turn, monopoly conduct leads to an inefficient economic *performance*. Thus, the fewer the firms, the worse the economic performance must be. The structure-conduct-performance view is refuted by empirical analyses that do not find such a systematic relationship. Game theory also rejects the structure-conduct-performance causation since market structure itself is determined by the strategies of firms.

2. Entry barriers created by government tend to be more permanent and harder to surmount. These include exclusive franchises or licenses, technology standards, subsidies, and tariff and non-tariff barriers to international trade.

3. Pradeep K. Chintagunta and Dipak C. Jain, "Empirical Analysis of a Dynamic Duopoly Model of Competition," *Journal of Economics & Management Strategy* 4 (Spring 1995), pp. 109–131.

4. The concept of evaluating the company's strengths, weaknesses, opportunities and threats (SWOT) is due to Kenneth R. Andrews, *The Concept of Corporate Strategy*, rev. ed. (Homewood, IL: Irwin, 1980). Among the many precedents is Sun Tzu, who in the chapter titled "Weaknesses and

Strengths" wrote "if I concentrate where he divides, I can use my entire strength to attack a fraction of his," Sun Tzu, *The Art of War* (London: Oxford University Press, 1963), p. 98.

5. Over 2,500 years ago, Sun Tzu emphasized the importance of indirect methods in warfare. He observed that "those skilled in war subdue the enemy's army without battle." Ibid., p. 79.

6. This section draws upon Daniel F. Spulber, *The Market Makers: How Leading Firms Create and Win Markets* (New York: McGraw Hill, 1998), Chapter 8.

7. The account in this paragraph draws from Sam Walton with John Hewey, *Sam Walton: Made in America* (New York: Bantam Books, 1992).

8. Ibid.

9. Pankaj Ghemawat, "Wal-Mart Stores' Discount Operations," Case Study 9-387-018, Harvard Business School, 1986.

10. Patricia Sellers, "K-mart is Down for the Count," *Fortune*, January 15, 1996, 102–103.

11. Ibid.

12. These data are adapted from *The American Almanac, 1994–1995 Statistical Abstract of the United States*, table 1078 (U.S. Bureau of the Census, Census of Agriculture), table 1085 (U.S. Department of Agriculture, National Agricultural Statistics Office). According to the 1997 Census of Agriculture, there were 1.9 million farms in 1997.

13. Barnaby J. Feder, "Putting Farmers on a Contract," *New York Times*, Saturday, May 20, 1995, p. 19.

14. Jeffrey Williams, *The Economic Role of Futures Markets* (Cambridge: Cambridge University Press, 1986).

15. For Cargill revenues for fiscal year 2000, see www.cargill.com/about/index.htm.

16. See www.cargill.com/about/index.htm. On ConAgre, see John Kimelman, "Pack Up the Packing House," *Financial World* 163 (May 24, 1994), pp. 30–31. See for example Archer Daniels Midland Annual Report to Shareholders, 1997, on Archer Daniels Midland.

17. Gary Hoover, Alta Campbell, and Patrick J. Spain, *Hoover's Handbook of American Business* (Austin, TX: The Reference Press, 1994), pp. 624–625.

18. Don P. Blayney and Alden C. Manchester, "Large Companies Active in Changing Dairy Industry," *Food Review* 23 (May–August 2000), pp. 8–13.

19. Barnaby J. Feder, "Putting Farmers on a Contract."

20. This definition of entry barriers is due to George J. Stigler, *The Organization of Industry* (Homewood, IL: Irwin, 1968), p. 67.

21. The definition is commonly applied. See for example William J. Baumol and Robert D. Willig, "Fixed Cost, Sunk Cost, Entry Barriers and Sustainability of Monopoly", *Quarterly Journal of Economics* 95 (1981), pp. 405–431. See Daniel F. Spulber, *Regulation and Markets* (Cambridge: MIT Press, 1989), pp. 40–42, for additional discussion of barriers to entry.

22. There are consequences if incumbents fail to recover the initial investments made to enter the market. If the incumbent has incurred debt to

finance the entry costs, the firm can face bankruptcy, restructuring, and possibly acquisition. If entry costs are financed from the sale of equity, a failure to recover initial investments will adversely affect the firm's stock price and increase the firm's cost of obtaining additional financing. Even if an incumbent firm exits the industry due to financial difficulty, productive capacity may still remain employed.

23. *United States* v. *American Tel. & Tel. Co.*, 552 F. Supp. 131, 171 (D.D.C. 1982).

24. *United States* v. *Western Elec. Co.*, 673 F. Supp. 525, 538 (D.D.C. 1987).

25. AT&T paid more than $11 billion for McCaw. MCI Metro planned to invest at least $2 billion in fiber rings and local switching infrastructure in major U.S. metropolitan markets; "MCI Unveils Long-Range Vision: Network MCI, Opens Nation's First Transcontinental Information Superhighway," *PR Newswire*, January 4, 1994. Sprint acquired Centel, with local exchange and cellular operations, for $2.5 billion. See "Gambling on Thin Air," *The Economist*, August 21, 1993, p. 49. The cellular industry has made a cumulative investment of $13.9 billion through December 1993; CTIA, *The Wireless Sourcebook* 9 (Spring 1994). Ten competitive access providers (CAPs) reporting cumulative investment to the FCC had sunk $94 million by year-end 1993; Jonathan M. Kraushaar, "FCC, Fiber Deployment Update: End of Year 1993," Federal Communications Commission report, Table 15, May 1994.

26. The information about Nucor is based on Pankaj Ghemawat, "Commitment to a Process Innovation: Nucor, USX and Thin-Slab Casting," *Journal of Economics & Management Strategy* 2 (1993), pp. 135–161, and Pankaj Ghemawat, "Competitive Advantage and Internal Organization: Nucor Revisited," *Journal of Economics & Management Strategy* 3 (1995), pp. 685–717.

27. Michael E. Porter, *Competitive Strategy: Techniques for Analyzing Industries and Competitors* (New York: Free Press, 1982).

28. Kenneth Labich, "Is Herb Kelleher America's Best CEO?" *Fortune*, May 2, 1994, p. 44.

29. Ibid.

30. http://www.nestle.com/all_about/at_a_glance/index.html.

31. A start-up company named Redox founded by former employees of P&G purchased the brand with the intention of building a company out of brands divested by other companies; Kruti Trivedi, "Chips Off the P&G Block," *New York Times*, August 5, 2000, p. B1.

32. "Make it Simple," *Business Week*, September 9, 1996, pp. 96–104.

33. Ibid.

34. Ibid.

35. See Thomas C. Schelling, *Micromotives and Macrobehavior* (New York: Norton, 1978).

36. For a discussion of network effects as externalities, see Michael L. Katz and Carl Shapiro, "Network Externalities, Competition, and Compatibility," *American Economic Review* 75 (June 1985), pp. 424–440; Jeffrey Church

and Neil Gandal, "Complementary Network Externalities and Technological Adoption," *International Journal of Industrial Organization* 11 (1993), pp. 239–260; and Jeffrey Church and Neil Gandal, "Systems Competition, Vertical Merger and Foreclosure," *Journal of Economics & Management Strategy* 9 (Spring 2000), pp. 25–51. For a discussion of why network effects should not be considered externalities, see Stanley J. Liebowitz and Stephen E. Margolis, "Network Externality: An Uncommon Tragedy," *Journal of Economic Perspectives* 8 (1994), pp. 133–150; Stanley J. Liebowitz and Stephen E. Margolis, "Path Dependence, Lock-In and History," *Journal of Law, Economics, and Organization* 11 (1995), pp. 205–226; Stanley J. Liebowitz and Stephen E. Margolis, *Winners, Losers and Microsoft: Competition and Antitrust in High Technology* (Oakland, CA: Independent Institute, 1999).

37. Timothy Bresnahan and Shane Greenstein in their article "Technological Competition and the Structure of the Computer Industry," *Journal of Industrial Economics*, March 1999, look at the market structure of the computer industry over a 30-year period. They argue that entry barriers were low for firms if not for platforms.

38. "The tale of Candescent Technologies Corp. shows how two of the technology world's most cherished ideas are often just plain untrue. First, that antiquated old technologies are always sitting ducks for 'disruptive' new technologies that come along to challenge them. Second, that every engineering problem can be solved if you have the best and the brightest working on it"; "Mourn the Tech Company That Failed at Trying Big Things," *The Wall Street Journal*, May 13, 2002, p. B1. Other information about Candescent also drawn from this source.

39. Kim Girard, "The Palm Phenom," *Business 2.0*, April 3, 2001, pp. 74–81.

40. The Linux operating system is an open computer standard based on the Unix operating system. See William J. Holstein, "Big Blue Wages Open Warfare," *Business 2.0*, April 17, 2001, pp. 62–65.

41. "Why Java Won't Repeat the Mistakes of Unix," interview with Scott McNealy, *Byte*, January 1997, p. 40.

42. Brent Schendler, "Whose Internet Is It Anyway?" *Fortune*, December 11, 1995, p. 120.

43. Ibid.

44. Ibid.

45. Ibid.

46. Tom R. Halfhill, "Today the Web, Tomorrow the World," *Byte*, January, 1997, pp. 68–80.

47. Ibid.

48. Ibid.

49. Ibid.

50. Tom Sullivan, "Java under Siege: Eyeing the Horizon," *InfoWorld* 24, no. 12 (March 25, 2002), p. 1,46, Copyright Infoworld Media Group.

51. Calmetta Coleman, "Kohl's Retail Racetrack," *The Wall Street Journal*, March 1, 2001, p. B1.

52. According to David Bunnell (p. 184): "Cisco's advantage over its rivals is its history as a data networker and a builder of the Internet Infrastructure. It has a dominant market share in routers and is the leader in local-area-network and wide-area network switches." David Bunnell, *Making the Cisco Connection: The Story Behind the Real Internet Superpower* (New York: Wiley, 2000).

53. Ibid., pp. 147–148.

54. Paul Korzeniowski, "Jumpin Juniper," *Business 2.0*, May 15, 2001, pp. 24–25.

55. Bret Swanson, "For Cisco, It's Change or Perish," *The Wall Street Journal*, April 28, 2001, p. A22.

56. Edward Wyatt, "Why Fidelity Doesn't Want You to Shop at Schwab," *New York Times*, July 14, 1996.

57. Ibid.

58. See the company history at www.dole.com/company/about/timeline4.ghtml.

59. Larry Rohter, "Foreign Presence in Colombia's Flower Gardens," *New York Times*, May 8, 1999, p. C1.

CONCLUSION

Our journey through the principles of management strategy is at an end, but your management odyssey continues. Armed with the basic tools presented in this text, you should be prepared to help your company or consulting client formulate a strategy. The text has covered the five basic steps of strategy making. First, select goals for the company that maximize its market value. Second, perform an external analysis of the company's markets and an internal analysis of its organization, and adjust the company's goals accordingly. Third, identify sources of competitive advantage over the company's rivals. Fourth, devise competitive strategies that are best responses to the anticipated strategies of competitors. Fifth, modify the company's organizational structure to follow the needs of the company's strategy.

The company's strategy is a plan of action to achieve its goals. Having formulated the strategy, the manager also must put the plan into action. This means implementing the plan by acting through others in the organization. Implementation cannot succeed without a practical plan, but even the best plans require effective implementation. The manager is responsible for making sure that there are sufficient resources to execute the required tasks. The manager is responsible for communicating the plan clearly to members of the organization. The manager must ensure that the company provides the incentives and motivation for managers and employees to carry out the necessary activities with excellence and dedication. Experience and leadership skills, and further study of management are essential to administering strategy implementation.

As the manager implements the strategy, the strategy-making process begins anew. Given information about the firm's performance and its environment, the manager must evaluate whether the firm's goals are still appropriate. External analysis and internal analysis are used to update those goals and to determine whether the company's organizational skills are still well-matched with market opportunities. The manager considers whether the company has succeeded in achieving a competitive advantage and whether the process of innovation poses threats or offers opportunities. The manager examines how expectations about rival strategies have changed and determines how to update the company's competitive strategies. Finally, the manager considers the structure and performance of the organization to determine whether it is effectively carrying out the strategy and whether organizational adjustments are needed.

Clearly, the process of formulating strategy must be ongoing. No manager can simply fix it and forget it. The market environment is dynamic: customers, suppliers, competitors and partners continually change their behavior and expectations. Continual innovations in manufacturing processes, product designs, and transaction methods require creative responses. The manager must design plans that anticipate change and be ready to revise those plans.

Managers should perfect their ability to formulate strategy through experience and practice. There cannot be a specific business policy prescription that meets every situation; indeed, managers should be careful to avoid such recommendations. Having a coherent and consistent framework for strategy formulation allows the manager to see the forest instead of only seeing the trees. Managers need to develop broad skills that allow them to both anticipate and respond to rapidly changing business environments.

The principles of management strategy presented in this book are sufficiently general to meet the demands of dynamic markets. There will always be a need for a well-conceived plan of action, and that need is particularly acute in times of economic turmoil. Managers may need to think on the run, but success favors those managers who meet the competition with the most effective game plan.

COMPANY INDEX

NAME INDEX

SUBJECT INDEX